Praise for the *Handbook to Practical Disaster Preparedness for the Family...*

"A great introduction to disaster preparedness in a single volume, both scholarly and easy to read! Provides practical information for families seeking to organize their preparedness efforts."
— **James Talmage Stevens, Making the Best of Basics**

"Families want and need actionable practical plans when disaster strikes, and the Handbook to Practical Disaster Preparedness for the Family has it all in one volume!"
— **Shane Connor, ki4u.com**

"Whether you're a novice just starting out, a seasoned prepper, or a die-hard survivalist, this is one book that you should add to your must read list."
— **Keith Erwood, disasterpreparednessblog.com**

"An easy to read handbook packed full of very helpful information to cover any emergency one might encounter."
— **Marg Pollon, Founding Director of Bridges of Love Ministry Society**

"This book is an incredible compendium of knowledge in the area of disaster preparedness. In spite of the many pages, it is never daunting or overpowering but rather written with a comfortable conversational tone."
— **Kerry Lewis, Master Anjing Banfa**

"This handbook exceeded my expectations! It is a comprehensive presentation of ideas not only for disaster preparedness but also helpful for everyday living."
— **Dr. Lee Green, Dangerous Man Christian Conferences**

"This book offers a comprehensive run-down of household preparedness activities that is easy to follow and implement."
— **Jeff Bowers, Confluence Disaster Preparedness Consulting**

"Dr. Bradley's sincerity, hands on research, and organization makes this book a treasure. It should be read by someone in every family, and a copy kept in every home."
— **Linda Hoover, 912 Group Lead, Centennial, CO**

"What I appreciated most from this book is Bradley's focus on being prepared so that we can help those around us, and the unintimidating, conversational style in which he presents the information."
— **Cindy Baum, Books and Chocolate**

Handbook to Practical Disaster Preparedness for the Family

2nd Edition, by Arthur T. Bradley, Ph.D.

Handbook to Practical Disaster Preparedness for the Family
2nd Edition

Author: Arthur T. Bradley, Ph.D.

Email: inquiries@disasterpreparer.com

Website: http://disasterpreparer.com

Special thanks are extended to Siobhan Gallagher and Bridget Flanagan for editing the book's content, Marites Bautista for designing the layout, and Bryan Macabanti for illustrations.

Library of Congress Control Number: 2011909453

ISBN 10:   1463531109
ISBN 13:   978-1463531102

Printed in the United States of America

This book is dedicated to family,
both yours and mine.
May they all be kept safe no matter how
dark the skies.

CONTENTS

ABOUT THIS BOOK

When I first set out to write the "Handbook to Practical Disaster Preparedness for the Family," I had one overarching goal in mind: to help people put together practical, yet effective, disaster preparedness plans. I must confess that I had my doubts as to whether there would be an audience for my commonsense message. Based on other books in print, it seemed that readers were more concerned with the-end-of-the-world-as-we-know-it (TEOTWAWKI) senarios and weren't particularly interested in preparing for the real-world threats that were killing people every day around the world. Thankfully, I was proven wrong.

The first edition of this handbook was hugely successful, receiving praise from internet blogs, park rangers, disaster preparedness consultants, soccer moms, firefighters, emergency management professionals, and real people from every walk of life. It seems that my practical message was one that was sorely missing. Sure, it's exciting to think about what would happen if the world suddenly came to a crashing halt due to an unfortunate run in with an asteroid or other such doomsday event, but when push comes to shove, people want to know how to survive things that are actually happening around them.

There is no better example of this than the tragedy that struck Japan in March of 2011. When people around the globe witnessed the devastating effects of the earthquake, tsunami, and radiation release, they starting assessing their own family's readiness, not against asteroids or zombies, but against the real dangers lying in wait all around them. I received hundreds of letters from people concerned about radiation poisoning—a very real threat for those close to the event. My first recommendation was to go back and read the first chapter of the handbook, including the waving of hands above one's head while yelling, "The sky is falling!" This was not the time to panic but rather to think clearly and formulate an effective plan. With that understanding, I then explained how to make sense of radiation levels and offered practical steps to reduce exposure.

At its core, this second edition remains the same compilation of practical steps found in the handbook's original release. It has, however, been corrected and expanded to address shortcomings that readers brought to my attention over this past year. Additions include discussions of long-term food storage options; firearm selection, handling, and carry; national preparedness organizations; our nation's coming financial insolvency crisis; and specific preparations for the five deadliest types of natural disasters.

I wrote this book for my own family but am honored to share it with yours. I would ask that if it helps you along your path to preparedness, pass it on to a loved one when finished. Our country desperately needs to grow a community of people ready and able to meet the challenges that we will all inevitably face.

Be safe,

Arthur Bradley

FOREWORD

The world of late has been in upheaval: earthquakes, tsunamis, radiation leaks, terrorist strikes, pandemics . . . the deadly list grows with each passing day. It is as if we are being given a warning to ready ourselves for things to come. Whether that warning is from God or simply a result of our own empirical observations is up to every individual to decide. What is undeniable is that the world is openly demonstrating its ferocity.

Every generation faces its own unique challenges. Our grandparents and great grandparents weathered world wars, food shortages, deadly pandemics, and nearly total financial ruin. They did so with grit and determination and by coming together as communities—sharing in their collective need to survive. This connectedness has all but disappeared, save for the most rural areas of our country. The sad truth is that, today, many people don't even know their neighbors' names.

Recently, however, there has been an awakening in our country. People are realizing that there truly is strength in numbers. Churches, fraternal organizations, veterans groups, friends, and neighbors are discussing more and more how best to prepare for uncertain challenges. Food storage, water purification, backup electrical power, and community protection are all being considered.

Having learned much from my many roles in life, including marine, police officer, father, and priest, I can attest to the three things necessary for survival: knowledge, preparation, and determination. This book you hold in your hands represents the first of those three steps. The author has done his part to lay out commonsense approaches to meet your family's needs during nearly any crisis. The rest is up to you.

Now let's get to Chapter One. Time may be short!

Curtis A. Bradley
Shichidan, 7th Degree Black Belt
Tracy's Kenpo Karate

CHAPTER 1

INTRODUCTION

Right out of the gate let me tell you what this book is NOT. It is not a book about fighting off hordes of flesh-eating zombies, should they ever rise from the grave. Nor will it describe how to survive a shipwreck by feasting on coconuts and roasted iguana. Finally, it is not intended to help you survive our planet being sucked into the cosmic fireball affectionately known as our Sun. If you wish to prepare for those types of events, I respectfully suggest that you continue your search for a more suitable text.

This book is designed to help your family prepare for more commonplace, yet still potentially deadly, disasters. The list is long and varied and includes hurricanes, tornados, terrorist attacks, earthquakes, pandemics, financial collapse, widespread blackouts, and much more. My hope is that this handbook will accomplish three things: (1) motivate you to become better prepared, (2) illustrate how to prepare effectively, and (3) help you to realize your place in a larger movement.

If you are going to become a true "prepper," you should start by learning the disaster preparer's mantra. Let's all say it together—ready, here goes . . .

"The sky is falling! The sky is falling!"

It helps if you wave your hands wildly above your head for effect. Seriously, give it a try. It is best to get this silliness out of your system now; that way you won't succumb to it later. Besides, you might as well say it a few times because you are almost certainly going to be accused of thinking it—even if only by way of stage whispers and snooty grins.

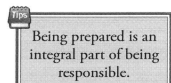
At the root of this incredulity lies a very basic question: *"Why?"* Why bother to prepare at all? Behind this question is the unspoken assertion that preparing for a disaster is unnecessary. It can be argued that most of us live in a fairly safe and stable world. What are the chances that you will ever need large stores of food or water? When will you actually use the carefully stocked first-aid kit that you keep in the car's trunk? Wouldn't your time, money, and energy be better served by focusing on life's "knowns" rather than its "unknowns?"

The answer to that question is a resounding YES! Disaster preparedness (sometimes abbreviated as DP) should never distract you from meeting life's other needs and responsibilities. Your kids will still need to go to college; your family will continue to benefit from the yearly getaway vacation; and you will undoubtedly have a better chance of advancing in your job if you remain vigilant at keeping your boss happy. People who neglect important areas of their life in order to focus on uncertain doom are losing sight of what's important—not to mention failing to see life's daily joys. With that said, it is still quite possible to live a full, rewarding life while preparing for hard times. Not only is it possible, but as the head of a household, I would argue that achieving this kind of balance is your responsibility.

Events have made me who I am, neither a hardcore survivalist nor an all-trusting soul. I am a Boy Scout in the truest sense of the word—dedicated to family, country, and good citizenship. As you may already know, an enabling element of those ideals is being prepared. I will never forget the helplessness that I felt in witnessing the horrific terrorist attacks of September 11, 2001. That event forced me to accept

Indian Ocean tsunami, 2004 *(Wikimedia Commons/David Rydevik)*

that our world is not as safe as we like to believe. Before I could even catch my breath, the world shook again with the tragic Indian Ocean tsunami of 2004. I remember sitting with my family that fateful Christmas, the holiday cheer forever interrupted by the suffering that hundreds of thousands of our brothers and sisters felt across the world.

Even as this book is going to print, the world is still recovering from the tragic events that unfolded in Japan, in March of 2011. Rocked by a powerful earthquake, deadly tsunami, and subsequent nuclear disaster, the country's citizens struggled to balance fear with their need to survive. The only positive outcome of the disaster was that people around the world paused to consider their own level of readiness. For a brief time, we stopped trusting that our governments would keep us safe and began to ask how we might ensure our own safety. Unfortunately, this newfound attention to the precariousness of life was quickly replaced with visions of the newest Apple products or gossip about the hottest celebrities. When the next global disaster strikes, most people will almost certainly be no better prepared than they were before the last calamity. In this world of comfort and consumerism, it has become human nature to resort to complacency and want.

What makes this book different from others is that it presents a practical approach to becoming prepared. As many books have demonstrated, it is easy to simply recommend that you keep everything you could ever need on hand. It doesn't seem to matter that it costs five times your annual salary, weighs a couple tons, and compromises many important aspects of your life. In my view, preparedness is best served by keeping all things in perspective. This premise will become clearer as you make your way through this book.

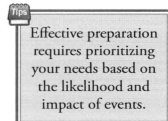

Effective preparation requires prioritizing your needs based on the likelihood and impact of events.

There are plenty of hardcore *preppers* who believe that everyone should be ready for Armageddon. These survivalist types routinely criticize some of my recommendations as not going far enough. Who knows for sure—they may be right! But I tend to take a more pragmatic approach. My motto is to **prepare for what makes sense.** It is highly unlikely the world will end tomorrow. Even if it does, it's equally unlikely that anything you do today will ensure your survival through the cataclysm. My opinion is that it's better to be prepared for the challenges you might actually face—harsh weather, power outages, loss of income, being stranded on the road, radioactive contamination, and so on.

But we still haven't answered the underlying question: "Why?" That single question is arguably the most important one in this entire book. If you don't know the answer, then you will likely lose your dedication to preparedness somewhere along the way. Also, without a clear rational answer that you truly believe in your heart, you will never win over family or friends—and establishing a network of fellow preppers is an important part of the readiness process.

The short answer is that the world is not as safe as we would like to believe. It is a chaotic system with limitless inputs and outcomes. **Anything can happen.** If you can keep these three simple words in mind, they will help you along your way. When people ask why you are going to such trouble to prepare for darker days, smile and say, "Anything can happen." Who can possibly contest that simple statement?

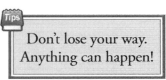

Don't lose your way. Anything can happen!

The point is that life is wholly unpredictable. Ninety-nine percent of the time everything works great. The world spins like a top, the skies are clear, and your refrigerator is full of milk and cheese. But know with certainty that the world is a dangerous place. Storms rage, fires burn, and enemies attack. No one is ever completely safe. Not you. Not your children. Not the richest man alive. We all live as part of a very complex ecosystem that, at its core, is unpredictable and willing to kill us without remorse or pause.

Don't believe me? Read on.

A SHOPPING LIST OF DISASTERS

There are countless ways in which the world can kill you. Many are simple accidents of nature, while others are the result of more malicious intent. Regardless of the cause, dead is still dead, and that eventuality is something we are all trying to avoid (or at least postpone). Remember the simple truism that nearly everyone who died as the result of a disaster fully believed they would live to see another day.

To help get everyone on the same page, as well as illustrate the point of how merciless our world can be, let's start by defining the word "disaster." One succinct yet informative definition is given below:[1]

disaster - *a calamitous event, especially one occurring suddenly and causing great loss of life, damage, or hardship, such as a flood, airplane crash, or business failure.*

Types of Disasters
➤ Natural
➤ Pandemic
➤ Man-made
➤ War/Terrorism/Crime
➤ Personal

The key words here are *calamitous event, suddenly, great loss of life, damage,* and *hardship,* none of which sound appealing, but particularly distressing when put together in a single sentence. When described in this way, it becomes clear why we are interested in mitigating the hardship caused by such a crisis.

The causes of each type of disaster may differ, but the impacts are often the same (e.g., loss of electricity, water contamination, food

Japanese town following earthquake and tsunami of 2011 *(U.S. Air Force)*

shortages, evacuations, mass casualties). The goal therefore is to come up with a comprehensive preparedness plan that addresses these impacts regardless of the specific cause. That is where this book comes in. Not only will it provide recommendations on how to prepare, but it will also show examples of real preparedness plans for every topic. By the time you finish the book, you should have enough advice and examples in hand to generate a DP plan that will meet your family's needs through nearly any crisis.

NATURAL

Weather events are often the first type of natural disaster that comes to mind. If you live in Colorado, you have experienced the isolation of heavy winter snowfalls. If you are on the Florida coast, you know

well the destructive forces of hurricanes. And if you have ever had to pack your family into a coat closet in rural Alabama, you can appreciate the very real threat that tornadoes pose. Many people don't realize that the United States experiences more severe weather than any other country in the world. Regardless of where you are in the world, dangerous weather events can and do affect you.

More generally, natural disasters are catastrophic events stemming from the dynamic nature of our universe. The world is constantly changing, and that grand-scale motion wreaks havoc. This category of disaster includes five deadly events that make the news almost daily: earthquakes, tsunamis, hurricanes, floods, and tornadoes. Specific preparations for these types of disasters are given in *Chapter 18: Five Horsemen of Death*. Natural disasters cause extensive property damage and unbelievable loss of life. A quick listing of some of the worst natural disasters in the last fifty years is sobering. [2,3,4,5]

Cyclone Nargis, 2008 *(NASA photo)*

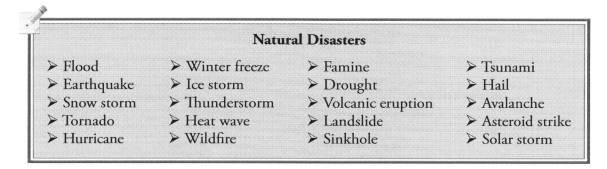

Natural Disasters

➤ Flood	➤ Winter freeze	➤ Famine	➤ Tsunami
➤ Earthquake	➤ Ice storm	➤ Drought	➤ Hail
➤ Snow storm	➤ Thunderstorm	➤ Volcanic eruption	➤ Avalanche
➤ Tornado	➤ Heat wave	➤ Landslide	➤ Asteroid strike
➤ Hurricane	➤ Wildfire	➤ Sinkhole	➤ Solar storm

Eruption of Mount Saint Helens, 1980 *(photo by Department of Natural Resources, State of Washington)*

EARTHQUAKES AND TSUNAMIS

Location	Date	Impact
Japan	March 2011	est. 30,000 dead, nuclear disaster
Haiti	January 2010	est. 230,000 dead
Sichuan, China	May 2008	est. 70,000 dead, 18,000 missing
Kashmir, Pakistan	October 2005	est. 80,000 dead, 3 million homeless
Sumatran Coast	December 2004	est. 230,000 dead
Bam, Iran	December 2003	est. 31,000 dead
Manjil-Rudbar, Iran	June 1990	est. 40,000 dead, 60,000 injured, 500,000 homeless
Tangshan, China	July 1976	est. 242,000 dead
Peru	May 1970	est. 66,000 dead

VOLCANIC ERUPTIONS

Location	Date	Impact
Mount Pinatubo, Philippines	July 1991	Blanketed 290 square miles with ash; more than 800 dead
Nevado del Ruiz, Columbia	November 1985	est. 25,000 dead
El Chichon, Mexico	April 1982	est. 2,000 dead
Mount Saint Helens, Washington	May 1980	Largest historic volcanic eruption in the 48 contiguous states; deposited ash across 11 states; 57 dead

DROUGHTS

Location	Date	Impact
North Korea	mid-1990's	est. 600,000 dead
Ethiopia	1984-85	1 million+ people dead
Uganda	1980	21% of the population (est. 50,000) dead
India	1965-67	est. 1.5 million dead from starvation and disease

HURRICANES, CYCLONES, AND FLOODS

Event, Location	Date	Impact
Cyclone Nargis, Burmese Peninsula	May 2008	est. 100,000 dead
Hurricane Katrina, U.S. Gulf Coast	August 2005	Massive coastal destruction; 80% of New Orleans flooded; est. 1,800 dead
Hurricane Mitch, Honduras and Nicaragua	October-November 1998	Unknown dead; 2.4 million homeless
Floods, Yangtze River, China	August 1975	Widespread famine; est. 85,000 dead
Floods, Hanoi, Vietnam	August 1971	est. 100,000 dead
Bhola Cyclone, Bangladesh	November 1970	est. 500,000 to 1 million dead

Even this cursory survey of recent natural disasters makes it clear that large scale catastrophes causing terrible loss of life are not uncommon events. Again, these are only the *major* natural disasters in the last half century. There are countless others of a lesser scale not listed.

PANDEMIC

A pandemic is loosely defined as a very widespread disease or illness—perhaps across a nation, perhaps around the world. For as long as I've been alive, and probably much longer, people have been predicting that a "superbug" will eventually destroy mankind (or at least set it back a few centuries). Given the list of potential candidates that make the daily news, including H1N1 Swine Flu, SARS, AIDS, Marbug virus, Ebola, H5N1 Avian Flu, and many others, it is certainly easy to understand their pessimistic predictions.

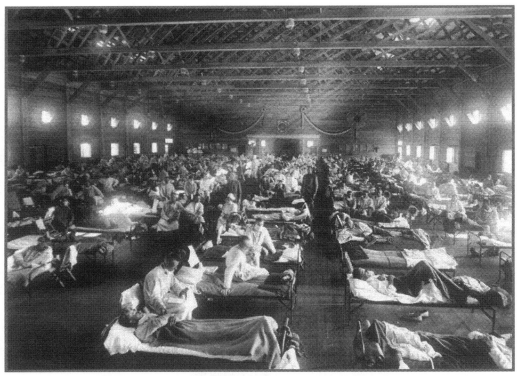

Spanish influenza, 1918 *(photo by National Museum of Health and Medicine)*

MAN-MADE

Man-made disasters are usually the result of things going wrong in our complex technological society. They include: blackouts, hazardous material spills, air pollution, house fires, radiation leaks, food or water contamination, industrial chemical releases, and oil spills. In the best case, they simply rob us of our modern conveniences for a brief time. In the worst case, they impact the entire ecosystem in which we live.

A profound example of a man-made calamity is the 1986 Chernobyl nuclear power plant accident that, according to the World Health Organization, exposed six million people to dangerous levels of radiation and left a portion of our planet nearly uninhabitable.[6] The recent events at the Fukushima nuclear plant have once again reminded us to reconsider the global impacts of our technological know-how.

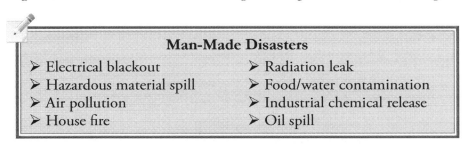

Man-Made Disasters
➤ Electrical blackout	➤ Radiation leak
➤ Hazardous material spill	➤ Food/water contamination
➤ Air pollution	➤ Industrial chemical release
➤ House fire	➤ Oil spill

Chernobyl nuclear accident, 1986 *(photo by Ben Fairless)*

WAR/TERRORISM/CRIME

Disasters caused by malicious intent are perhaps the most terrifying because their impact is limited only by the perpetrator's knowledge and available resources. Consider that humans are the only species to have ever existed who possess the ability to destroy the planet (in so far as to make it uninhabitable for mankind). Recently, there has been significant emphasis around the globe on combating the threat of terrorism. Some argue that such attention has improved our level of safety, while others quickly dismiss that view as naïve.

Terrorist attacks of 9/11 *(photo by FEMA News/Bri Rodriguez)*

PERSONAL

Sometimes a disaster is nothing more than an unfortunate turn of events in your personal life—perhaps an illness, loss of job, or unexpected financial burden. Typically, it causes hardship through financial pressures or physical disability. Serious personal misfortunes like these must also be included in your emergency plans since they disrupt your life in much the same way as other disasters.

DEFINING "BAD" AND "REALLY BAD"

All things are relative. What is considered to be a disaster by one person, might be thought of as little more than an inconvenience by another. To sort this out, it is helpful to have a general method of ranking disaster events. There are no recognized rating systems, so one will be arbitrarily created for the sake of discussion. This admittedly unscientific system is designed solely to illustrate the point that the term "disaster" may be used to describe a wide range of events.

Let's rate disasters by three important metrics: *area affected, duration,* and *severity.* Let's further rank each metric from 1 to 4, where 1 represents the best case, and 4 represents the worst case. Table 1-1 puts values to the categories. Again, keep in mind this is just an arbitrary rating system. But it gets your head in the right place—thinking about how broad disasters can be, how long they can last, and what impacts they might have on you and those you love.

To make sense of this, consider two examples—one at each end of the disaster spectrum.

Example 1:

Your community is hit by a winter ice storm. Power goes out due to fallen tree limbs; other services are available and unaffected. A few grocery stores have managed to stay open, but the roads are treacherous to drive on. Within a few days, the ice melts and things return to normal. Using our rating system, we would define this disaster as 1-1-1 (Area Affected: Cat. 1), (Duration: Cat. 1), and (Severity: Cat. 1). All in all, this is a mild disaster with limited impact.

Think this sounds impossible? Do a little research on the super volcano lying under Yellowstone National Park, affectionately named "The Beast" by geologists.[7]

Example 2:

Now, let's consider something more devastating. A large volcano erupts without warning in the center of the United States. It spews ash up to six feet thick across the entire country. The weather becomes unpredictable, daylight is limited, and there are no services (medical, fire, police) or utilities (phone, power, water, gas) available. Roads are completely impassable, so escaping the worst hit regions is not an option. The government is overwhelmed and quickly unable to function. Financial institutions worldwide are crippled. Worst of all, the effects of the disaster could be felt for a decade or more. This disaster would rank as 4-4-4 (Area Affected: Cat. 4), (Duration: Cat. 4), and (Severity: Cat. 4). Serious indeed!

Again, the key point of this rating exercise is to recognize that disasters come in many shapes and sizes, not all of which require the same preparations.

Table 1-1 Arbitrary Disaster Ranking

Category	Area Affected	Duration	Severity
1	Local	<5 days	- Limited disruption to services/utilities - Food/water/fuel available - Government/financial institutions intact
2	Regional	5-14 days	- Limited disruption to services/utilities - Limited food/water/fuel - Government/financial institutions intact
3	National	15-60 days	- No services/utilities available - No food/water/fuel available - Limited disruption to government/financial institutions
4	Global	>60 days	- No services/utilities available - No food/water/fuel available - Collapse of government/financial institutions

PREPARING VERSUS *PREPARING!*

Just as there are various degrees of disasters, there are also many different levels of preparedness. This brings us to one of the most important points in this book: ***Know what you are preparing for.***

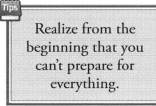

Realize from the beginning that you can't prepare for everything.

You can't prepare for everything. Instead, try to draw a line in the sand based on what you see to be the greatest threats to your family. Then target your efforts to mitigate the impact of those threats. Do you want your family to be capable of riding out a particularly bad winter storm? Or are you preparing for something much worse? Only you can answer these questions. And your priorities may change over time. Perhaps you will begin by simply reviewing and updating your homeowner's insurance and filling the pantry with extra canned goods. Later, you might feel compelled to take a more exhaustive look at things. Taking that next step might lead you to develop emergency preparedness skills (e.g., first aid, home maintenance, gardening, canning, self defense), establish a network of fellow preppers, and more closely monitor world events that might portend of an impending disaster.

The point here is that you need to maintain a realistic understanding of how well your preparations will carry you through different situations. As a Green Beret soldier once advised me, "Know your capabilities, but more importantly, know your limitations." Accept that being prepared does not necessarily require planning for the end of the world. A better starting place is to focus on meeting your family's needs during the most likely disaster events.

WILDERNESS SURVIVAL

People often confuse disaster preparedness with wilderness survival. Many books treat them as one and the same, and that is a critical shortcoming. Let's distinguish between the two. Disaster preparedness is having a wood-burning stove for keeping your family warm on cold winter nights when the power fails. Wilderness survival is building a lean-to using branches and shoestrings. Both are important and can save your life. Even though there will be times when the line gets blurred, there should be no illusion about it. This book is focused on disaster preparedness, *not* wilderness survival. There are two reasons for this.

First and foremost, preparing for likely disasters will serve the vast majority of people much better than honing their "live off the land" skills. Perhaps some, such as avid hikers or pilots, might fall outside this rule, but for most of us it is far more likely that we will be facing a power outage than a charging moose.

The second reason is that disaster preparedness is something that can be taught using a book. Regardless of the number of ways someone describes how you clean a squirrel with a pocketknife, it requires hands-on instruction and a heck of a lot of practice in order to become proficient. Much of what is contained in this book, on the other hand, is simply designed to help you identify the needs you may experience during hard times.

This is not to suggest that wilderness survival is unimportant. On the contrary, it just might save your life one day. Those blessed few who truly possess the skill to survive in extreme wilderness conditions are to be saluted.

Wilderness survival (U.S. Navy)

AVOIDING PITFALLS

There are dozens of disaster preparedness (DP) books available. Some titles are unique and informative, while others do little more than rehash the Air Force Survival Guide. Even with the breadth of choices, there remain many unanswered questions and a great deal of misinformation. An overriding objective in writing this handbook was to pull together a collection of useful information while avoiding the short-comings encountered in other DP books.

Some of the biggest shortfalls are discussed below. They are listed here only to illustrate what you should watch out for when conducting your own research.

Back to Basics—Much is written on the need to become self-sufficient by generating your own power (a.k.a. "getting off the grid"), growing your own food, retrieving your own water, and essentially checking out of modern society. This approach is not particularly practical or even desirable for most people.

List Driven—Preparing is often conveyed by exhaustive lists of tools, clothing, food supplies, and other gear. Unfortunately, this hoarding can cost tens of thousands of dollars with a questionable return on the investment. The approach tends to be popular because, let's face it, we all love lists! They itemize what you need in a format that allows checking a box—thereby getting one step closer to readiness. The assumption with exhaustive stockpiling is that the world's supply chain will collapse, forcing your family to live for a year or more on only what you have stockpiled.

> *Tips*
> Unless all civilization breaks down, you don't have to be self-sufficient to be prepared.

If history is to be our guide, however, this doomsday assumption is very unrealistic. Most of the recommended supplies would prove costly and un-necessary. You would likely end up with a jumble of stuff collecting dust in a "junk" closet until a spring cleaning forced them to be sold at a garage sale.

> *Tips*
> Try not to get overly excited by lists. Preparing is not just about collecting "stuff."

Food Focus—Countless pages in dozens of DP books have already been dedicated to food topics, including: basic recipes, bulk storage, canning techniques, sprouting, gardening, and grinding your own wheat. All of this I personally find very interesting, and if you share that interest, I encourage you to investigate further. However, as you will discover, the food plan presented in this book is built around very different goals.

Hearsay—Unfortunately, much of the information circulated in the DP community is unsubstantiated—based as much on folklore as fact. I am, and advise that you be, very skeptical of unreferenced information. People *think* they know lots of things, but more often than not, the knowledge is unproven and untested. Always ask yourself, "How do I know this is true?" The best answer is to put it to the test yourself. Second to that, is to locate a well-documented report of someone else testing the idea.

> *Tips*
> If ever there was a time to be a skeptic, it is when your life is on the line.

Scare and Prepare—Scare tactics are often employed as motivators for disaster preparedness. Worst-case "what if" scenarios are employed—another nation threatening a preemptive nuclear strike, an earth-destroying asteroid heading our way, the latest plague mutating and become airborne—you get the

A worst case scenario *(photo by US Department of Energy)*

idea. Once an author has scared the pants off you, he may spend another fifty pages emphasizing that you can survive anything if you are adequately prepared.

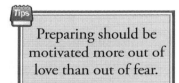

Preparing should be motivated more out of love than out of fear.

I do not engage in this *scare and prepare* method. The fact that you are reading this book tells me that you already know that the world is a dangerous place and want to prepare. Preparation motivated by fear is ill-advised. Remember, you already dealt with this when you ran around waving your hands wildly, shouting, "The sky is falling, the sky is falling." Besides, I believe that preparing should be motivated more out of love than out of fear. Things done out of concern for others almost always yield better results.

All of this discussion of shortcomings is not in any way to suggest that this handbook is without flaws. Certainly there will be unintentional omissions and errors. I do hope, however, that the book's organization and no-nonsense approach prove helpful when developing your own family's disaster preparedness plan.

TOME OF ALL KNOWLEDGE

It would be easy to write an 800-page book on disaster preparedness. There are scores of topics that could be studied, many of which would be of secondary value at best. This book aims toward a basic nuts and bolts approach rather than being the "tome of all knowledge."

My objective was to write a book that you can share with your family and friends. In fact, my intent was to create a book to give to *my* children and friends, and all those whom I care enough about to help them make it through tough times. In the end, it is not quite Strunk and White's *The Elements of Style,* but at least it doesn't rival the girth of Margaret Mitchell's *Gone with the Wind.*

This book is intended to be a study guide—something that can be read as a whole or quickly referenced by individual sections. While the organization is similar to many other DP books, the content is a compilation of personal recommendations based upon my own research. The ultimate goal is simple: to ensure that you and your family are more confident, prepared, and secure in this very unpredictable world.

Let's be clear about one thing: **Disaster preparedness is all about you.** If you choose to read this book and take no action, then fine, you are a little smarter than you were at the onset—although arguably no better prepared. On the other hand, if you find yourself loading up the supermarket cart with some extra cans of tuna, or tossing a gas can and blanket into your trunk "just in case," then it was all worth it.

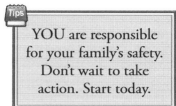

YOU are responsible for your family's safety. Don't wait to take action. Start today.

CANDOR

Years ago I worked as a college professor, and one thing I came away with is that teaching is most effective when done with empathy and candor. For that reason, I am writing this book with the same tone and honesty that I would want to see if I were reading it. Try to think of us as two friends out in the shop changing the spark plugs while drinking cold ones (Coke or Coors, your choice). We will work through the discovery process together. I will read through the Chilton's guide, and you turn the wrench. One thing that may not be obvious yet is that a significant part of surviving a disaster is leaning on your fellow citizens, sharing resources, and finding strength in numbers. We might as well start out that way.

ORGANIZATION OF BOOK

The bulk of this handbook is organized around the basic needs that must be met in order to survive. At the beginning of each chapter is an example scenario designed to help you consider your current level of readiness. At the end of the chapters are quick summaries of the important points for easy future reference.

After the summaries are brief lists of recommended supplies. The lists are intentionally kept short and limited to actual needs, not "nice to haves," and are crafted with generality in mind—focusing more on the general need than specific items. You should tailor the lists such that your supplies are practical and effective in meeting your specific needs. Taking an individual approach to preparation will help you pull together a DP plan that works for you and your family.

Also, you will likely be better served if you start thinking like Mac-Gyver—learning to make the most of what you have on hand. Even with a house full of supplies, there will undoubtedly be challenges that require you to adapt and use what you have in ways you never before considered. Recognizing how to use things for other than their intended purpose may in fact be the most valuable DP skill you can develop.

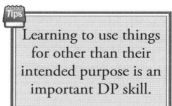

Learning to use things for other than their intended purpose is an important DP skill.

In addition to the chapters that discuss basics needs are four that focus on other topics. The first addresses special needs, specifically those of children, pregnant women, the elderly, people with disabilities, and pets. This section is a must read for most families. After that is a chapter on the

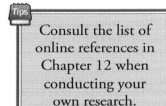

Worksheets

➤ Food storage list
➤ Telephone numbers
➤ Important papers
➤ Home assessment
➤ Home hazards checklist
➤ List of medications/allergies
➤ Leave behind note
➤ Personalized DP plan

importance of establishing a support network. You will be stronger and more capable in a group of like-minded families. Next is a chapter with information about preparing for five common types of natural disasters: earthquakes, hurricanes, tornadoes, floods, and tsunamis. This chapter was added to the second edition to help families better prepare for threats that make the news almost daily. Finally, the last chapter of the book encourages you to test your level of readiness by rehearsing several scenarios in order to test your plan *before* an actual disaster occurs. In my experience, this is the best way to reveal any preparations that you may have overlooked.

At the end of this handbook are a number of worksheets designed to help you organize your family's information. Feel free to photocopy them for easier use.

WEB REFERENCES

I have endeavored to research the covered topics extensively and limit recommendations to only those areas where sound conclusions could be drawn—eliminating any guesswork that might get you killed.

Consult the list of online references in Chapter 12 when conducting your own research.

One advantage of living in the 21st century is that nearly any kind of information can be found from one's own living room via the internet. This convenience speeds up research to be sure, but unfortunately it also leaves a fleeting trail for others to follow. The question then becomes how does a writer reference his work to sources that might change on any given day?

My answer to this challenge is to do three things:

1. Primarily cite websites and papers from large private organizations (e.g., American Red Cross, WebMD) and government services [e.g., Centers for Disease Control and Prevention (CDC), Federal Emergency Management Agency (FEMA)]. Websites for large institutions such as these are typically better managed than those of smaller organizations and change less frequently. When changes do occur, there is often a link left behind to redirect you.

2. Maintain a complete listing of the references at *http://disasterpreparer.com*. The webpage is periodically updated to reflect any known changes in the reference web links. A provision is also provided on the website to report a dead link or changed content.

3. Cite more than one reference whenever possible. Duplicate sources are useful in locating supporting information in the event of a reference becoming unavailable.

For more information about this book or the author, see *http://disasterpreparer.com*.

CHAPTER 2

STAYING ALIVE

Effective preparation of any kind requires a clear understanding of the goals you are trying to achieve. In the case of disaster preparedness, the primary goal is to minimize the impact of a crisis—whether it's a house fire, earthquake, or flood. At an absolute minimum, this means you want to survive, but in most cases, you also expect to maintain a reasonable quality of life. A comprehensive DP plan should help you and your family to accomplish both.

THE CARDINAL RULE

The cardinal rule to surviving a disaster is, whenever possible, to get out of its way! Whether you're being threatened by a tsunami, radiological contamination, or wildfire, getting out of the direct zone of impact is often paramount to survival. Preparedness begins with being alert and ready to take action. Obviously, some disasters occur suddenly and without warning and may not allow you to avoid their impact. Others, such as dangerous weather events, may provide warning but are not easily avoided. These are all situations in which your preparations will pay huge dividends, not only in helping you to avoid or ride out the event but also perhaps saving your life. Avoidance and preparation go hand in hand on the path to survival.

Remember the cardinal rule—get out of the way! *(photo by U.S.D.A. Forest Service)*

CORNERSTONES TO PREPARING

> **Cornerstones**
> 1. Leverage existing safety nets.
> 2. Stock extra consumables.
> 3. Collect tools and supplies.
> 4. Develop useful knowledge and skills.
> 5. Establish a support group.

Disaster preparedness requires much more than just stocking up on food and water. A thorough approach is developed around five cornerstones (see tip box). As you work through the discussion of the 14 basic and supporting human needs, consider each of these cornerstones when deciding how to establish the most effective preparedness plan.

Of the five, developing useful knowledge and skills is arguably the most important. As you will see throughout the book, many of the recommendations emphasize what you need to *learn,* rather than what you need to *buy.* The truth is most of us already have plenty of stuff. Rather, it's our know-how that is seriously lacking. That is where this book comes in—to teach you critical skills, such as how to store food, purify water, make your house safer, generate electricity, find disaster information on the internet, protect your family, and much more.

WHERE TO BEGIN

If the thought of pulling together a personalized DP plan feels a bit overwhelming, don't be discouraged. It's hard enough to prepare for a single catastrophe let alone ready your family for dozens of possible scenarios. You already know that your family will need food, water, and shelter, but how do you pull together something more comprehensive?

Preparing for many different disaster scenarios requires careful planning.

There are two possible methods. The first is to list as many disaster situations as you can think of, and then identify what your family would need in order to survive them. The problem with this approach is that there is significant overlap of needs from one disaster to another. For example, the preparations you make to prepare for a blizzard might overlap those needed for a food shortage. Both are likely to leave you depending on your food stockpile. This approach also introduces the risk that you might overlook a specific scenario, leaving your family vulnerable.

The second approach is to identify your family's fundamental needs that must be met in order to survive nearly any situation. You then tailor the specifics of that "need list" to the type and severity of the disaster. This is the preferred approach because it results in a more general solution, one without the tedious repetition.

HUMAN NEEDS

An excellent discussion of human needs was presented by the humanistic psychologist Abraham Maslow. As a humanist, he proposed that people not only try to survive but also prosper and live meaningful lives. In our current society, this trait often manifests in professional achievement or the pursuit of

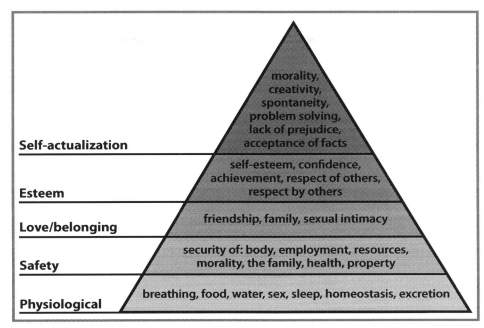

Maslow's hierarchy of needs

advanced consciousness and wisdom. The assertion is that people wish to be "fully functioning." Under most conditions, humans are not only interested in meeting their most basic physical needs but also in living full and consequential lives. As conditions deteriorate, however, people are forced to turn their attention to meeting only their most fundamental needs, the ones that help keep them alive. This necessary retrogression is frequently seen when a major disaster strikes as people forage for food and water or resort to less than optimal methods of sanitation. We do what we must to survive.

Maslow established a hierarchy of needs, often illustrated using a pyramid.[7] In his hierarchy, he identified the most basic survival needs as physiological. Included in these are food, water, air, and sleep. Above those are needs associated with safety and prosperity. Even farther up the hierarchy are more personal needs, such as friendship, self-esteem, and morality.

As far as disaster preparedness, you need to concern yourself primarily with *physiological, safety,* and to a lesser extent the *love/belonging* needs. The abstract needs above those are better suited to self-improvement studies. If you consider Maslow's hierarchy and apply a little common sense, you can quickly identify eight basic needs that must to be met in a disaster situation: food, water, shelter, light, heating/cooling, air, sleep, and hygiene/sanitation.

This is a short list compared with all the perceived necessities you might require in the course of your normal daily life (e.g., fashionable clothes, premium coffee, a luxury sedan, or a comfortable bed). But unlike those desires, the sustained loss of any one of these eight basic needs will

Basic Needs
➢ Food
➢ Water
➢ Shelter
➢ Light
➢ Heating/cooling
➢ Air
➢ Sleep
➢ Hygiene/sanitation

Supporting Needs
➢ Medicine/first aid
➢ Communication
➢ Electrical power
➢ Financial security
➢ Transportation
➢ Protection

almost certainly lead to your demise. The goal of preparation, therefore, is to meet these needs regardless of the crisis. There are also six secondary needs that directly or indirectly support your survival and help to maintain a reasonable quality of life. They include: medicine/first aid, communication, electrical power, financial security, transportation, and protection.

Together these two lists comprise fourteen areas that should be addressed by a comprehensive preparedness plan. It seems logical therefore to organize this handbook around these needs. By the end of the book, you should have a good understanding of these subjects. And with that knowledge, you will be ready to tailor your own personalized DP plan.

YOUR PERSONALIZED DP PLAN

What exactly is a disaster preparedness plan, and how do you go about making one? Despite the wealth of information published on the subject of emergency preparedness, you will find it difficult to find a direct answer to this question. It's also equally unlikely that you will find an actual example of what a comprehensive DP plan might look like. Rest assured that this question will be tackled head on. Not only will you understand what a DP plan should include, but you will also have examples to guide you in creating a customized plan for your family.

A disaster preparedness plan is simply a personalized summary of how you will meet your family's needs—those listed above, or others that you identify. There are many ways to organize the details, but the simplest is to create a table that pairs the dangers your family may face to the steps you will take to mitigate their impact. Keep in mind that your family's DP plan will be unique because it is built around their specific needs as well as the methods you choose to meet those needs.

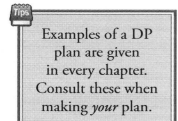

Examples of a DP plan are given in every chapter. Consult these when making *your* plan.

A blank DP Plan worksheet is included in the *Appendix* to help you better organize your approach. Feel free to make as many copies as necessary to detail your preparations—perhaps using one page for your water plan, another page for your food plan, and so on.

Each chapter also includes an example of what a corresponding DP plan might look like. Feel free to create your own format if you don't like the one provided. It's also not a bad idea to document your plan on the computer since doing so allows you to make changes quickly and easily. Remember, a DP plan is nothing more than a manner in which to get your preparations down on paper. There is no one right way.

SURVIVAL IN A BUCKET?

We all love one-stop shopping. It's easy, and there's little thought required. Capitalizing on that line of convenience thinking, several companies now offer prepackaged disaster preparedness kits. Most are stored in airtight buckets or easy to carry backpacks—both good ideas. If you read the retailer websites, you might be convinced that preparing involves nothing more than forking over $99 and finding a shelf on which to store the bucket of goodies.

Let's take a look at one such disaster preparedness kit meant for a family of four, all conveniently stored in a five-gallon potty bucket:

> Only stock up on useful high-quality supplies.

- 12 energy bars
- 12 8-oz. water boxes
- 4 space blankets
- 4 light sticks
- 4 thin plastic ponchos
- 4 small packs of tissues
- 1 tube tent
- 4 lightweight dust masks
- 1 pair work gloves
- 12 plastic bags
- 1 small first-aid kit (adhesive bandages, gauze, tweezers, antibiotic cream)

- 1 pair disposable medical gloves
- 5 five-hour emergency candles
- 1 whistle
- 50 water purification tablets
- 1 GI-style can opener
- 50 feet of nylon cord
- 1 small Swiss Army knife
- 50 waterproof matches
- 1 solar radio with flashlight
- 1 package of toilet chemicals
- 1 roll of duct tape

Wow, that seems like quite a list! Now, let's imagine being in a real life situation and see how well the supplies hold up.

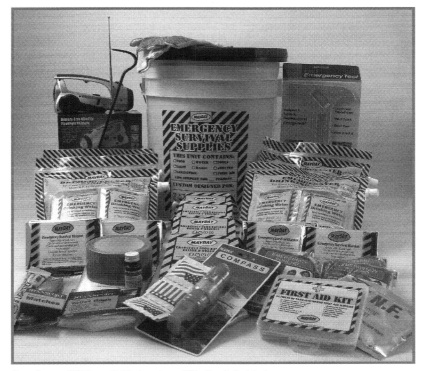

One of many DP kits available *(courtesy of MayDay Industries)*

Example scenario:

Assume that your family of four lives on the East Coast and a hurricane blows through. Thousands of power poles are downed, leaving you without electricity for two weeks. Gasoline is in very short supply, supermarkets are unable to restock, and local water reserves are contaminated due to flooding. The temperature is about 85 degrees during the day and 50 degrees at night. There are subsequent bouts of heavy rain. Worst of all, a large tree has fallen through the roof of your house. And, though not prevalent, there are scattered incidents of looting.

Let's test this kit against your family's fourteen fundamental needs:

- **Food**—The energy bars will keep your family alive for an additional day or two over having nothing at all. However, the less than appetizing meals certainly won't help the morale of your hungry family. The GI-style can opener will open any canned food, but it's slow, difficult to use, easy to lose, and won't hold up for long.
- **Water**—The purified water will last your family of four about six hours (assuming a minimum of one gallon per person/day). After that, you will have to either boil or treat the city water in an effort to nullify whatever contaminants have entered the water supply.
- **Shelter**—There are no tarps or tools in this kit to patch the hole in your roof, so your shelter is slowly being destroyed by the deluge of rain.
- **Light**—The kit's single solar radio with flashlight will be inadequate for your family to function at night. The five emergency candles will likely only last a single evening. After that, your family will be forced to hunker down as soon darkness arrives.
- **Heating/Cooling**—The temperatures are mild enough that you can survive without heating or cooling. Daytime temperatures will make it uncomfortably warm, as well as increase the need for drinking water. Evening temperatures will be chilly, forcing you to depend on the blankets you already have to keep everyone warm. Space blankets are best suited for short-term wilderness survival (see *Emergency Blankets* in *Chapter 8: Heating/Cooling*). They are typically small, easily torn, hard to wrap yourself in properly, and difficult to use without practice.
- **Air**—In this scenario, there are no dangers to your air supply. However, if there were, the kit offers only lightweight dust masks that would be ineffective against most airborne threats.
- **Sleep**—Sleeping in your home is still possible for a while, but conditions will quickly become unhealthy because of sanitation issues.
- **Hygiene/Sanitation**—The few packages of tissues and the potty bucket are inadequate for your family's personal hygiene. Without an ample supply of water, your family faces risk of disease from poor sanitation. Also, the inability to maintain an accepted level of cleanliness will drastically drive morale down.
- **Medicine/First Aid**—The first aid supplies provided are completely insufficient for anything beyond a scraped knee. If anyone in your family becomes injured or ill (both very possible), you will be faced with a very serious situation.
- **Communication**—There is nothing in the kit to help contact family or friends, no evacuation plan, no games or activities to keep children entertained, and no information about emergency services.
- **Financial Security**—The kit does nothing to help your financial security. In particular, it lacks home inventory sheets to help itemize property damage.

- **Transportation/Navigation**—The small radio might offer information to help decide if evacuation is necessary. However, with this kit alone, you will have no spare gasoline, no alternate means of transportation, and no maps of the surrounding area.
- **Protection**—The kit offers no personal protection beyond a small pocketknife. Should looters invade your home thinking it empty, your family will be helpless to defend themselves.

The bottom line is that, upon further analysis, the bucket DP kit falls far short of meeting your family's post-hurricane needs. Test this kit against other scenarios, such as a winter storm, terrorist strike, or widespread blackout. No doubt you will agree that it does little to improve your chances of survival, let alone make the situations more tolerable.

This criticism, however, isn't directed toward any company offering such kits. Having a bucket of supplies is better than having nothing at all. So in one sense, they are to be applauded for identifying a very real need. It is, however, strongly recommended that you take a more practical approach to disaster preparedness.

> Tips
> The best DP kit is not the one with the *most* supplies but the one with the *right* supplies.

Preparing will require you to do significantly more than just order a bucket of supplies. It will require you to take numerous actions, including assessing your family's needs, stocking supplies, shoring up your shelter, mitigating potential hazards, and preparing for likely threats. At the end of your efforts, you will be rewarded with knowing that your family is far better prepared to handle real world challenges.

Now let's get on with it.

BUTTER VS. BULLETS

When asked the difference between preppers and survivalists, some might say that survivalists focus more on guns, while preppers focus more on food. Though this isn't always true, it does bring up an important distinction between two very different preparedness philosophies. There are those who believe that personal protection should be foremost in all preparations and others who will argue that food should be at the heart of any disaster preparedness plan.

Both camps can provide very good arguments for their position, and therefore, both opinions are to be respected. I tend to steer clear of taking sides by advocating that people take a balanced approach to preparedness. What is most important is to develop a complete disaster preparedness plan, one that considers *all* your family's needs.

Perhaps learning to use a firearm and stocking ammunition fall into your plan for personal protection. That's fine, but you should also consider food, water, electricity, first aid, and many other important needs. Remember, you can't eat bullets. Then again, you can't fight off a mob with canned peaches. Preparedness is about being ready for a wide range of threats.

BECOMING A STOCK-GYV-ALIST

In the event of a major disaster, who would you rather be:

- The Stockpiler—someone with a wide assortment of supplies, but very little knowledge of how to actually *do* anything,
- The MacGyver—someone who can jury rig anything with duct tape, a pencil, and a pack of chewing gum, or
- The Survivalist—someone who can find dinner in an old stump, and make a heater using toilet paper and a rusty coffee can?

Clearly there are advantages to each type of person. But the greatest benefit can be had by combining the traits of all three and becoming a true "stock-Gyv-alist." A stock-Gyv-alist is someone who has carefully stockpiled critical supplies (e.g., food, water, batteries), taken the time to learn how things work and more important how to make them work, and developed the mindset necessary to survive nearly any encounter. *This* is who you want to become.

THIRTEEN BASIC STEPS

Most people who purchase this handbook intend to read it from cover to cover, carefully jotting down notes on how best to put together their personal DP plan. That's exactly what you should do. However, I am frequently asked, "Can you give me a few basic steps to help me get started?" After all, busy schedules have a way of keeping good intentions from ever turning into real actions.

For this reason, I've included a list of some basic steps that you can start taking immediately, even before your full DP plan is complete. This is in no way meant to summarize the book or act as a complete set of preparations but rather to serve as a starting point from which you can build a more comprehensive plan for your family.

1. Start paying attention. Get a weather radio. Monitor local and national events. Be more aware of your surroundings and things that may affect your family. Stay Alert = Stay Alive!
2. Make a simple list of dangers that you are most likely to face, many of which are dependent on where you live. Next, assess the shortages or hardships that these dangers might cause, such as loss of electricity, water contamination, or inability to travel the roadways. Finally, make a few basic preparations to mitigate the effects of those hardships (perhaps keeping a generator in your garage or installing a water purifier).
3. Stock up on consumables that might end up in short supply: food, water, candles, batteries, generator fuel, ammunition, diapers, etc.
4. Shore up your shelter. Take time to inspect your dwelling to make sure it is in good repair and capable of protecting your family.
5. Plan your possible evacuation. Identify where you will go, at least two ways to get there, and what supplies or valuables you will take with you.
6. Put together a small emergency kit for your automobile.
7. Review your insurance policies and adjust or supplement them to have an adequate safety net in place.

8. Have ready a properly-sized backup heating system (if appropriate to climate).
9. Establish an emergency fund that can be quickly accessed when a financial hardship occurs.
10. Learn first aid, and put together a well-stocked family kit.
11. Maintain a minimum 30-day supply of important medications and medical supplies.
12. Create a network of like-minded individuals committed to working together to survive dangerous events.
13. Consider the special needs of those within your household, including children, the elderly, those with disabilities, and pets.

All of these actions (and many others) are discussed in detail throughout this book. My advice is to take your time pulling together your family's DP plan. Try to let the preparedness mindset soak in before rushing out to stock your cupboards. Disaster preparedness is as much a frame of mind as it is a final destination. Enjoy the journey.

CHAPTER 3

FOOD

Challenge

A series of coordinated terrorist attacks on the country's food transportation system have left grocery stores unable to restock their supplies. Shelves are quickly emptied, and it may take up to three weeks for shipments to be resumed. How will you feed your family?

When preparing for a disaster, food is often the first need that comes to mind. A good illustration of this is when people rush to purchase milk and bread at the first warning of a bad storm. Admittedly, some people also turn their attention to batteries, generators, and bottled water, but those needs are usually considered after their stop at the grocery store.

Your access to food can be cut off in two different ways. The first is local in nature when stores are forced to close due to an immediate crisis—most often inclement weather. The second situation is one in which the food supply itself is disrupted. What is the likelihood of that? Not as remote as you might think. Minor disruptions occur all the time, often caused by weather events or contamination. For example, recently there was a winter freeze that ruined much of California's citrus crop, and an outbreak of *E. coli* poisoning that restricted much of Europe's produce. More significant disruptions can be caused by many incidents, including a prolonged power outage, transportation problems, major weather events, or an attack on the food supply system.

Did you know that most supermarkets only stock about a week's supply of food? This "just in time" supply system serves everyone well under normal circumstances. The supermarket owner doesn't suffer the overhead of having to hold large stockpiles of goods, and the consumer is guaranteed fresh goods that haven't sat on the shelves for a long time.

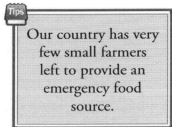

Our country has very few small farmers left to provide an emergency food source.

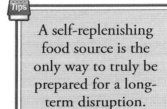

A self-replenishing food source is the only way to truly be prepared for a long-term disruption.

Ask yourself what would happen if the grocery stores were forced to shut down, or if the food supply was disrupted and stores could no longer be resupplied in a timely manner. Where would your family get food? There are very few small farmers left in the country, so local farmers' markets wouldn't be much help. Likely it would fall to the government to provide emergency rations, and let's face it; our government hasn't always shown itself to be efficient or timely when it comes to disaster relief.

The only way to truly be prepared for a long-term disruption of the food supply system is to have a self-replenishing food source, such as a farm or livestock. Unfortunately, this is not a practical option for most people. Landlords just wouldn't understand cows in the apartment! Many experts suggest doing the next best thing—stockpiling large quantities of non-perishable food. This book challenges that premise. Instead, the recommendation here is for a modest food storage plan that will meet your family's needs in all but the very worst disaster scenarios. The recommendations presented here regarding food are intentionally contentious because they are meant to challenge you to think about the practicality of your preparations. You do not have to agree with *any* of the recommendations in this handbook. What is important is that you have a clear understanding of what your preparations are and why you made them. Said another way, your preparations have to make sense to *you*.

There are four fundamental questions to consider when preparing a well thought out food storage plan. As it turns out, the answer to the first question determines the answers to the other three.

Fundamental Food Questions
1. How much food should you store?
2. What types of food?
3. How should you store it?
4. How long will the stored food last?

HOW MUCH FOOD SHOULD YOU STORE?

The question of how much food to store ties directly back to the more fundamental question posed in Chapter 1. What are you preparing for? The answer ultimately determines the quantity of food and every other type of supply that you will need. Short-term food interruptions, such as a snowstorm that prevent you from getting to the store, are unlikely to require more than a week's supply. Whereas a terrorist attack on the food supply system might cause shortages lasting weeks or even months—albeit likely limited only to certain food types.

Making preparations requires you to draw the line somewhere. Some preppers gravitate towards extreme, "end of the world" food storage plans but fail to consider the likelihood of such events or the costs associated with that kind of preparation. The recommendations in this book strive to be more practical and pocketbook friendly.

When it comes to food storage, a reasonable approach is to keep a minimum of a 30-day supply. Perhaps you are shaking your head in disbelief. You're recalling the advice of numerous experts who suggest having an entire year's supply of food stockpiled. Certainly having a year's supply will leave you more ready than having only a month's. However, that level of food storage is impractical, wasteful, and almost always unnecessary.

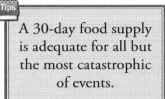

A 30-day food supply is adequate for all but the most catastrophic of events.

Let's start with *impractical*. Consider that the average American consumes about 2,000 pounds of food per year. Now imagine a family of four storing *four tons* of food in a location that is both temperature and humidity controlled. Not only storing it, but rotating it, and keeping it from spoiling or being ruined by pests. A challenge indeed!

Storing such a large quantity of food would likely be very *wasteful* as well. You would need to master techniques to keep food from becoming contaminated—perhaps using nitrogen-purged air-tight containers. Even when you did become skilled at food storage, you would still experience loss due to spoilage, rodents, and insects. This means that you would be required to regularly inspect a huge stockpile of food, discarding and replacing things that were no longer consumable. This would be a costly and time consuming process. To minimize waste, you could choose to store foods with a very long life, such as Meals Ready-to-Eat or freeze-dried products. However, those are very expensive and not particularly appetizing for daily consumption outside of a crisis. Such products are best suited for food caches that are infrequently accessed—see *Long-Term Food Storage*.

Finally, ask yourself which disasters would require you to have a one-year food cache. Go back and examine the *Shopping List of Disasters* in Chapter 1. Of all the disasters listed there, only a few truly catastrophic scenarios would require you to have more than a month's food supply (e.g., asteroid strike, world-wide drought, global war—a truly world changing event). The bottom line is that the sacrifices incurred from storing large quantities of food would very likely be *unnecessary*.

Consider the impact of every family in America following this recommendation to keep a minimum of a full month's supply of food. No one would ever have to race to the store when weather threatens. Catastrophic disasters would still cause worry, but the 30-day supply would give families time: time for the nation's emergency management services to provide relief, time for people to rally with friends and family to pool resources, and time to evacuate to areas where shortages are not as severe.

Also, people don't realize how much food is required for a 30-day supply. As a simple example, assume that everyone in your family eats cereal in the morning. How long would it take for them to finish a box? A day or two? Certainly my family of six can finish a box of cereal in a morning. This means that establishing a 30-day supply would require having thirty boxes of cereal! Not something that most people have in their cupboards. The example is obviously too simplistic since everyone likely wouldn't eat the same thing, but hopefully you get the point: a 30-day food stockpile is much larger than most people imagine.

As with every aspect of preparedness, however, your food storage plan is just that—yours! If you decide that a month's supply isn't enough to protect your family, then by all means store more. Again, the point in making the recommendation for a modest food plan was to force you to put justifications behind your preparations. Understanding why you take every preparation will go a long way to helping you pull together the most effective, balance DP plan, as well as save your pocketbook from unnecessary withdrawals.

WHAT TYPES OF FOOD?

Once you've decided how much food to store, the next step in creating a food plan is deciding what to store. Table 3-1 is an example of what a few entries in a food storage list might look like. A blank food storage worksheet is given in the *Appendix*. Make a copy of the worksheet, and then fill it in with the food items your family enjoys eating. You may find it helpful to use different color highlighters to color code the foods, indicating if the item is stored in the pantry, refrigerator, or freezer.

Keep the food storage list and a pencil near your pantry, perhaps taped to the door. As you consume items, change the "Qty Needed" to reflect what needs to be restocked. When you take your next trip to the supermarket, bring the worksheet with you as a shopping list. Don't forget to replace the old list with a new blank one.

Feel free to vary it up each week by making food substitutions, such as replacing canned peaches with raisins, but try to stay true to your general dietary goals—discussed in detail later in this chapter.

Far too often, food storage plans consist mostly of bland, easy-to-store food. This is usually a result of trying to store massive quantities of food where shelf life and getting enough calories in the diet are the driving requirements. Unfortunately, little thought is put into how unhappy the menu will make the people having to eat it.

Shying away from a very large food plan frees you up to store a wide variety of commercially available food. This type of food plan can revolve around three requirements: (1) nutrition, (2) taste, and (3) shelf life. Your family will thank you for adopting a consumer view in your preparations. No more dehydrated banana chips or molasses-flavored cookies.

Table 3-1 Sample Food Storage List

Food Category	Type	Full Stock		Current Stock	Qty Needed
		Qty	Size		
Vegetables	Green Beans	4	12 oz. can	1	3
	Frozen Corn	2	12 oz. bag	2	-
Rice and Beans	Brown rice	2	16 oz. bag	1	1
	Pinto beans	3	12 oz. bag	1	2
Dairy	Milk	2	Gallon	1	1
	Milk, dry	1	16 oz. box	1	-

To stock up, go no further than your local grocery store or discount warehouse. Storing foods with a very long shelf life, such as military Meals Ready-to-Eat (MREs) are thankfully no longer necessary. If you have ever eaten MREs (or their predecessor, C-Rations), you likely understand. A meal or two can be fun, but eating them as your three squares a day gets old very quickly. *I say this as an ex-Army grunt who has eaten his fair share of both.*

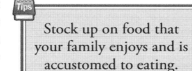

Stock up on food that your family enjoys and is accustomed to eating.

There is a line of thinking in the DP community that preparing is synonymous with getting "back to basics." Many books have lengthy discussions on gardening, sprouting, canning, and dehydrating. Any activity that has you eating healthier is certainly to be encouraged, but understand that these activities aren't necessary to pull together a reasonable food storage plan. Therefore, they are not discussed at length here. If you are interested in learning about these food topics, you should have no trouble finding numerous books specializing in them.

Gardening, sprouting, canning, and dehydrating are all very interesting, but not necessary for establishing a modest food cache.

When deciding what types of food to store, consider the USDA's new "Choose My Plate" recommendations.[8] The dietary guidelines were recently updated, and though you may still find them flawed, they represent well-researched guidance to proper nutrition. Contrary to popular belief, creating a stockpile does not mean that you have to store food oddities—don't go out and buy a huge tub of beef jerky thinking that it would make great "survival food." Stick to what your family eats and enjoys.

Said another way . . . ***store what you eat, and eat what you store.*** This makes rotation a snap; you put the newest food to the rear, and eat that which is oldest.

When selecting canned goods, opt for those with pull tops. This eliminates the need for can openers, and does not affect the shelf life.

Canned and boxed foods will be perfectly fresh 30 days after purchase. Breads and raw meats must be frozen to maintain a 30-day supply. Canned meats are also readily available (e.g., Spam, tuna, chicken). As far as dairy, many items have adequate refrigerator life, such as cheese, yogurt, and butter. Shelf stable milk and hard cheeses are options that don't require refrigeration. Shelf stable milk is created using ultra high temperature pasteurization, and has a shelf life of several months. It is offered through several manufacturers,

Shelf stable milk *(courtesy of Borden)*

including Borden, Nestle, and Parmalat, and is readily available at grocery and warehouse stores. For even longer storage, you can keep powdered products, even if just as a backup.

Be careful not to stock up too heavily on frozen or refrigerated foods because this leaves your food supply susceptible to the loss of electricity. There are, however, a few tricks to keeping food cool during a power outage (up to a few days):

- If available in your area, put blocks of dry ice in the refrigerator and freezer.
- Keep the refrigerator and freezer stocked full with little empty space, even if it is just filled with containers of frozen water, such as 2-liter bottles.
- Minimize the times you open the refrigerator or freezer.

If you do find yourself with extra food that will go to waste due to thawing, invite the neighbors over for a big cookout. Goodwill toward others, especially in times of crisis, can go a long way toward making lifelong friendships. Even if you don't like your neighbors and are not driven by "the common good," your generosity may well lead to a prudent exchange of resources.

CHOOSE MY PLATE

You already know what your family likes to eat, and your food stock-pile should definitely be tailored to keep everyone happy. With that said, it wouldn't hurt to review a little expert guidance to ensure that you pull together a balanced, healthy food storage plan. Much of

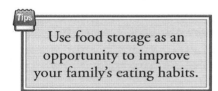

Use food storage as an opportunity to improve your family's eating habits.

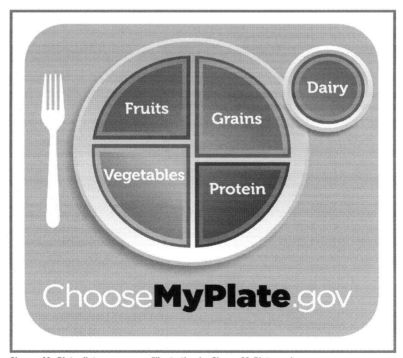

Choose My Plate dietary program (illustration by ChooseMyPlate.gov)

what follows might seem better suited to a guide on nutrition, but an important part of being prepared is stocking your cupboards, refrigerator, and freezer with the right kinds of food.

In 2011, the United States Department of Agriculture (USDA) released a significant update to their dietary recommendations—replacing the Food Pyramid with Choose My Plate. The new recommendations are illustrated by showing relative proportions of the five different food types: grains, vegetables, fruits, dairy, and proteins. Though not part of their primary illustration, recommendations are also provided for the consumption of fats and oils.[8]

Recognizing that everyone is different, the USDA has created an online tool that allows you to create your own personalized food plan. Recommendations are based upon age, sex, size, and activity level. There are also many other useful features, including menu planners, meal tracking worksheets, nutrition tips, food ratings, and more. Go to *www.choosemyplate.gov* to try the interactive tool. An example of a USDA personalized food plan is shown below. In this case, it recommends consuming 10 ounces of grains, 3½ cups of vegetables, 2½ cups of fruits, 3 cups of milk/dairy, and 7 ounces of low-fat meat each day. The specific quantities will vary based on weight, age, and activity level, but the general profile will remain very much the same.

Sample dietary plan *(illustration by ChooseMyPlate.gov)*

GRAINS

The foundation of the Choose My Plate diet is built on grain products, such as breads, cereal, pasta, tortillas, popcorn, oatmeal, and rice. When possible, choose foods made with whole grains over those with refined or enriched grains. The program recommends that at least half of your grains be whole grains.

Whole-grain foods are made with the entire grain kernel (i.e., bran, germ, and endosperm) and as a result offer more fiber and vitamins, and have been shown to reduce the occurrence of several diseases.[9,10,11] Refined grains are processed such that the bran and germ are removed. This is done to

Courtesy of Country Living Productions

give the grain a finer texture and extend the shelf life. The drawback is that removing them also removes much of the fiber and B vitamins. Whole-grain foods are thus healthier, and since they are higher in fiber, they also keep your digestion on track while helping you to feel full longer.

It is also possible to store raw wheat for grinding, but this requires a grain mill (either electric or manual). For very large food stockpiles, a grain mill becomes particularly important because raw wheat lasts longer than ground wheat. For the 30-day plan however, it is not required.

Don't forget to store baking items (e.g., baking powder, baking soda, salt, yeast, sugar, corn starch) if planning to bake your own foods. Home-baked goods are considered healthier because they contain less preservatives. Also include spices and flavorings, such as vanilla, cinnamon, cocoa, and chili powder. As always, the bottom line is to store what you know how to use.

FRUITS AND VEGETABLES

You have heard it a thousand times: Americans don't eat enough fruits and vegetables. People use a host of excuses ranging from not liking the taste, to restaurants not offering enough choices. Excuses aside, everyone knows that fresh produce contributes to better health, and is thus deserving of a more concerted effort. Americans in general would be better served by moving away from meats and sweets and toward fruits and vegetables.

You can get your recommended servings of fruits and vegetables many different ways—fresh, cooked, frozen, canned, dried, or even as juice (not the sugar-loaded stuff). Nutritionists recommend eating a variety of fruits and vegetables with different colors and textures. The reason for this is that each food has unique properties that contribute to good health. Try using this excursion into disaster preparedness to take two actions:

1) **Try new foods**—Try those fruits and vegetables that you have passed by so many times in the grocery store fearing that you wouldn't like the taste. You may stumble upon a gem while

contributing to your own good health. Besides, having a broad appreciation for a variety of foods is a valuable DP skill since it will help you adapt to dietary changes introduced by food shortages.

2) **Eat a more balanced diet**—Consume more fruits and vegetables, and try some of the whole-grain products available today. Eating more fruits and vegetables will likely help you lose any unwanted pounds, thereby improving your health and helping you to be more physically active. Physical ability is often important when facing the challenges associated with a disaster, such as repairing your roof, hiking down to the river for water, or clearing the roadway of fallen trees.

Fruits and veggies *(USDA photo)*

DAIRY

It is no surprise that the Choose My Plate diet recommends primarily consuming low-fat dairy foods. They are significantly lower in calories and saturated fat than the full-up versions. A diet low in saturated fat is believed to help maintain low cholesterol, which is an important part of heart health.[12]

Dairy products can include such things as milk, cheese, yogurt, pudding, and ice cream. For those who are lactose-intolerant, there are many lactose-free dairy products now available.

Lactose-free dairy *(USDA photo)*

PROTEINS

Protein sources include a wide variety of meat, poultry, fish, beans, eggs, nuts, and seeds. The USDA recommends opting for lean meats, fish, and poultry with the skin removed. Try to avoid processed meats since they often contain added sodium and other ingredients that can best be described as "miscellaneous animal parts."

Protein can also be found in a wide variety of beans and peas. Eggs are another good source of protein, but be aware that the yolks are high in cholesterol (containing about 70% of your recommended daily allowance). A healthier choice is to eat only the whites if you are going to eat eggs often.[13]

As far as nuts and seeds, the medical community has recently recognized that eating 1½ ounces of nuts per day may reduce the risk of coronary heart disease.[14] Keep in mind that nuts are high in calories, and an ounce is only a small handful (about ¼ cup).

Proteins *(NCI photo)*

FATS, OILS, AND SWEETS

Fats and oils *(NCI photo)*

Most people have no trouble getting enough fat in their diet. From butter, to mayonnaise, to cooking oil, to animal fat, to candy bars, fat is everywhere! You can't escape it.

Oils are simply fats that are liquid at room temperature. Most are high in monounsaturated or polyunsaturated fats, low in saturated fats, and have no cholesterol. Notable exceptions are coconut and palm kernel oil, which are high in saturated fats—very undesirable. Cooking oils include canola oil, corn oil, cottonseed oil, olive oil, safflower oil, soybean oil, and sunflower oil.

Solid fats are just that—fats that are solid at room temperature. This includes food items such as butter, margarine, vegetable shortening, beef fat (tallow, suet), and pork fat (lard). Solid fats come from many animal products and can also be made from vegetable oils through a process called hydrogenation. The American Heart Association (AHA) recommends that you consume fats and oils with less than 2 grams of saturated fat per tablespoon, such as canola, olive, and soybean oils.[15]

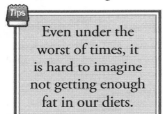

Tips

Even under the worst of times, it is hard to imagine not getting enough fat in our diets.

The AHA also recommends limiting dietary fat to 30% of total calories. How easy is this to do? Consider that whole milk gets about 48% of its calories from fat, peanut butter—75%, and steak—52%.[16] It's quite easy to see why people routinely exceed the 30% calories-from-fat recommendation. Short of having to live on lean, wild rabbits, it is difficult to imagine a case where inadequate fat intake would become a health concern.

Sweets and sweeteners fit best into this category too. Sweeteners include such things as sugar, brown sugar, honey, maple syrup, corn syrup, and artificial sweeteners. Consider stocking a collection of sweeteners for fresh-baked goods, as well as a small supply of your family's favorite sweets (e.g., candy bars, brownie mix). They can help make life a bit more pleasurable, which can offset some of the hardship.

FOOD SUMMARY

The message here is that *now* is a good time to introduce your family to a balanced, nutritious diet. The new Choose My Plate recommendations can help you craft a healthy eating plan. As far as food storage, increase your stockpile of foods to a minimum of 30-days supply, being careful not to rely too heavily on frozen or refrigerated foods. Also, remember to store what you eat and eat what you store—no need for "survival food." This strategy keeps waste to a minimum. Perhaps this food storage plan lacks the adventuresome feel of storing a bunker full of military rations. It is, however, practical and would serve your family well in hard times.

One final thought on food storage—of all the steps in the DP process, stockpiling food will provide you with the most unique sense of security. Every time your family asks, "What's for dinner?" and you have dozens of choices at your fingertips, you can't help but feel better prepared. Try it, and you will quickly discover the comfort in having well-stocked cupboards.

HOW DO YOU STORE IT?

The difficulty of storing food is directly proportional to how much you decide to store. Larger storage plans present significant challenges, including storage space, rotation, infestation, and spoilage. If you are on board with the 30-day food plan, then the biggest challenge is simply finding enough space.

To help you better appreciate just how much food is required for 30 days, go to your cupboards and count up how many full meals you have (breakfast, lunch, and dinner). If you are like most families, your current stockpile is somewhere around five to seven days, with lots of little extras that don't necessarily fit into a specific meal. Take a moment to imagine what 30 days of food might look like. Would it even fit in your cupboards? Almost certainly not. More than likely you will have to be a bit creative and free up some closet space or perhaps keep extra food in out of the way places. Many people find that the wasted space under a stairway can be easily converted to a convenient food pantry.

Canned and boxed foods are best stored in dry conditions where temperatures are kept fairly constant, ideally between 40°-60°F.[17] Keep food away from appliances or heating vents where undue heat could cause spoilage. Storage places could include a basement, under your stairs, closets, a utility room, kitchen cupboards, or even under your bed. Due to wide temperature swings, storing extra food in the garage or attic is probably a bad idea. Try to keep all but canned foods off the ground to minimize potential problems with bugs, rodents, or moisture. If you are tight on space, check out *www.shelfreliance.com*, a provider of slanted shelves that allow efficient storage of canned foods, and make rotation a snap.

If you have opted for a larger food storage plan, then you will need to do your homework and learn important storage methods; most of which depend on storing bulk food in airtight containers. Plastic buckets and mylar-foil bags are effective at keeping out air, bugs, and light—all of which can compromise your food cache. Don't store food in garbage bags because the bags may contain pesticides and are made of plastic resins that are not recommended for contact with food.[18] A good deal of advice regarding large food plans can be found in Spigarelli's *Crisis Preparedness Handbook.*[19]

Food storage shelves *(courtesy of Shelfreliance.com)*

Refrigerator thermometer *(courtesy of Taylor)*

When storing refrigerated foods, the temperature should be kept between 34°-40°F.[20] Cool temperatures help slow the growth of bacteria, extending the shelf life of foods. To ensure that you have the proper refrigerator temperature, check it in several places using a small refrigerator thermometer. Make sure that the highest reading is no more than 40°F. Here's your first big investment—a $10 thermometer!

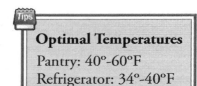

Optimal Temperatures
Pantry: 40°-60°F
Refrigerator: 34°-40°F
Freezer: 0°-5°F

Frozen foods should ideally be kept at 0°F but never above 5°F.[20] Once again, you can use the refrigerator thermometer to ensure you have the proper setting. As a rule of thumb, if your freezer can't keep a block of ice cream brick solid, then it is set too high. It is a good idea to put a freeze date on your freezer foods if they aren't already dated by the store. Also, freeze foods in moisture-vapor-proof packages, such as zippered freezer bags, or airtight freezer containers. Poorly sealed freezer packaging can lead to freezer burn.

HOW LONG WILL STORED FOOD LAST?

You might think that if you limit your food supply to 30 days, then spoilage won't be of much concern. While it is true that a smaller storage plan does remove much of this worry, there are still several reasons why every prepper should be aware of the shelf life of different foods:

1. During a crisis, food poisoning can be potentially life threatening. Knowledge of food safety is therefore critical.
2. Even with a modest 30-day food plan, spoilage can still affect some of your supplies (e.g., thawed meats, fresh produce, dairy).
3. In situations where you are forced to scavenge for food, you will need to assess whether that food is safe to eat.
4. If you ever decide to adopt a larger food storage plan, shelf life is critical to planning your stockpile and establishing a rotation schedule.

The first thing to know about shelf life is that it is affected by many factors and is therefore impossible to precisely predict. Food can degrade and become dangerous to consume due to either microbial or non-microbial causes. Some forms of microbial growth are easily seen, such as the fuzzy mold that grows on food that sits in your refrigerator for too long. Other types of bacteria are more difficult to detect. Eating contaminated food is dangerous because it can cause a variety of different foodborne illnesses, a discussion of which is given in the next section. The time it takes for microorganisms to affect foods depends on the particular methods used in production as well as storage conditions.

Non-microbial spoilage includes moisture gain or loss, chemical effects leading to changes in color or flavor, light-induced rancidity or vitamin loss, and physical damage (e.g., bruising of fruits/vegetables, denting of cans).[21] Beyond spoilage, food can also be ruined by rodents or insects.

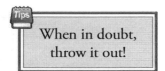

When in doubt, throw it out!

Foods can develop an unpleasant odor, flavor, texture, or appearance due to bacterial spoilage. Don't initially taste test food to see if it's spoiled. Rather, begin by checking it for bulging, bubbles or foam, mold, or cloudiness. If it looks okay, check to see if it smells sour, cheesy, fermented, or putrid. Next, touch the food to see if it has become slimy. Finally, if it passes the visual, smell, and touch tests, taste a small quantity to see if it tastes okay. If the food has become sour, bitter, chalky, or mushy, spit it out immediately.[19]

If any of the above characteristics are noted, the food should be immediately discarded. Your senses are trying to warn you! Food can also be mishandled, causing bacteria to grow rapidly. A well known example is that leaving hot dogs sitting in the sun for several hours at a picnic will cause them to spoil prematurely.

Food that is properly handled and stored at 0°F will remain safe almost indefinitely. Only the quality of food (e.g., tenderness, flavor, aroma, juiciness, and color) suffers from lengthy freezer storage.[23] This rule applies to freezer burn as well. Freezer burn occurs when air reaches the food's surface and dries the product, usually due to the food not being wrapped in air-tight packaging. Freezer-burned food *is* safe to eat, but for optimal taste, it is recommended that you cut away the affected areas before or after cooking.

> **Tips**
> Food kept frozen at 0°F can remain safe almost indefinitely.

For additional information on food safety, check out the Department of Health and Human Services website, *www.foodsafety.gov*. It contains useful information regarding food safety, food poisoning, and current recalls.

If you should ever become ill and suspect it is due to food contamination, save the packaging materials. Label any remaining food as "Dangerous," and freeze it for future examination by health officials. If the food is a meat or poultry product, call the USDA Meat and Poultry Hotline at 1-888-674-6854. If it is something other than meat or poultry, or is from a restaurant, contact your local health department. To locate the phone number for your local health department, consult the Health Guide USA website.[24]

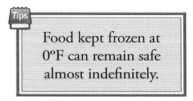

> **Tips**
> If you become ill due to food contamination, contact the USDA or local health department.

FOODBORNE ILLNESSES

In times of crisis, you definitely don't want illness added to your list of troubles. Traveling to a doctor may prove difficult or even impossible, and medical facilities are likely to be overwhelmed. Having to rough it out during a disaster can be bad enough, but doing so while hovering over the toilet can make it nearly intolerable. The following sections provide brief descriptions of seven important foodborne illnesses found in the United States (and throughout the world). The purpose of reviewing them is to become familiar with their symptoms and likely causes. This information is drawn from several references.[25,26,27,28,29,30] Later, some simple but effective ways to prevent food poisoning are discussed.

SALMONELLA

How you get it: *Salmonella* is a rod-shaped motile bacterium that occurs widely in animals, especially poultry and swine. *Salmonella* is frequently contracted by eating food contaminated with animal or human feces (a.k.a. fecal-oral contamination). Contamination often occurs when the food preparer doesn't

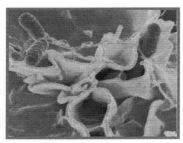
Salmonella *(NIAID photo)*

wash his hands after using the toilet. Other possible sources include undercooked eggs, poultry, or meat; and unpasteurized milk or juice.

What it causes: Onset of salmonellosis typically occurs within 1 to 3 days. Acute symptoms can last a few days to a week and may include nausea, vomiting, abdominal cramps, diarrhea, fever, and headache. Additionally, arthritic symptoms may follow 3 to 4 weeks after the onset of acute symptoms. The Food and Drug Administration (FDA) estimates that 2-4 million cases of salmonellosis occur annually in the United States.

CAMPYLOBACTER JEJUNI

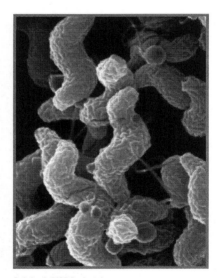
C. jejuni *(USDA photo)*

How you get it: *Campylobacter jejuni* is a slender, rod-shaped bacterium that causes campylobacteriosis. Most cases result from eating raw or undercooked poultry, or from cross-contamination. The bacteria are often found in healthy cattle and birds. Surveys show that 20-100% of retail chickens are contaminated. However, properly cooking chicken *will* kill the bacteria—see Table 3-2. *C. jejuni* is also contracted from contaminated water and unpasteurized milk.

What it causes: Onset of symptoms usually occurs 2 to 5 days after ingestion of the contaminated food or water. Symptoms can include diarrhea, fever, abdominal pain, nausea, headache, and muscle pain. The illness generally lasts 7 to 10 days, but relapses are also common. *C. jejuni* may also spread to the bloodstream and cause a life-threatening infection. Surveys have shown that *C. jejuni* is the leading cause of bacterial diarrheal illness in the United States. The FDA estimates that there are at least 2-4 million cases annually.

CLOSTRIDIUM PERFRINGENS

How you get it: *Clostridium perfringens* is a sporeforming, rod-shaped bacterium that is widely distributed in the environment, and can be found in decaying vegetation, marine sediment, and the intestines of humans and domestic animals. Consumption can cause *C. perfringens* food poisoning. Eating of improperly prepared meat or meat dishes, such as gravy or stew, is the main source of infection. Food may be either undercooked or prepared too far in advance of consumption. This bacterium is often referred to as the "cafeteria germ" because most outbreaks occur at institutional kitchens in hospitals, school cafeterias, prisons, and nursing homes.

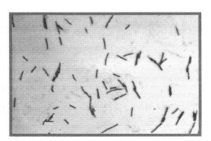
Clostridium perfringens *(CDC photo)*

What it causes: Symptoms usually start 8 to 22 hours after consumption and include abdominal cramps and diarrhea. The illness is usually over within 24 hours, but less severe symptoms may persist for 1 to

2 weeks. Symptoms may last longer in the elderly. *C. perfringens* is the third most common foodborne illness in the United States with an estimated 250,000 people affected annually.

E. COLI

How you get it: *Escherichia coli* are a large and diverse group of bacteria. Some strains are harmless, while others can cause diarrhea, urinary tract infection, or respiratory illness. The most commonly identified Shiga toxin-producing *E. coli* (STEC) is *E. coli O157*. When you hear reports about outbreaks of *E. coli*, they are usually referring to this particular strain. Infections are caused by fecal-oral contamination (i.e., ingesting small quantities of human or animal feces). This can be caused by consuming contaminated food or water, drinking unpasteurized milk, or coming into direct contact with human or animal feces. Other possible sources include undercooked hamburger and unpasteurized milk or juice.

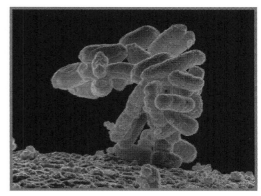
Escherichia coli *(USDA photo)*

What it causes: Symptoms of STEC infections include severe stomach cramps, diarrhea (sometimes bloody), vomiting, and a mild fever. Most people recover within a week. Onset of symptoms may begin anywhere from 1 to 10 days after infection. Very young victims may develop hemolytic-uremic syndrome (HUS), which may cause acute kidney failure. The FDA estimates that 70,000 cases of *E. coli O157* occur each year in the United States.

SHIGELLA

How you get it: *Shigella* are non-sporeforming, rod-shaped bacteria that cause shigellosis. Shigellosis is primarily a human disease, not found in animals except monkeys and chimpanzees. There are several different types of *Shigella* bacteria, but Groups B and D account for almost all the cases in the United States. The organism is frequently found in water polluted by human feces. The disease is most often spread through fecal-oral contamination. It is particularly common with toddlers who are not fully toilet-trained. Shigellosis can also be contracted by eating contaminated food. Uncooked salads (e.g., potato, tuna, shrimp, maca-

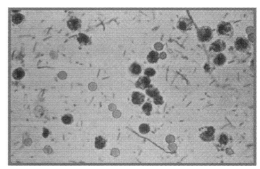
Shigella *(CDC photo)*

roni, and chicken), raw vegetables, milk and dairy products are all foods that may be contaminated. Water can also become contaminated if sewage runs into it, or if someone infected with shigellosis swims in it.

What it causes: Symptoms usually occur 12 to 50 hours after ingestion and may include abdominal pain, cramps, diarrhea, fever, vomiting, blood/pus/mucus in the stool, and difficulty defecating.

Shigellosis typically resolves in 5 to 7 days. A severe infection can occur in children under age 2 resulting in high fever and seizures. An estimated 300,000 cases of shigellosis occur annually in the United States.

CLOSTRIDIUM BOTULINUM

How you get it: *Clostridium botulinum* is a spore-forming, rod-shaped bacteria that produces a potent neurotoxin. Seven types of botulism are recognized (A-G), of which types A, B, E, and F cause human botulism. Food botulism is a rare but severe type of food poisoning caused by the ingestion of foods containing the potent neurotoxin formed during growth of the organism. The toxin is heat-sensitive and can be effectively destroyed if heated to 176°F for a minimum of 10 minutes. Sausages, meat products, canned vegetables, and seafood products are the most frequent vehicles for

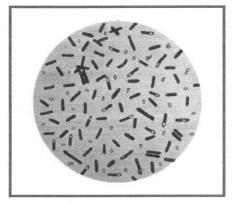

Clostridium botulinum (CDC photo)

human botulism. Most of the outbreaks in the United States are associated with inadequately processed home-canned foods, but commercially produced foods have also been involved (e.g., Castleberry's hot dog chili in 2007).

What it causes: The botulism toxin causes paralysis by blocking motor nerve terminals. The flaccid paralysis progresses downward, usually starting with the eyes and face, moving to the throat, chest, and extremities. When the diaphragm and chest muscles become paralyzed, respiration is impaired and death from asphyxia can result. The incidence of this type of poisoning is very low, but the mortality rate is high. Annually, there are on average 10 to 30 outbreaks of food botulism in the United States.

STAPHYLOCOCCUS AUREUS

How you get it: *Staphylococcus aureus* is a spherical bacterium capable of producing a highly heat-stable enterotoxin that causes illness in humans. The most common way of contracting Staphylococcal food

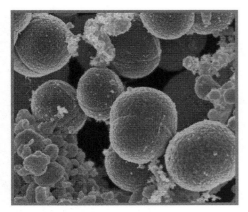

Staphylococcus aureus (NIAID photo)

poisoning is through contact with food workers who carry the bacteria, or by eating contaminated foods. It is usually associated with meat products, poultry and egg products, homemade salads (e.g., egg, tuna, chicken, potato, macaroni), cream pastries and pies, and dairy products.

What it causes: Onset of symptoms is usually rapid and may include nausea, vomiting, retching, abdominal cramping, and exhaustion. In severe cases, headache, muscle cramping, and changes in blood pressure and pulse rate may occur. Recovery may require a few days. The incidence rate of staphylococcal food poisoning is unknown due to poor reporting of the illness and frequent misdiagnosis.

AVOIDING FOOD POISONING

The CDC estimates that 325,000 hospitalizations and 5,000 deaths occur every year in the United States due to foodborne illnesses.[27] To help combat food poisoning, a partnership between the USDA Food Safety and Inspection Service and the FDA Center for Food Safety and Applied Nutrition established the *Fight BAC!* campaign.[31] Their recommendations are obvious, yet important. Millions of people get food poisoning each year from not following these simple steps:

Avoiding Food Poisoning
➢ Clean
➢ Separate
➢ Cook
➢ Chill

1. **Clean**—Wash hands, kitchen utensils, and surfaces with hot, soapy water. Many foodborne illnesses occur due to fecal-oral contamination. This is often caused by someone with "dirty" hands touching the food, a utensil, or a kitchen surface. In addition to washing, a weak bleach solution (1 tbsp bleach to 1 gallon water) can also be used to sanitize cutting boards or utensils. Always rinse fresh fruits and vegetables with tap water, scrubbing lightly.
2. **Separate**—To prevent cross-contamination, separate raw meat, poultry, and seafood from ready-to-eat foods. Use separate cutting boards and knives. Never place cooked or ready-to-eat foods on a plate or surface that had raw meat on it previously.
3. **Cook**—Cook foods to safe internal temperatures. Table 3-2 gives minimum safe internal temperatures for various food types. Use a clean thermometer to test temperatures. Color is *not* a reliable indicator of doneness. Also, keep cooked foods hot until they are eaten. Bring sauces, soups, and gravies to a boil when reheating.

Table 3-2 Minimum Internal Temperatures[30,32]

Category	Food	Temp. (°F)
Ground Meat & Meat Mixtures	Beef, Pork, Veal, Lamb	160
	Turkey, Chicken	165
Fresh Beef, Veal, Lamb	Steaks, Roasts, Chops	145
Poultry	Chicken, Turkey, Duck, Goose	165
Pork, Ham	Fresh Pork	160
	Fresh Ham	160
	Precooked Ham	140
Eggs & Egg Dishes	Eggs	Firm
	Egg dishes	160
Leftovers & Casseroles	Leftovers	165
	Casseroles	165
Seafood	Fish	145
	Shrimp, Lobster, Crab	Flesh opaque

4. **Chill**—Refrigerate leftovers and takeout foods within 2 hours. Refrigerate perishables as quickly as possible after purchase. Keep cold foods cold (34°-40°F for refrigerated, 0°-5°F for frozen). Never defrost food at room temperature. Most food can be thawed in one of three ways: in cold water, in the microwave, or in the refrigerator. Food thawed in the microwave or in cold water should be cooked immediately.

Having your family follow these four simple steps will significantly reduce their chances of contracting food poisoning. A few additional suggestions that will further reduce the likelihood of contamination are listed below:

- Make sure that poultry is adequately cooked. The meat juice should run clear, not pink.
- Don't eat raw or undercooked eggs; or homemade products with raw eggs, such as salad dressings, mayonnaise, ice cream, cookie dough, and frostings.
- Don't consume unpasteurized milk or juices. Nearly everything you buy at the store is pasteurized.
- Avoid swallowing water from swimming pools, lakes, ponds, or streams.
- If your tap water has been declared unsafe, cook or peel all fresh produce.
- Follow strict hygienic practices when home canning, especially with low-acid foods, such as green beans and corn. Boil all home-canned foods for 10 minutes before eating.

PRODUCT DATING

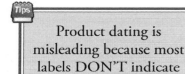

Product dating is misleading because most labels DON'T indicate food safety dates.

Except for infant formula and some baby food, product dating is generally not required by federal regulations. State food dating requirements vary. If calendar dating is used (a.k.a. "open dating"), it will include both month and day. Also, a phrase must be included to indicate the meaning of the date, such as "use by" or "best before."

Fortunately, manufacturers have voluntarily labeled the vast majority of foods with freshness dates. However, there are several different types of phrases associated with these dates, and this can cause confusion about what is safe to eat. Take a moment to go through your cupboard and locate the dates on various types of food products. They are not always easy to find—sometimes hidden under flaps or stamped in inconspicuous places. An understanding of the following food dating terms will help you put meaning to the dates.[23,33,34]

"USE BY" OR "BEST IF USED BY" OR "BEST BEFORE" OR "BEST" OR "USE" DATE

This is the last date a product is deemed to be at its best. This date recommendation is based solely on flavor or quality. It is *not* a food safety date, however, highly perishable foods may present a safety risk if consumed after this date.

"SELL BY" DATE

This date indicates the last day the product can be sold. The "sell by" date tells the retailer how long to display a product. The date is quality driven, and is *not* a food safety date.

"EXPIRATION" DATE

The expiration date indicates the last date a food should be eaten or used. This *is* a food safety date.

Foods that are not already labeled by the manufacturer should be manually labeled following the guidelines given in shelf life tables (see next section).

SHELF LIFE TABLES

Many published tables detail the expected shelf life of pantry foods. The tables presented here are based on input from several sources.[20,25,30,35,36,37] Keep in mind that shelf life tables generally represent conservative estimates, and in many cases, food will last longer. For example, there are tales of people eating canned food after 20 or more years of storage. That's not something to be recommended, but you get the point.

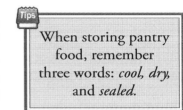

When storing pantry food, remember three words: *cool, dry,* and *sealed.*

The most important recommendation regarding pantry food is to keep it cool, dry, and tightly sealed. Heat and moisture are bacteria's best friends and will speed up the spoilage process. Air, on the other hand, will dry out exposed food and cause flavor and color changes due to oxidation.[38]

Another way to extend the shelf life of food from the grocery store is to buy the freshest products available. Check the dates on packages whenever in doubt. Dusty cans or icy deposits may indicate old stock. Don't purchase dented or bulging cans as this may indicate spoilage. Also, don't purchase frozen goods whose boxes are discolored due to having been wet as this might indicate that they were inadvertently thawed during transport or during stocking. Feel free to cherry-pick the foods from the back of the shelf to find those that are freshest.

Table 3-3 Cupboard Storage Shelf Life Table

Food	Recommended Storage Time
Baking powder	18 months
Baking soda	2 years
Beans/peas, dried	1 year (airtight container)
Biscuit, brownie, muffin mixes	9 months (airtight container)
Bouillon cubes, granules	1-2 years
Bread crumbs (dry)	6 months
Bread	3-5 days
Cakes	
- purchased or prepared	1-2 days
- mixes	1 year

Food	Recommended Storage Time
Canned foods, unopened	2+ years
Catsup, chili sauce	
- unopened	1 year
- opened, sealed well	1 month (refrigerate)
Cereal, ready to eat	
- unopened	6-12 months
- opened, sealed well	2-3 months
Cheese, grated Parmesan	
- unopened	10 months
- opened, sealed well	2 months (refrigerate)
Chocolate	
- unsweetened	18 months
- semi-sweet	18 months
- syrup, unopened	2 years
- syrup, opened, sealed well	6 months (refrigerate)
Cocoa	Indefinitely
Cocoa mixes	8 months
Coconut, shredded or canned	
- unopened	1 year
- opened, sealed well	6 months (refrigerate)
Coffee	
- cans or bags, unopened	2 years
- cans or bags, opened	2 weeks (airtight container)
- instant, unopened	1-2 years
- instant, opened	1-2 months (airtight container)
Coffee creamers, dry	
- unopened	9 months
- opened, sealed well	6 months
Cookies	
- homemade	2-3 weeks (airtight container)
- packaged	2 months
Corn syrup	up to 3 years (refrigerate to extend life)
Cornmeal	1 year (airtight container)

Food	Recommended Storage Time
Cornstarch	18 months (airtight container)
Crackers	3 months
Fish, canned	3-5 years
Flour	
- white	6-8 months (airtight container)
- whole wheat	6-8 months (airtight container)
Frosting	
- canned, unopened	3 months
- mix, unopened	8 months
Fruit	
- dried	6 months (airtight container)
- fresh	3-5 days (longer in refrigerator)
Fruit juice	
- canned	9 months
- juice/drink boxes	9 months
Gelatin	18 months
Grits	1 year (airtight container)
Honey	1 year - if crystallized, warm opened jar in hot water.
Hot roll mix	18 months
Hot sauce	2 years
Jellies/jams	1 year (refrigerate after opening)
Marshmallows	2-3 months (airtight container)
Mayonnaise	
- unopened	4-6 months
- opened	2 months (refrigerate)
Milk	
- condensed, unopened	1 year
- evaporated, unopened	1 year
- nonfat dry	6 months (airtight container)
- shelf-stable, unopened	2-3 months
Molasses	
- unopened	2 years
- opened	6 months (refrigerate)

Food	Recommended Storage Time
Mustard, prepared yellow - unopened - opened	 2 years 6-8 months (refrigerate)
Nuts - in shell, unopened - vacuum can, unopened - vacuum can, opened	 4-6 months 1-3 years 3 months
Pancake syrup - unopened - opened	 18 months 3-4 months (refrigerate)
Pasta - spaghetti, macaroni, etc. - egg noodles	 2 years (airtight container) 6 months (airtight container)
Peanut Butter - unopened - opened	 6-9 months 2-3 months Natural peanut butter must be refrigerated after opening.
Pie crust mix	8 months
Pies and pastries	2-3 days Refrigerate whipped cream, custard fillings.
Popcorn, unpopped	2 years (airtight container)
Potatoes - instant mix - fresh	 6-12 months (airtight container) 2-4 weeks (keep in dark place)
Powdered drink mix	18-24 months
Pudding mix	1 year
Rice - white - brown, wild - flavored or herb	 2 years 6-12 months 6 months
Salad dressings - bottled, unopened - bottled, opened - made from mix	 10-12 months 3 months (refrigerate) 2 weeks (refrigerate)

Food	Recommended Storage Time
Shortenings (solid)	8 months
Sauces and gravy mixes	6-12 months
Soft drinks	6 months
Soup mixes, dry	1 year

Soy sauce
  - unopened
  - opened, sealed well
	
	3 years
	9 months

Spices and herbs	6 months-2 years depending on type (airtight container)

Sugar
  - brown
  - confectioners
  - granulated
  - artificial sweeteners
	
	4 months (airtight container)
	18 months (airtight container)
	2 years
	2 years

Syrup
  - unopened
  - opened
	
	18 months
	1 year (refrigerate)

Tea
  - bags
  - instant
  - loose
	
	18 months (airtight container)
	3 years
	2 years (airtight container)

Toaster pastries	2-3 months (airtight container)

Vanilla (and other extracts)
  - unopened
  - opened, sealed well
	
	2 years
	1 year

Vegetables, fresh
  - onions
  - potatoes
  - sweet potatoes
	
	2 weeks
	2-4 weeks
	1-2 weeks

Vegetable oils
  - unopened
  - opened
	
	6 months
	1-3 months

Vinegar
  - unopened
  - opened
	
	2 years
	1 year

REFRIGERATOR / FREEZER STORAGE

Just as there are storage guidelines for shelved foods, there are also recommendations regarding the usable lifetime of refrigerated and frozen foods. The storage times listed here assume that food has been stored in appropriate airtight containers, and that temperatures are maintained properly (i.e., 34°-40°F for refrigerated and 0°-5°F for frozen). Note that the freezer storage times are provided for optimum quality only since food can be stored almost indefinitely in a properly cooled freezer.

Table 3-4 Refrigerator and Freezer Storage Shelf Life Table

Product	Refrigerator	Freezer
Eggs		
- Fresh, in shell	4-5 weeks	Don't freeze
- Raw yolks or whites	2-4 days	1 year
- Hard-boiled	1 week	Don't freeze
- Egg substitutes, unopened	10 days	1 year
- Egg substitutes, opened	3 days	Don't freeze
- Mayonnaise	2 months	Don't freeze
TV Dinners, frozen casseroles	Keep frozen until ready to eat	3-4 months
Deli & vacuum-packed products		
- Egg, chicken, macaroni salads	3-5 days	Don't freeze
- Stuffed pork, lamb, or chicken	1 day	Don't freeze
- Store cooked convenience meals	3-4 days	Don't freeze
- Vacuum-packed dinners	2 weeks	Don't freeze
Raw meats and poultry		
- Ground beef, turkey, pork, veal	1-2 days	3-4 months
- Steaks	3-5 days	6-12 months
- Pork chops	3-5 days	4-6 months
- Roasts	3-5 days	4-12 months
- Tongue, kidneys, liver, heart	1-2 days	3-4 months
- Chicken, turkey—whole	1-2 days	6-12 months
- Chicken, turkey—parts	1-2 days	6-9 months
- Giblets	1-2 days	3-4 months
- Bacon, store-packed	7 days	1 month
- Bacon, vacuum-packed	1-2 months	6-12 months
- Sausage, raw, store-packed	1-2 days	3 months
- Sausage, factory-packed		
unopened	1-3 months	6-12 months
opened	1-2 days	3 months

Product	Refrigerator	Freezer
Cooked meats and poultry		
- Poultry		
Fried chicken	3-4 days	4 months
Chicken nuggets	1-2 days	1-3 months
Cooked poultry dishes	3-4 days	4-6 months
- Beef, pork, veal		
Cooked meat	3-4 days	2-3 months
Cooked meat dishes	3-4 days	2-3 months
- Gravy and meat broth	1-2 days	2-3 months
- Soups and stews	3-4 days	2-3 months
- Ham, fully cooked		
whole	7 days	1-2 months
half	3-5 days	1-2 months
sliced	3-4 days	1-2 months
canned	6-9 months	Don't freeze
Fish and shellfish		
- Lean fish	1-2 days	6 months
- Fatty fish	1-2 days	2-3 months
- Canned fish, opened	3-4 days	2 months
- Shellfish	1-2 days	3-6 months
- Cooked fish	3-4 days	4-6 months
- Smoked fish	14 days	2 months
Hot dogs and lunch meats		
- Hot dogs		
opened	1 week	1-2 months
unopened	2 weeks	1-2 months
- Lunch meats		
opened	3-5 days	1-2 months
unopened	2 weeks	1-2 months
Breads, pastries, and cakes		
- Bread/rolls		
unbaked	2-3 weeks	1 month
baked	Don't refrigerate	2-3 months
- Cookies		
dough	3 months	3 months
baked	Don't refrigerate	6-12 months

Product	Refrigerator	Freezer
Breads, pastries, and cakes *(continued)*		
- Cakes		
batter	Don't refrigerate	1 month
baked, unfrosted	3 days	2-4 months
baked, frosted	3 days	1 month
- Fruit pies		
unbaked	1-2 days	2-4 months
baked	2-3 days	6-8 months
Cheese		
- Cottage, ricotta	1-2 weeks	4 weeks
- Soft cream, opened	5 days	Don't freeze
- Hard and wax-coated	1-2 months	6-8 months
- Sliced	2 weeks	Don't freeze
- Parmesan, Romano—grated	2 months	Don't freeze
- Processed	3-4 weeks	6-8 months
Fruits and vegetables	Varies, based on type and freshness	6 months (blanching may be required)

LONG-TERM FOOD STORAGE

With all this talk of storing what you eat and eating what you store, you might think that there is never a need for food with an extended shelf life. On the contrary, long shelf life foods are useful for stocking food pantries that are infrequently accessed. A good example of this is food that might be stored by your disaster preparedness network to serve those who find themselves underprepared. Another example is a food bank established by a church or civic organization to be used for community emergency relief. Food caches such as these are not likely to be accessed under normal circumstances but could prove lifesaving when a serious disaster occurs.

Distributing emergency food rations *(U.S. Navy)*

One of the primary reasons individuals give for creating an infrequently-accessed food cache is to prepare for "the-end-of-the-world-as-we-know-it" (TEOTWAWKI) scenarios. These scenarios center around a high-impact, low-frequency (HILF) event occurring that completely breaks down modern civilization, leaving families to fend for themselves. Certainly these events are possible. A short list might include: a large solar flare, electromagnetic pulse, deadly pandemic, asteroid strike, supervolcano eruption, or nuclear war. Any one of these could certainly set our planet back a century or more. However, by definition, HILF events are highly unlikely. Therefore, it can be argued that TEOTWAWKI preparations should be made only *after* preparations are in place for more commonplace disasters. Also, it's a good idea for families to periodically eat from their emergency cache, allowing everyone to become familiar with the food before being forced to rely on it.

Regular canned food, such as vegetables, soups, or potted meats, can be used for an emergency food bank. However, you may find it difficult to store in bulk due to weight and size. Also, the shelf life of canned products may not be adequate for very long term storage. Over time, food deteriorates in quality and nutrients even when left unopened. Long-term storage is best achieved with highly-stable food products. With that said, canned food may still be edible for many years past its expiration date, albeit with reduced nutritional content. There is no one right answer as to how long a food will last because it is a strong function of the particular food type, storage conditions, and canning conditions. Given that canned foods typically have a shelf life of at least a few years, why bother with specialty items such as MREs, dehydrated, or freeze-dried products? The answer is found by better understanding these important emergency foods.

MEALS, READY-TO-EAT (MREs)

The military first delivered MREs to soldiers back in 1981 to replace the existing C-Rations. The MRE I was the first incarnation of a "meal in a bag" that has become the staple field ration not only for the military services but also for emergency relief activities. Meals, Ready-to-Eat have improved significantly since their first release and now include flameless heaters, freeze-dried coffee, Tobasco sauce, shelf-stable bread, biodegradable spoons, and even high-heat-stable chocolate bars. Every year, the least accepted menu items are discontinued, and new recipes are tried out. Not only are MREs convenient and tasty, they also contain about 1200 calories and meet the Office of the Surgeon General's nutritional requirements. This along with their stringent durability requirements, including surviving airdrops and extreme temperatures, make MREs a truly remarkable emergency food.

Meals, Ready-to-Eat *(US Army)*

Contrary to misinformation found on the internet, MREs have a shelf life of *three* years (assuming a storage temperature of 80°F). Many retailers claim that MREs have a shelf-life of 10+ years, but that is not true for modern MREs unless they are stored at very low temperatures. Studies have shown that in many cases MREs *are* safe

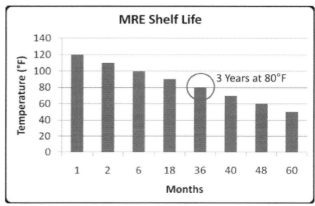

Shelf-life of MREs

Contents of a modern MRE *(U.S. Army)*

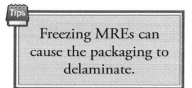

Tips

Freezing MREs can cause the packaging to delaminate.

to eat beyond the three year expected life. However, every food type deteriorates differently, so it is impossible to make a sweeping statement like *"MREs will last x years before making you sick."* With all food, let your senses be your guide. If the food smells or tastes bad, is discolored, or has changed consistency, toss it out.

Over the past few years, cases of MREs have been equipped with time-temperature indicators (TTIs) to help consumers know if the food has reached the end of its expected shelf life. The indicators are very effective at reducing waste and ensuring consistent food quality. The TTI starts off with a dark outer reference ring and a bright red inner circle (a.k.a. the bulls-eye). The center darkens over time, with the color change occurring more quickly at elevated temperatures. For example, if the MREs are stored at 80°F, the circle will take 36 months to darken to match the outer ring. Likewise, if the MREs are stored at a cooler temperature, the circle will take longer to darken. This time-temperature behavior makes the TTI an absolute reference for determining the freshness of the MREs. The lighter the inner circle, the fresher the food. When the inner circle matches the color of the outer ring, the MREs are considered to be at the end of their shelf life. If the inner circle is darker than the outer ring, the MREs are past their shelf life and should be carefully inspected prior to eating. Keep in mind that if you purchase individual MRE pouches or a case of MREs without a TTI, there is no way to determine the remaining shelf life of the food.

While there is still some question as to the legality of selling MREs to individual consumers, they are readily available. Besides getting them from friends or family members in the military, you can also find them at army surplus stores, guns shows, and on eBay. Beware that many of the MREs for sale are past their expiration date and may not be equipped with TTIs. Limit your buying to cases equipped with TTIs that clearly indicate the product's freshness. Many times, sellers on eBay will state the "inspection date" of the MREs. This is simply the projected three-year shelf life date and assumes storage at 80°F. If the food was stored in hotter conditions, the end-of-shelf-life will arrive sooner than this date. Likewise, if it was stored in cooler conditions, the end-of-shelf-life will arrive after the inspection date. For this reason, it is better to use the TTI to determine the freshness of the MREs rather than a projected expiration date. Also, be aware that many vendors sell non-military, prepackaged meals under the guise of being MREs. Some of these

(Fresh)   (At Shelf Life)   (Past Shelf Life)

Time-temperature indicators

products are perfectly fine for long term storage, but their food, packaging, and contents do not necessarily meet the standards set forth by the military.

Personal aside: I was serving as an Army paratrooper in 1981 when MREs were first introduced. It didn't take long to conclude that carrying heavy-duty pouches of food was a really good idea. For one thing, they could easily be carried in cargo pockets without bruising up your leg every time you dove to the ground. They also came with chewing gum and a clean eating utensil, and they didn't require using the notoriously slow P38 can opener. As for the food in those early MREs, it could best be described as tolerable.

DEHYDRATED

Many people have experience with dehydrated foods, whether it be from snacking on dried fruits or using dehydrated beans and vegetables in recipes. Dehydration is the process of removing most of the water from food, leaving it lighter, smaller, and with a much longer shelf life. The food can be eaten

THRIVE dehydrated food in one gallon cans *(courtesy of Shelfreliance.com)*

directly or rehydrated by allowing it to sit in hot water for several minutes. Keep in mind that storing dehydrated or freeze-dried foods is of little benefit unless you also have access to sufficient potable water. Don't make the common mistake of focusing heavily on food storage while neglecting to also have an adequate backup water plan.

The texture of dehydrated food is typically kind of chewy because the dehydration process is completed slowly and at warm temperatures. If you wish to make your own dehydrated food, home-based systems are readily available and easy to use. Most dehydrated food is made as a single item, such as mushrooms, apple slices, or beans. This is because the dehydration process does not lend itself well to more complex one-package meals. By removing most of the water from the food, the shelf life is significantly increased—perhaps 10-15 years when stored unopened at room temperature.

FREEZE-DRIED

Freeze drying is a process in which food is flash frozen, the ice evaporated away, and the food sealed in a vacuum package. This rapid freezing and sealing process requires sophisticated equipment and is not able to be done easily at home. Since nearly all of the water is removed, the food is very lightweight, and the shelf life is significantly improved (e.g., 5-7 years for pouches, 10-25 years for larger cans). Freeze-dried products are quickly rehydrated by adding hot water.

Since freeze drying is a fast, uniform process, storage of more complex foods is possible. Entire meals are often freeze dried in a single package, such as spaghetti, clam chowder, beef stroganoff, and chicken

Mountain House freeze-dried food pouches

Dehydrated and freeze-dried strawberries *(courtesy of Faith E. Gorsky of AnEdibleMosaic.com and Freeze Dried Food Suppliers)*

stew. This meal-in-a-package functionality gives freeze-drying the advantage over dehydration when storing for emergency purposes.

Freeze-dried food tends to keep its original appearance better than dehydrated, and that leads many people to say that it looks more appetizing. As a simple example, the figures above show a side-by-side comparison of dehydrated and freeze-dried strawberries.

A more complete comparison of MREs, dehydrated, and freeze-dried products is given in Table 3-5. Each food type has its respective advantages and disadvantages. Meals, Ready-to-Eat are convenient and do not require water but have the shortest shelf life. Dehydrated food is the cheapest option but doesn't offer meal-in-a-package functionality. Freeze-dried food has a very long shelf life and food complexity

Table 3-5 Comparison of MREs, Dehydrated, and Freeze-Dried Food

Metric	MREs	Dehydrated	Freeze-Dried
Food complexity	Full meals	Single food	Single food or full meals
Rehydration time	Not required	Longer	Shorter
Appearance	Natural	Shriveled	Natural
Texture	Processed	Chewy	Crispy/Soupy
Cost	Most expensive	Least expensive	More expensive
Weight	Heaviest	Medium	Lightest
Typical shelf life	3+ years	10 to 15 years	7 to 25 years

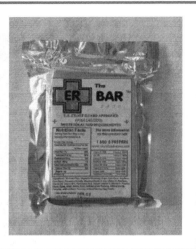

Taste test samples

but requires rehydration. Regardless of what long-term food type you select, be sure to have a clear understanding of its respective place in your overall food storage plan.

Personal aside: One thing that is often overlooked when selecting emergency rations is just how edible the food is. Any of the options discussed above will certainly keep you alive. But remember, your goal is not only to survive but also to maintain a reasonable quality of life. Your family has to be willing to eat the food that you store. Keeping this in mind, my family conducted a very unscientific comparison of emergency foods by taste testing an assortment of military MREs, freeze-dried dinners from Alpine Aire Foods, and an emergency food bar from Vita-Life Industries. Admittedly, the food bar was not meant to be eaten except during extreme situations (e.g., stranded at sea), but we thought it would be fun to throw in. The food was then ranked according to three metrics: appearance, taste, and consistency. A summary of the results is provided in Table 3-6.

Table 3-6 Unscientific Taste Test of Emergency Foods

Food Item	Appearance	Taste	Consistency
Freeze-dried Meals	GOOD - soupy; less processed looking than MREs	GOOD - flavor varies significantly by meal type	GOOD - similar to a thick soup or chili; controlled by amount of water added
MREs	FAIR - similar to canned food; very processed looking	FAIR - flavor varies significantly by meal type	GOOD - like canned raviolis
Emergency Food Bars	POOR - like a solid block of Soylent Green	POOR - single taste; slight vanilla-lemon flavor	FAIR - dry and crumbly; uniform throughout

FOOD SAFETY AND TERRORISM

The terrorist attacks of September 11, 2001, brought renewed focus on shoring up the country's defenses. In 2002, Congress and the president together enacted the Public Health Security and Bioterrorism Preparedness and Response Act.[39] The law is worth reading and can be found at the FDA website on Food Defense and Terrorism.[40]

The Bioterrorism Act is divided into five sections:

- National Preparedness for Bioterrorism and Other Public Health Emergencies
- Enhancing Controls on Dangerous Biological Agents and Toxins
- Protecting Safety and Security of Food and Drug Supply
- Drinking Water Security and Safety
- Additional Provisions

There are numerous goals and provisions outlined in the document. How well they will actually protect the population against a terrorist attack on the food, water, or drug supply is anyone's guess.

Radioactive Effects on Food

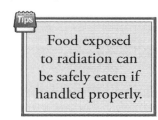

Food exposed to radioactive fallout requires cleaning prior to eating. Radioactive particulates should be rinsed off canned foods before opening. Boxed foods can also be cleaned off and consumed if the food is sealed inside an airtight bag. Produce should be washed and peeled to remove any radioactive fallout. Foods that can't be thoroughly cleaned, such as opened cans or breads, should be discarded.[41] Radioactive contamination can also be passed through animals into the food supply system, such as cows eating contaminated grass or seed and then passing on the contamination through their milk.

Food exposed to radiation can be safely eaten if handled properly.

MISCELLANEOUS

SPECIAL NEEDS

When stockpiling food, it is important to consider any special needs that your family may have. This includes baby food, pet food, and food for those with dietary restrictions. See *Chapter 16: Special Needs* for more information.

MULTIVITAMINS

There tends to be two camps of people: those who swear by multivitamin supplements, and those who believe they are a waste of money. The people in favor of daily multivitamins will argue that the diet of the average American is sorely lacking in important nutrients. Those opposing the use of vitamins will cite studies that show that high levels of some vitamins and minerals have been shown to cause health problems as well as interfere with medications.[42,43,44,45]

Fortunately, almost everyone seems to agree that daily multivitamins should be taken during times of bodily stress, such as when pregnant, confined indoors for an extended period, or eating a restricted diet. Therefore, in situations where food quantity and selection might be limited—such as during a disaster— taking a multi-vitamin supplement seems both reasonable and prudent.

PETS

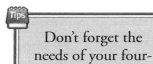

Don't forget the needs of your four-legged friends.

For most people, pets are members of the family. Don't forget to keep the same 30-day supply of food for your pets. For dogs and cats, this isn't usually difficult—stocking a spare bag of dry food in the cupboard, or rotating canned food. Of course, if you run out of pet food, it may be possible (depending on the type of animal) to improvise by feeding your pets "scraps" from your meals. It might not be the healthiest diet, but it beats starving. Fortunately, most animals quickly adapt to scavenger mode when the need arises.

A much more complete set of recommendations regarding pets is given in *Chapter 16: Special Needs.*

NON-FOOD ITEMS

There is certainly no need to go out and purchase a new set of kitchen supplies as part of your preparations, but you should at least do a quick inspection of your utensils and pots and pans. During a disaster would be a poor time to realize that your only manual can opener is being held together by zip ties.

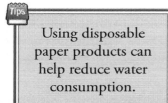

Using disposable paper products can help reduce water consumption.

Also, take a good inventory of the consumable non-food items in your kitchen, making sure that you have an adequate supply. This list would include such items as aluminum foil, plastic wrap, paper products, plastic utensils, plastic storage bags and containers, and napkins.

Most families have a shelf lined with cookbooks. It would be wise to select one of your favorites (perhaps an all-purpose cookbook from Betty Crocker, Better Homes, or Mark Bittman) and familiarize yourself with different recipes. If you are stranded at home for an extended period, you may need to improvise by pulling together dishes using only the particular ingredients you have on hand. Learning a few shortcuts and substitutions now means you won't find yourself staring at a cupboard full of supplies during a crisis wondering how best to use them. Also, knowing how to cook using your microwave or barbeque grill can be especially handy for situations with limited electrical power (see *Cooking* in *Chapter 8: Heating/Cooling*).

DP PLAN EXAMPLE

Table 3-7 Sample DP Plan Entry - Food

Need: Food			
Danger	**Goals**	**Needs**	**Implementation**
Shortage	Feed family for 30 days without resupplying	Sufficient food for three meals a day for a family of five	Stock a minimum thirty breakfasts, lunches, and dinners for five people, along with fruit juices and shelf stable milk.
		Additional food to supplement a needy elderly neighbor	Create a pantry by installing shelving in closet under stairs.
			Keep a few days of food in the kitchen cabinets, and the bulk of pre-packaged and canned foods in the pantry.
			Keep a food storage list inside pantry door.
			Rotate food with each weekly purchase.
		Pet food for household pets	Stock two 20-lb bags of dry dog food.
	Feed those in network who find themselves underprepared	200 spare meals with a long shelf life	Stockpile 200 MREs in local church basement.
Loss of refrigeration	Don't compromise food supply with power outage	Bulk of food stockpile that doesn't require refrigeration	Limit refrigerated or frozen food to no more than 7 of the 30-day supply.
			Keep additional frozen foods in the garage freezer.
			Use water bottles as necessary to keep refrigerator and freezers full.

Quick Summary - Food

➢ Store a minimum of 30 days of non-perishable food.

➢ Consult the USDA Choose My Plate dietary guidelines when deciding what to store.

➢ Store what you eat, and eat what you store.

➢ Rotate your food by placing the newest to the back.

➢ Always keep pantry food cool, dry, and tightly sealed.

➢ Set your refrigerator 34°–40°F and your freezer from 0°–5°F for maximum shelf life.

➢ Sweeteners, oils, and seasonings are not critical to survival but can ease the hardship by making food more enjoyable.

➢ Consult product labels and shelf-life tables to determine how long foods will store safely.

➢ Consider food with a longer shelf life when stocking emergency pantries that won't be accessed frequently.

➢ Before stocking up on emergency foods (i.e., MREs, dehydrated, or freeze dried products), have your family conduct taste tests.

➢ Always buy MREs by the case so that you can inspect the time-temperature indicators.

➢ Don't forget to stock up on non-food and special need items.

➢ Food poisoning can turn a bad situation into a deadly one. Practice the four steps (clean, separate, cook, and chill) to avoid food poisoning.

Recommended Items - Food

❑ A stockpile of food
   a. Minimum of 30 days of food that generally satisfies the guidelines of the USDA New Choose My Plate program and does not rely heavily on refrigeration
   b. Any special needs foods, including baby food, pet food, and dietary restricted foods
   c. (Optional) Food with extended shelf life for DP network food pantry

❑ Refrigerator/freezer thermometer to properly set the temperatures

❑ Non-food items (e.g., aluminum foil, paper products, plastic utensils, plastic storage bags, napkins, cooking utensils)

❑ General all-purpose cookbook

Challenge

Heavy rains have flooded the nearby water treatment facility, introducing two dangerous pathogens (Giardia and Shigella) into the water supply. Local authorities have issued an order to use bottled water or boil all tap water. The rains are expected to continue for the next five days. How will you provide clean drinking water for your family? Do you understand the risks that these pathogens pose?

Never underestimate the importance of having clean, drinkable water. A useful saying is that humans can live three minutes without air, three days without water, and three weeks without food. Given that air quality is often not a problem in many disasters, it leaves water as the primary need—certainly much more important than food for short-term survival. Also, take a moment to consider that even in the best of times, over a billion people on this planet don't have access to clean drinking water.[46]

There are two approaches to making sure that you have water in a crisis. You can either maintain a permanent stockpile of water, or you can have empty containers ready to fill when a disaster is approaching. The obvious advantage of the permanent stockpile is that you are always ready. The disadvantage is that water is heavy, bulky, and can be a mess if not handled correctly. Also, unless treated with a water preserver, it must be poured out and refilled about every six months—see *Storing Water* further in this chapter.

Regardless of your approach, one thing holds true. If a disaster is imminent, store as much water as possible. If you don't have enough water containers, fill bathtubs, buckets, pots, barrels, and anything else you have available. Remember that water is not only used for drinking and cooking, but also hygiene and sanitation. Don't neglect to account for these important needs. As discussed in *Chapter 3: Food,* many of the worst bacteria-related illnesses are a result of fecal-oral contamination. Keeping yourself and your environment clean is extremely important in times of crisis.

Finally, don't forget to consider the needs of your pets. If you have a couple of cats, they probably won't have much impact on your water consumption. However, if you have two German Shepherds, three cats, and a donkey, you should definitely determine their water usage and budget accordingly!

SANITATION

When water is in short supply, the toilet is going to be your biggest enemy. The amount of non-potable water (i.e., water not fit for drinking) needed depends on how old your toilet is and how frugal you are with your flushes. If your toilet was made prior to 1982, it probably takes 5 to 7 gallons per flush. That is a lot of water. Newer toilets require only about 2 to 3 gallons per flush. This conservation is an excellent reason to upgrade at least one toilet if you happen to live in an older home. If your budget doesn't allow the upgrade, consider putting a few heavy glass jars or bottles in the tank to displace some of the water—thereby reducing the amount used with each flush.

When water service is no longer available, there are four obvious choices for sanitation:

1. Dig a hole or trench outdoors. This can get old in a hurry, as well as be a source of disease.
2. Use a portable toilet with disposable liners; smelly but manageable with the correct supplies.
3. Use a self-contained composting toilet; an excellent, but expensive alternative (see *www.lehmans.com*).
4. Ration your water, and continue to use your conventional toilet; the least impact to your family, but one that requires access to a significant amount of water.

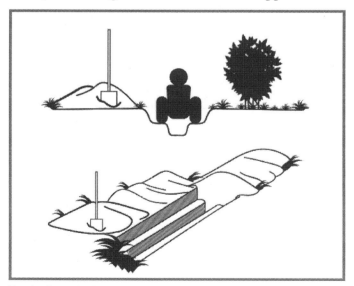

Trench toilet—a back-to-nature experience

Digging a hole or trench is certainly the easiest in terms of preparations. All you need is a shovel and a roll of toilet paper (newspapers or magazines will do in a pinch). But this back-to-nature approach is not without its drawbacks. If you've ever used an outhouse, you are familiar with the problems, namely the offensive smell and the flies.

To help with these issues, you can sprinkle the

waste with a little lime or lye, but be careful not to get any on your skin as they will burn you. If you don't have either of those available, you can substitute sawdust, wood ash, or peat moss. Another obvious drawback to a hole or trench is that you have to travel outdoors some distance from your home to relieve yourself. This can be dangerous in the middle of the night, not to mention very inconvenient.

Portable toilets are definitely a step up from trench toilets. Not only are they more comfortable and convenient, they are also less stinky. It's important to stock up with the necessary supplies, including plastic liners and spray disinfectants. There are also products available that make waste easier to dispose of, such as "Poo Powder."

Self-composting toilets turn human waste into compost. They do a good job of venting the smell away from the home and use very little, if any, water. Some require electricity to operate internal fans (reduces odors) and heaters (aids in waste decomposition), while others use passive venting. Bulking agents, such as peat moss or sawdust, are frequently added to further aid in the fluid adsorbtion, odor management, and waste decomposition. The biggest advantage of self-composting toilets is that they don't require much water. Their biggest drawback is that they are expensive (perhaps $1,500 for a quality unit). Units are often found in cabins and rural locations where water is in short supply.

The final sanitation option is to simply continue using your home's toilet. However, if you opt to do this, you will need to store or have access to enough non-potable water for at least one flush per person per day. The idea is to flush the toilet only after bowel movements. A great way to remember this is to keep in mind the saying, "If it's yellow, let it mellow. If it's brown, flush it down." A tad vulgar perhaps, but you won't forget it.

There are two ways to flush a toilet when water has been disconnected. The first is to cut off the incoming water valve (usually just behind the toilet), pour water into the back of the tank, and flush as usual. This works fine, but can be a little messy. The second method is to pour the water directly into the bowl. If unfamiliar with plumbing, you might think that the toilet would overflow. However, as the water level rises, a partial vacuum is created as water spills over the dam in the back of the toilet boil. This vacuum pulls the water out of the bowl and down into the sewage pipe.

One final note about operating toilets with external water sources: once you are finished flushing, add a little water to the toilet bowl. If the water level is too low, it may allow sewer gas to enter the home.

Tips

A five-gallon bucket, two boards, and a garbage bag can serve as a makeshift portable toilet.

GO Anywhere Portable Toilet® (courtesy of Cleanwaste)

Sun-Mar composting toilet

Flushing the toilet with a water bucket

SEWER BACKFLOW

One other topic that fits in the category of sanitation is sewage backflow. Your sewer or septic system is designed to remove sewage from your home, but that same piping can inadvertently bring sewage back up into your home. This most frequently occurs when flood water flows into the sewer system and floats raw sewage up through a home's toilets, tubs, and sinks—disgusting to be sure!

Install a backflow valve to prevent sewage from backing up into your home from flooded sewer systems.

The surest way to prevent sewage backflow is to install a backflow valve on your sewage line. The backflow valve allows sewage to flow in only one direction—that is, *out* of your home and not back into it. If you are a handyman with a bit of plumbing experience, you can probably do this job yourself. Otherwise, contact your local plumber. If possible, have the backflow valve installed somewhere readily accessible. This way, if you ever have a clog associated with the backflow valve, you can easily clear it.

HYGIENE

This might be a good place to emphasize the importance of maintaining good personal hygiene during times of crisis. Simply put, you must keep your hands clean of fecal matter and other contaminants that might make you sick. Many serious infections, including salmonellosis and *E. coli*, can be the result of contamination from tiny amounts of fecal matter entering your body through the mouth, nose, or eyes. These bacterial infections can be especially deadly when access to medical care is limited.

When you use the toilet or touch anything else that might be contaminated (e.g., a sick person, garbage can, raw meat), you must wash your hands thoroughly. Likewise, before handling food, you should always assume your hands are dirty and wash them. For these reasons, budget a gallon of water for hygiene per person per day. This recommendation exceeds those of many other DP books, but hygiene is critical to preventing illness and should not be shortchanged.

Washing Hands

1. Use warm water if possible.
2. Lather soap into a thick foam.
3. Scrub hands thoroughly for at least 20 seconds.
4. Rinse thoroughly.
5. Air dry, or use a disposable towel.
6. Use towel to turn off faucet.

Washing has one primary goal; to remove the contaminants from your skin. The soap foam bonds to the contaminants, and water rinses them away. Teach your children the proper way to wash their hands (see tip box).[47]

Hand sanitizers with 60% or more alcohol are an excellent alternative to hand washing when water is not available. They do a great job of killing pathogens but don't remove waste, blood, or dirt from your skin. As with soap and water, most sanitizers *don't* provide extended protection. They only kill what is currently on your hands. There are a few lotion-based sanitizers that claim to provide several hours of protection, but it is not clear that they are an adequate substitute for periodic hand washing.

Disposable baby wipes can also be used to clean your hands and body, but most wipes are not alcohol-based and don't clean as well as soap and water. Using larger, disposable bathing wipes can be an excellent temporary replacement for taking showers or baths. Campers have used bathing wipes for years, and they can really help you to feel refreshed.

It is also a good idea to have plenty of heavy-duty garbage bags and twist ties on hand. Plastic bags are handy for getting rid of food remnants, medical waste, and contaminated clothing. They can also be used to seal leaks, gather water, serve as a rain poncho, act as a toilet, and much more.

Finally, if you need to clean a hard surface, such as a countertop, door knob, or toilet, and are out of Lysol or other germ-killing cleansers, you can call upon bleach to serve as a powerful disinfectant. Simply mix 1 part bleach to 9 parts water.[48] Just be careful not to spill the mixture on carpet or clothing since it will cause whitening. If the surface will later be used for preparing foods, it should be rinsed first with clean water.

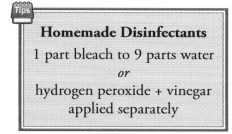

Homemade Disinfectants
1 part bleach to 9 parts water
or
hydrogen peroxide + vinegar
applied separately

An alternative to using bleach is to spray the infected surface with 3% hydrogen peroxide (i.e., the standard drug store concentration), and then again with white vinegar. This combination has been shown effective at killing *E. coli, Salmonella,* and *Shigella,* and is safe to use on countertops and cutting boards without additional rinsing.[49] Don't combine the vinegar and hydrogen peroxide in the same bottle. Also, keep the hydrogen peroxide in an opaque bottle since light will degrade the solution.

ADDING IT ALL UP

To determine your family's daily water needs, simply add up the water required for drinking and cooking (1 gallon per person), hygiene (1 gallon per person), and sanitation (2 to 7 gallons per person depending on your toilet type). Drinking, cooking, and hygiene needs must be met with potable water, but sanitation can be handled with non-potable water. It is recommended that you store (or have access to) enough water to support your family for a minimum of 14 days.

For a family of five, this corresponds to: 5 people × 2 gallons per day × 14 days = 140 gallons of potable water. And if toilets are to be used (assuming 3 gallons per flush): 5 people × 3 gallons per day × 14 days = 210 gallons of non-potable water.

It should go without saying that storing 350 gallons of water is something that requires prior planning. If you are fortunate enough to have access to a large body of water, such as a swimming pool, stream, or lake, you can draw upon it for your non-potable needs. Using outside sources like these significantly reduces your storage requirements.

The average American household uses about 94,000 gallons of water each year.[50] Using the census estimate of 2.59 people per household, this consumption figure converts to about 100 gallons per person per day. Compare this to the 5 gallons of water recommended per person (2 gallons of potable, 3 gallons of non-potable). Suffice it to say, your family should be prepared to live on much less water than they are accustomed to. The days of lounging in the hot shower will be a thing of the past.

The average American uses about 100 gallons of water per day.

STORING WATER

Potable water in jerry cans and a large tank

The shelf life of water is difficult to predict because it is based on many factors, including the purity of the source, storage temperature, lighting, and type of storage container. In some conditions, water can be stored safely for several years, while in others, it can become contaminated in a matter of months. Follow the water storage guidelines (in tip box) for achieving the best shelf life.

It is a good idea to have both large and small containers for your water storage. The large containers are efficient at storing sizable quantities of water in a small space. But if you are forced to relocate, or simply have to go out in search of additional water, you will need containers light enough to transport. A gallon of water weighs approximately eight pounds, so portable containers shouldn't be over about six gallons in size (less if you are unable to handle 50 pounds). Keep two sets of portable containers. Use one set for retrieving potable water and the second set for retrieving non-potable water. By keeping them separate, you eliminate the risk of cross-contaminating your containers.

Prepare your potable water containers in advance of any disaster. Containers should be cleaned following these simple steps:

1. Mix 1 tablespoon of unscented liquid household bleach (i.e., 5-6% sodium hypochlorite) into one gallon of water.
2. Pour the solution into the container, rubbing or brushing it on the threads and mouth.
3. Shake it around in the container, and then let it sit for 10 minutes.
4. Rinse the container thoroughly with clean water.

Water storage containers *(courtesy of Baytec and Reliance Products)*

Water Storage Guidelines

➤ Use FDA-approved DOT #34 opaque containers
➤ Pre-treat if it comes from untreated sources (e.g., well)
➤ Store out of light and away from pesticides, gasoline, paint, or chemicals
➤ Keep warm enough to prevent freezing
➤ Cycle every 6 months unless treated with a preserver

TREATING WATER

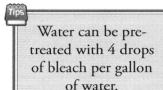

Water can be pre-treated with 4 drops of bleach per gallon of water.

Unscented household bleach can also be used to pre-treat water prior to storage by adding 4 drops per gallon and mixing well. However, pre-treating water with bleach is only recommended if the water comes from an untreated source, such as a well or rural township water. Most cities pre-treat their water before it is piped to customers, so further pre-treating is unnecessary and won't increase the shelf life of the water. Additional pre-treating won't cause any harm, but it may leave an unpleasant odor and taste.

Commercial concentrated water preservers, such as 7C's Safety & Environmental's Water Preserver, can be added to water to significantly increase the shelf life. These products are made of stabilized sodium hypochlorite, which is designed to prevent microorganism contamination for a minimum of five years. If you have large water storage containers, such as 55-gallon drums or 250-gallon Super Tankers, then using a water preserver makes sense. However, if you are just using jerry cans or smaller containers, it isn't difficult to simply cycle your water every six months.

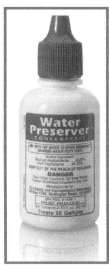

Courtesy of 7C's Safety & Environmental, Inc.

CONTAMINATION

The Environmental Protection Agency (EPA) has estimated that 90% of the world's fresh water is contaminated and unsuitable for drinking without some form of purification.[50] The days of kneeling down on a hike and sipping from the stream are long gone. Don't make the mistake of thinking that just because water looks clear or tastes good that it is free of contaminants. That includes frozen water, which can house hepatitis A, *Salmonella*, and *Cryptosporidium* for months.[51] The CDC estimates that 88% of the world's cases of diarrhea are the result of unsafe water, inadequate sanitation, and poor hygiene. Water-related illnesses rank as one of our planet's deadliest killers, resulting in the death of 1.5 million people annually, most of them children.[52]

The EPA classifies water contaminants into six categories (see tip box). Based on these categories, the agency publishes a long list of contaminants and sets their maximum allowed levels in drinking water.[53] Likewise, the National Sanitation Foundation International (NSF) certifies that water filters remove those contaminants to a specified level of effectiveness. Contaminant removal is tested by adding an influent and then measuring the effluent. In other words, they start with clean water, add a contaminant, pass the water through a filter and see how much of the contaminant has been removed. For a filter to be certified as effective against a contaminant, it must meet or exceed the NSF requirements.[54]

Water Contaminants
➤ Pathogens
➤ Organic chemicals
➤ Inorganic chemicals
➤ Disinfectants
➤ Disinfection by-products
➤ Radionuclides

It is instructive to compare the EPA safe-water requirements to the list of contaminants removed by NSF-certified water filters. For example, the EPA requirements dictate that drinking water contain less than 0.015 mg/L of lead, and the NSF certification process requires that water contaminated with 0.15 mg/L of lead be reduced down to a maximum of 0.01 mg/L (about 93% removal)—just within the EPA limit. It becomes clear that EPA requirements and NSF certification align closely to first dictate, and then test, water quality.

PATHOGENS

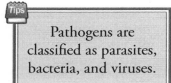

> **Tips**
> Pathogens are classified as parasites, bacteria, and viruses.

Pathogens are microscopic organisms that include protozoa, bacteria, parasitic worms, fungi, and viruses.[55] They can generally be classified into one of three broad categories: parasites, bacteria, and viruses. For detailed information about many different water contaminants, see the Center for Disease Control and Prevention's online listing.[56]

PARASITES

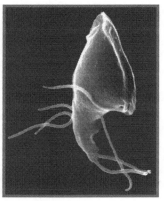

Giardia *(CDC photo)*

Parasites include worms, lice, and protozoa. All can be transmitted through water, but for modern societies, protozoa represent the greatest waterborne parasitic threat. Protozoa are single-celled organisms that may have more than one nucleus. They are generally found in water as microbial cysts and cause serious gastrointestinal illness when ingested. Their source is human or animal fecal waste, and they typically range from 2 to 30 microns in size. A micron (a.k.a. micrometer) is 1/1,000,000 of a meter. Fortunately, parasites of this size are large enough to be easily removed by quality water filters (i.e., those with an absolute pore size of less than 1 micron).

BACTERIA

Salmonella *(CDC photo)*

Bacteria are single-celled microorganisms without a nucleus, generally measuring from 0.2 to 4 microns in size. They are found in every habitat; growing in soil, seawater and fresh water, and even deep in the earth's crust. Many bacteria are not harmful, and some are even beneficial to your health. There are approximately 40 million bacterial cells in a gram of soil and 5 million in a single teaspoon of fresh water.[57] Some of the bacteria found in water are the result of transfer from fecal waste. As with parasites, ingestion usually results in severe gastrointestinal illness, along with other associated infections. Larger bacteria are effectively removed by conventional water filters, while other smaller bacteria are more difficult to remove.

VIRUSES

Viruses are essentially raw genetic material with a protective coating called a capsid. They infect healthy cells and inject their genetic characteristics into them. In turn, those contaminated cells quickly replicate.

Consuming virus-contaminated water can cause gastrointestinal illness, weakness, fever, liver disease, and paralysis. Viruses are very small in size, measuring from 0.004 to 0.3 microns in size. Because of their minute size, most water filters will *not* effectively remove viruses. However, there are a few filters that do remove them by taking advantage of the fact that viruses will attach to larger particles under certain conditions.

For a conventional filter to effectively remove a contaminant of any type, the absolute (not average) pore size of the filter membrane must be smaller than the contaminant. Table 4-1 provides a quick comparison of general filter requirements to remove several common pathogens.

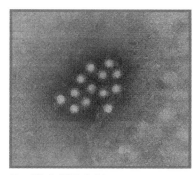

Hepatitis A *(CDC photo)*

Depending on where you look, you will find conflicting data regarding the size of various microorganisms. That's because microorganisms can be many different sizes depending on the conditions in which they grow. How then do you decide if a filter will remove a particular pathogen? The best way is to review the test data for the specific filter in question. Certified filters will provide performance information such as "removes 99.99% of *Cryptosporidium*." Much more information about selecting a water filter is given later in this chapter.

Table 4-1 Filter Absolute Pore Size Recommendations

Microorganism Type	Microorganism	Recommended Absolute Filter Pore Size (microns)
Protozoa	Giardia lamblia	1
	Cryptosporidium	
	Entamoeba histolytica	
Bacteria	Escherichia coli	0.2-0.4
	Vibrio cholera	
	Shigella	
	Campylobacter	
	Salmonella	
Virus	Hepatitis A	N/A—see note
	Norovirus	
	Poliovirus	
	Rhinovirus	

Note: Due to the very small size of viruses, most filters don't adequately remove them. However, there are a few exceptions, as noted later in the chapter.

ORGANIC AND INORGANIC CHEMICALS

> Distillation and reverse osmosis are the best ways of removing chemical contaminants.

The EPA tests drinking water for a long list of organic and inorganic chemicals.[53] These include chemicals discharged from various types of factories (e.g., petroleum, plastic, metal, coal-burning, pulp), runoff from herbicide treatments, corrosion from plumbing, and erosion of natural deposits. Consumption of chemical contaminants can cause cancer as well as damage to the body's basic systems and organs.

Most conventional membrane filters do a poor job of removing chemicals. However, distillation and reverse osmosis systems have been shown to be effective treatment methods.

DISINFECTANTS AND BY-PRODUCTS

Disinfectants are chemicals added to water during the purification process to kill microbes. These additives include chloramines, chlorine, and chlorine dioxide. By-products created by the water treatment process include bromate, chlorite, haloacetic acids, and total trihalomethanes. Consuming high levels of these increases the risk of cancer as well as liver, kidney, and neurological disorders.

Fortunately, these additives and by-products are often able to be removed with a quality, carbon-based water filter.

RADIONUCLIDES

Drinking water in the United States has very low levels of radioactive contamination (a.k.a. radionuclides). Most of what is present in water occurs naturally as result of erosion of natural deposits of certain minerals. Radioactive contamination can also be the result of the introduction of human-made nuclear materials, such as from a nuclear accident, accidental spill, or improper disposal practices. Examples of radioactive contamination include alpha/beta particles, uranium, and radium 226/228. Long-term exposure to high levels of radionuclides in drinking water causes an increased risk of cancer. Exposure to uranium in drinking water also causes kidney damage.[58]

Conventional membrane filters will not remove radionuclides. Three methods have been shown to reduce radionuclides: ion exchange devices used to remove water hardness, reverse osmosis systems, and lime softening.[59]

LOCAL LEVELS

If you are interested in learning about the levels of contaminants in your local water supply, go to the EPA's website, *www.epa.gov/safewater,* and find the link for your area. If your water provider doesn't post its findings on the EPA website, you may have to contact your provider directly. It is both informative and well advised to understand the water quality issues facing your community.[60]

Even in the best of times, maintaining the purity of your drinking water requires careful attention. You may find that your local drinking water falls short of EPA guidelines. During disasters, water quality can quickly degrade, or worse, can be completely shut off. To adequately prepare for contamination or shortage, you will need a well thought-out water management plan.

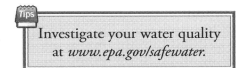

> Investigate your water quality at *www.epa.gov/safewater.*

PURIFICATION

Unless you have a large storage tank of clean drinking water, or can draw water from an uncontaminated source, such as an underground well, you will likely need some method of purifying water from non-potable sources. Let's start by quickly examining the water purification process that large treatment facilities use.

The purification process consists of six fundamental steps:[61]

- **pre-treatment**—screen to remove large debris, and perform limited softening and chlorination
- **pH adjustment**—adjust alkaline/acidic level
- **coagulation/flocculation/sedimentation**—clarify to remove particles
- **filtration**—remove smaller particles
- **disinfection**—kill pathogens using chemicals and UV light
- **additional treatments**—fluoridate, hard/soft condition, and remove radium

Strict replication of these steps on a small scale is not practical. However, the basic goals of purification still apply—remove the debris and kill or remove the impurities. Unfortunately, there is a great deal of misinformation about how to make non-potable water safe. Even within the DP community, confusion abounds. The information presented here will make clear the efficacy of different methods.

There are six primary techniques that can be used independently or in conjunction to purify water (see tip box). Each method has its

Purifying Water
- ➢ Boiling
- ➢ Filtering/purifying
- ➢ Chemical disinfection
- ➢ Distillation
- ➢ Reverse osmosis
- ➢ Ultraviolet light

Water treatment facility *(FEMA photo/Manny Broussard)*

respective advantages and disadvantages. Let's begin with a discussion of each the various techniques, followed by several recommended approaches to purifying water.

BOILING

Boiling is the optimal way of killing microorganisms. It is simple and effective at neutralizing all types of pathogens, but it will not improve the taste of water. Nor will it remove particulates or chemical impurities. Tests have shown that microorganisms are destroyed by the time water reaches the boiling point (212°F). Boiling the water for extended durations (e.g., 10 to 30 minutes) as suggested in some texts is unnecessary.[50] The generally accepted guidance is to bring the water to a rolling boil for one minute. After that, let the water cool naturally, providing a final measure of safety by ensuring that the water remains at high temperatures for an extended time.

Boiling is an excellent method of purifying already clear water, such as tap water, that has become infected with a pathogen. The main drawback is that it is slow and requires a heat source. Boiling is also not optimal for purifying water from natural sources, such as a lake, since it doesn't remove any particulates or debris and won't improve the taste.

FILTERING/PURIFYING

Filtering can be as simple as passing water through a handkerchief or as sophisticated as forcing it through a nearly solid substance that allows little else but water molecules to pass through. Water coming from natural sources, such as a lake or river, is often filtered in stages—a coarse filter first to remove the dirt and debris, followed by a much finer filter to remove pathogens. The idea is to get as much crud out of the water as possible before trying to purify it. This approach greatly helps to prevent clogging of the filter.

The terms *filter* and *purifier* are often used interchangeably by retailers even though they refer to very different things. *Filter* is a general term describing any device that can be used to remove contaminants from water. This could be as crude as a simple coffee filter, although typically it refers to a device that contains some sort of fine membrane. Most general purpose water filters remove protozoa and select bacteria. Contrary to popular belief, they *do not* remove minerals, heavy metals, or salt.

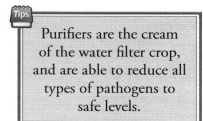

Tips
Purifiers are the cream of the water filter crop, and are able to reduce all types of pathogens to safe levels.

A water *purifier* is a special type of filter, one with a very specific definition. The EPA defines a purifier as something that reduces *all* pathogens to safe levels—exceeding log-6 protection for bacteria (99.9999% removed), log-3 for *Cryptosporidium* (99.9% removed), and log-4 for viruses (99.99% removed).[62] When you see something described as a "certified purifier," you should recognize that the device offers broad protection from waterborne contaminants.

Water filters can be grouped into one of three categories: point-of-entry (POE), point-of-use (POU), and portable. A POE filter attaches to the main water line, filtering water throughout the entire house.

A POU filter sits on the countertop or under the sink and filters only a single tap. Finally, portable units are lightweight pump-driven or gravity-fed devices often used when camping or traveling. Pitchers and other filtered dispensers can also be categorized as portable units.

Having a quality POU or POE filtering system will ensure that you can very likely continue to use your city's water even if it becomes contaminated. This provision relieves the burden of trying to find clean or treated water. Having a POU or POE filter obviously does not help in the event that water service is completely shut off. For those situations, you will need to locate an alternate water source and use a pump-driven or gravity-fed water filter as necessary.

Whether choosing a POU or POE system, select one with an absolute pore size of 1 micron or smaller. This rating indicates that the filter will not pass anything larger than 1 micron in size. A filter with 1 micron absolute pore size will safely remove all protozoa (e.g., *Cryptosporidium, Giardia*) and provide some protection from bacteria (e.g., *E. coli* and *Salmonella*). However, it is unlikely that the filter will protect you from viruses since they are significantly smaller than 1 micron.

Katadyn's Pocket filter

Fortunately, there are a few select water filters that have been independently verified to eliminate all forms of pathogens (i.e., bacteria, protozoa, and viruses) without the use of chemical disinfectants. These filters are classified as true water purifiers, and offer a simple one step process to removing pathogens (discussed in *Selecting a Water Filter*).

CHEMICAL DISINFECTION

Two widely used halogen chemicals for killing waterborne pathogens are iodine and chlorine (along with their respective derivatives). Popular products include Micropur MP1 chlorine dioxide tablets and Potable Aqua's titratable iodine tablets. These tablets are easy to use and have shelf lives of at least four years if unopened—one year if opened.[63,64] Another product, Polar Pure iodine crystals, is a little more difficult to precisely administer but is much less expensive and offers the advantage of having a nearly indefinite shelf life.[65]

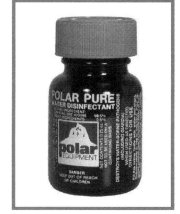

If you have no other means, you can use household bleach or 2% tincture of iodine to disinfect water. These products are as effective as the commercial chemical water disinfectant products (e.g., MP1), but not as convenient to use. Also, bleach has a much shorter shelf life—requiring replacement every six months to maintain potency.[66]

Polar Pure

MSR's MIOX

Halogen disinfectants (a.k.a. electronegative chemicals) of this type are considered effective against bacteria, somewhat effective against viruses, and of limited value against protozoa. The reason for this limitation is that protozoa cysts have protective coatings around them. *Cryptosporidium* in particular is resistant to halogen treatment.[67]

Finally, there is one device (Mountain Safety Research's MIOX) that uses salt, water, and electricity to create a foamy brine of oxidizing agents. When added to water, the brine will neutralize protozoa, bacteria, and viruses. Disposable test strips are used to verify that the water has been adequately treated. Independent tests have shown MIOX meets the EPA Guide Standard and Protocol for Testing Microbiological Water Purifiers (a.k.a. EPA Guide Standard).[68] Suffice it to say that this level of purification is exceptional—far better than iodine or chlorine. One noticeable disadvantage of the mixed oxidant method is that it requires a four-hour treatment time for *Cryptosporidium*.

DOSING

> **Chemical Disinfection**
>
> 1. Add halogen per Table 4-2.
> 2. Mix the water thoroughly.
> 3. Splash the disinfected water onto the container's threads.
> 4. Wait 30 minutes. Increase the wait time to 2 hours for very cold water.

If using commercial chemical disinfectants, such as Polar Pure or Micropur MP1, the dosages are clearly marked on the package. However, if you are using bleach or iodine to disinfect water, you will need to measure your own dose.

Table 4-2 gives the recommended dosages for disinfecting clear and cloudy water using bleach or iodine. The recommended dose for bleach is 2 drops per quart of water, which translates to approximately 5-6 parts per million (assuming a bleach solution with 5-6% sodium hypochlorite). Similarly, the recommended dose for iodine is 5 drops per quart of water, corresponding to approximately 5 parts per million (assuming a 2% tincture of iodine solution). After treating with either chemical, let the water sit covered for 30 minutes, giving time for the halogen to work. Treated water should have a detectable chlorine or iodine odor.

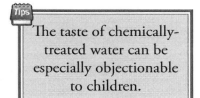

> The taste of chemically-treated water can be especially objectionable to children.

In general, the more turbid the water, the more disinfectant you need to add.[72] Consequently, most references suggest doubling the concentration for cloudy water—as indicated in Table 4-2. Although effective, the increased dose introduces a strong chemical taste. An alternative approach is to filter the water first to clear it up, and then use the standard dose of halogen. A second alternative is to double the dose of disinfectant and then filter using a small-pore filter to remove some, although not all, of the unpleasant chemical taste.

Very cold water (i.e., water below 40°F) slows the reaction time of the halogen, leading to the suggestion once again to double the normal dose of disinfectant. An alternative to this is to allow the water to

Table 4-2 Ratios for Purifying Water with Bleach or Iodine[69,70]

Water Quantity	Clear Water	Cloudy Water
1 Quart/Liter	Bleach - 2 drops Iodine - 5 drops	Bleach - 4 drops Iodine - 10 drops
1 Gallon	Bleach - 8 drops (1/8 tsp) Iodine - 20 drops	Bleach - 16 drops (1/4 tsp) Iodine - 1/2 tsp
5 Gallons	Bleach - 1/2 tsp Iodine - 1 tsp	Bleach - 1 tsp Iodine - 2 tsp
10 Gallons	Bleach - 1 tsp Iodine - 2 tsp	Bleach - 2 tsp Iodine - 4 tsp
55 Gallons	Bleach - 5½ tsp Iodine - 11 tsp	Bleach - 11 tsp Iodine - 22 tsp

Notes:
1. *1 drop = 0.05 mL*
2. *Water that has been disinfected with iodine is not recommended for pregnant women, people with thyroid problems, those with known hypersensitivity to iodine, or continuous use for more than a few weeks at a time.[71]*

warm before treating it. A second alternative is to permit the halogen to work for two hours instead of 30 minutes when disinfecting very cold water.[73]

One drawback of using chemical disinfectants is the objectionable smell and taste of the treated water. There are several things that can help with this:

- Avoid using higher doses by first clarifying the water or allowing the disinfectant to work longer.
- Use a filter subsequent to the chemical disinfection process.
- Mix in a powdered fruit drink mix, such as lemonade or Kool-aid, that contains ascorbic acid (vitamin C). The ascorbic acid helps convert the chlorine and iodine to tasteless chloride and iodide.
- Allow the treated water to air out for a couple hours before drinking.

If you plan to use a chemical disinfectant, it is advisable to try out several methods to determine which yields the best-tasting solution for your family.

Personal aside: In a very unscientific taste test of chemical treatment methods, my own family concluded that iodine-treated water was by far the worst smelling and tasting, bleach-treated was second, MIOX-treated water third, and water treated with Micropur MP1 ready-to-use tablets was the least objectionable.

DISTILLATION

Distillation is a process of boiling water and then collecting the water vapor as it condenses. This method of purifying water is extremely effective at removing all types of pathogens (i.e., bacteria, protozoa,

Water distillation system *(courtesy of Nutriteam)*

and viruses). Distillation and reverse osmosis are also the only methods discussed that do an excellent job of removing chemical contaminants.[74]

A disadvantage of using distillers is that they are slow, typically taking about six hours to yield one gallon of water. They also require electricity to operate, which given the amount of time they have to run, might be a significant problem during many disasters.

REVERSE OSMOSIS

Osmosis is the process of flowing from low concentrate level to high. It is the reason drinking seawater can kill you. The seawater in your stomach draws water out of your body trying to dilute the high concentrate of salt, eventually leading to dehydration and death.

Reverse osmosis (RO) is the process of flowing from high concentrate to low. Pressurized water is forced through a very fine membrane while discharging excess water and concentrate (pollutants in this case). Reverse osmosis systems are usually constructed of several stages, including a pre-filter, semi-permeable membrane, pressurized storage tank to hold the treated water, and carbon adsorption post-filter.

Reverse osmosis systems do an excellent job of removing all forms of pathogens as well as chemical impurities. However, they require high water pressure (typically > 40 psi), and therefore may necessitate the use of an electric pressure-boost pump in some homes. They also waste a great deal of water, turning out only 5-10% of the incoming water. The remaining water is flushed away with the pollutant. To create 1 gallon of purified water might require 8 to 18 gallons of incoming water.[75] Finally, reverse

Reverse osmosis system *(courtesy of Watts)*

osmosis systems require periodic maintenance, including replacing the pre- and post-filters annually and the membrane every few years.

REVERSE OSMOSIS VERSUS DISTILLATION

Be forewarned that there are two very vocal groups of water filter marketeers—those who sell distillation devices, and those who sell reverse osmosis systems. Salesmen for distillation devices will claim that reverse osmosis removes important minerals. Likewise, those selling reverse osmosis systems will claim that distillation causes the water to taste flat. Both salesmen will claim that the competing technique doesn't remove contaminants as well as the one they are selling.

There is a little bit of truth in both arguments. Yes, distilled water can taste flat. And yes, reverse osmosis does remove minerals. But both issues are not really problematic. The flat taste can be eliminated by aerating the water—simply pour the water back and forth between containers and allow it to sit for a few hours. Also, the loss of minerals from reverse osmosis is not usually considered a health concern since Americans get most of their minerals from the foods they eat. The good news is that both types of systems filter water exceptionally well, removing all forms of pathogens and many chemical contaminants.

ULTRAVIOLET LIGHT

It is also possible to disinfect water using ultraviolet (UV) light. This treatment method has been used for years by large water treatment facilities, but it has only recently been available for the home user, traveler, or hiker. Two units are widely available: Meridian Design's AquaStar and Hydro-photon's SteriPEN. The

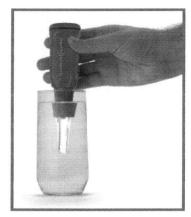

Courtesy of Hydro-Photon

AquaStar includes a heavy plastic bottle with an enclosed UV light source. The SteriPEN is a small, portable UV light pen used to stir water in a glass or bottle. Both claim to have been independently certified to meet the EPA Guide Standard.[76]

Using either of the portable UV devices is quick and easy (taking only 30-80 seconds) and has been shown to be very effective against all types of pathogens. The UV light effectively disrupts the microbes' DNA, preventing them from multiplying. Unlike chemical disinfectants, ultraviolet light is also effective at neutralizing protozoa such as *Giardia* and *Cryptosporidium*.[77]

However, neither device will remove impurities (e.g., dirt, particulates, chemicals) or help improve the taste of the water. There are also some concerns over the effectiveness of the devices in turbid water. For example, independent testing showed that the SteriPen did a poor job of purifying muddy water.[78] For this reason, it is recommended that water be clarified prior to using a UV light purification device.

Those limitations aside, portable UV purifiers do offer a convenient method of purifying small quantities of water—especially when away from home.

SUNLIGHT

When no other methods are available, sunlight can be used to decontaminate water. This method of solar water disinfection (a.k.a. SODIS) takes advantage of the ultraviolet emission of the sun to kill waterborne microorganisms and is used in developing countries around the world.[79]

The SODIS method is easy to follow:

1. Wash the bottles thoroughly with soap and water.
2. If necessary, perform basic filtering to clarify the water.
3. Fill the bottles with water, and close the lids.
4. Expose the filled bottles to direct sunlight for a minimum period of time.
5. Drink directly from the bottles, or pour the water into a clean cup.

The method is simple, however, you must use colorless, transparent, polyethylene terephthalate (PET) bottles no larger than 2 liters in size. Glass and other plastic materials can block the ultraviolet light needed to purify the water. Fortunately, most convenience-sized beverage bottles sold in the United States are made of PET materials.[80]

At a water temperature of 86°F, a minimum of six hours of direct, summer sunlight is needed for mid-latitude regions. Colder water requires longer, as do cloudy conditions. For example, on days that are 50-100% overcast, water decontamination requires a full two days of exposure.[79]

The SODIS method has been proven to reduce cases of diarrhea caused from *some* waterborne pathogens. However, it has not been shown to kill *all* waterborne pathogens. Also, it is not clear that SODIS

SODIS *(courtesy of EAWAG)*

will kill trace contamination on the threads of the bottle. For these reasons, it should be considered a last resort for water decontamination.

RECOMMENDED METHODS

With all the advantages and disadvantages in hand, it becomes easier to select the optimal purification methods. Remember the two goals of purification: remove debris/particulates, and neutralize the pathogens. Given the methods discussed, to remove anything requires either filtration, distillation, or reverse osmosis. Fortunately these methods also improve the taste of water—something that can be important to many finicky drinkers. Refer to Table 4-3 for a comparison of water purification methods.

For general in-home use, an EPA-certified purifier, water distiller, or reverse osmosis system are all perfectly adequate. However, for disaster preparedness, a certified purifier is preferred over a distiller or a reverse osmosis system. This preference is because the purifier doesn't require electricity like distillation, and doesn't waste water like reverse osmosis. Table 4-4 summarizes the benefits and drawbacks of the three preferred POU and POE methods.

When it comes to portable systems, the only single-step solution that removes particles and neutralizes pathogens is a certified purifier. If you are willing to combine two methods, then several other options exist. For example, combining boiling, UV light, or the MIOX with a standard filter can neutralize pathogens, clarify the water, and make it more palatable.

Table 4-3 Comparison of Purification Methods[74]

Method/ Device	Neutralizes Pathogens	Removes Particulates	Removes Chemicals	Affects Taste	Portable	Notes
Boiling	All	No	No	Leaves water tasting flat	Yes	The best method of killing pathogens.
Filter (1 micron pore size)	Protozoa – All Bacteria – Limited Viruses – Limited	Yes	Some	Improves	Yes	Filter effectiveness varies greatly based on pore size and membrane technology.
Certified Purifier	All	Yes	Some	Improves	Yes	Purifiers are proven to remove all forms of pathogens.
Chemical Disinfectant	Protozoa – Limited Bacteria – All Viruses – Limited	No	No	Introduces chemical taste	Yes	Slow to neutralize protozoa.
MIOX	All	No	No	Introduces salty taste	Yes	Slow to neutralize protozoa.
Distillation	All	Yes	Yes	Improves, but may leave water tasting flat	No	Requires electricity; very slow.
Reverse Osmosis	All	Yes	Yes	Improves	No	Wasteful; requires high water pressure; may require electricity for booster pump; removes minerals.
Ultraviolet Light	All	No	No	No	Yes	Does not work as well in cloudy water.

Table 4-4 Comparing Preferred POU and POE Methods

Method	Neutralizes all Pathogens	Improves Taste	Needs Electricity	Wastes Water
Purifier	Yes	Yes	No	No
Distiller	Yes	Yes	Yes	No
Reverse Osmosis	Yes	Yes	Maybe	Yes

SELECTING A WATER FILTER

When selecting a water filter, you should know that all water filters are not created equal. Many manufacturers promise "clean water" using vague claims about their products that are impossible to fully understand. When it comes to filters, it is definitely *let the buyer beware*. Don't assume anything. When buying a used car, you want to see the Carfax report to verify that things are as the seller claims. In the case of water filters, you need to see the impurity removal test data.

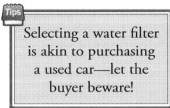

Selecting a water filter is akin to purchasing a used car—let the buyer beware!

If you can't find independent technical data showing how well the filter removes a certain type of contaminant, then you should assume that the filter falls short in that area. Remember, manufacturers want to convince you that their product is the best, so if they are not showing you the data then they likely have something to hide. Table 4-5 lists several misleading terms that you may see in water filter advertisements. In general, these claims indicate nothing about the capabilities of the filter.

There is also a great deal of confusion over the difference between filters and purifiers. As discussed previously, for a product to be described as a water purifier, it must be highly effective at removing, killing, or inactivating *all* forms of pathogens.[81] The word "filter" is a broader term describing any device used to remove contaminants, regardless of its effectiveness.

In reality, there are very few filters that can be classified as true water purifiers. This is largely due to the difficulty in filtering out viruses. Two exceptions worth noting are the General Ecology Seagull IV series (sink mount), the General Ecology First Need (portable). These purifying systems have been independently certified to meet the EPA Guide Standard without the use of chemical purification. Other filters may outperform these purifiers for specific contaminants, in particular *Cryptosporidium,* but the noted models offer a unique one-step solution to purifying water without the use of chemicals, UV light, or boiling. General Ecology also offers an emergency preparedness conversion kit (EPK) that converts their Seagull IV POU purifier to a portable, pump-driven system.

Table 4-5 Making Sense of Filter Claims

Claim	Meaning
Registered with the EPA	All filters that use a chemical disinfectant must be registered. It is not an indication of their effectiveness.
Tested/Approved by the EPA	The EPA does not test or approve filters.
Nominal pore size of *x* microns	"Nominal size" indicates the average pore size, not absolute. The filter could have pores significantly larger than the average size—making it ineffective at filtering smaller pathogens.
Effective against *Giardia,* etc.	Claiming to be effective without providing specific data is meaningless.

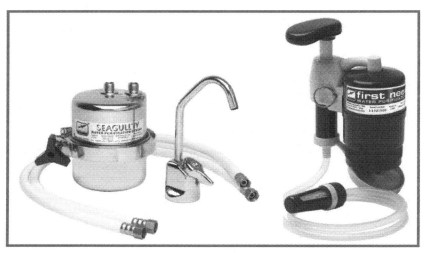

Water purifiers *(courtesy of General Ecology)*

CERTIFICATION

Filter certification can also be confusing because there are three sources of official water filter certification, all of which are accredited by the American National Standards Institute (ANSI):

- National Sanitation Foundation International (NSF)[82]
- Underwriters Laboratories (UL)[83]
- Water Quality Association (WQA)[84]

Each of the organizations has its own certification program (e.g., WQA Gold Seal, UL Water Quality Mark), but all require testing to the same ANSI/NSF standards.

Standards set the requirements for each of the four different purification methods: filtering, UV light, reverse osmosis, and distillation. These standards require that careful testing verifies that the filter reduces specific contaminants by a certain amount (e.g., 99.9% removed). Construction, product labels, and sales literature are also inspected. Periodic re-certification and inspection are required. The key ANSI/NSF standards relating to point-of-use (POU) water filters are listed below.[85]

- Standard 42: *Drinking Water Treatment Units - Aesthetic Effects*
- Standard 53: *Drinking Water Treatment Units - Health Effects*
- Standard 55: *Ultraviolet Microbiological Water Treatment Systems*
- Standard 58: *Reverse Osmosis Drinking Water Treatment Systems*
- Standard 62: *Drinking Water Distillation Systems*

To verify that a product was tested by an official lab, go to the NSF, WQA, or UL website and search their database of certified products. Testing by any of these three accredited organizations is expensive. As a result, many companies elect to have independent testing done by universities or other outside labs. Accordingly, they may state that their

WQA Gold Seal

product is "tested to the NSF/ANSI standards by an independent lab." Most independent test labs are reputable, and the results are considered accurate. Regardless of who does the testing, the results should be published and available for the customer's review.

Given the list of ANSI/NSF standards, it would seem that selecting a POU water filter would be as easy as finding one that is certified to Standards 42 and 53. Or, if using alternative technologies (e.g., ultraviolet light, reverse osmosis, distillation), you would ensure that the product meets the associated standard (i.e., Standards 55, 58, or 62 respectively). In general, this selection criteria is sound. However, there is one important point to note: a significant shortcoming of Standard 53 is that it has no criteria to check effectiveness against bacteria or viruses. Broadening of the standard to include bacteria and viruses has been discussed for many years. Until that effort is complete, Standard 53 is of limited value in selecting a filter system.

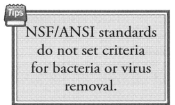

NSF/ANSI standards do not set criteria for bacteria or virus removal.

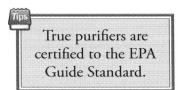

True purifiers are certified to the EPA Guide Standard.

The NSF has recognized this shortcoming and has been working on standards that address all pathogens.[86] As a stop gap, NSF currently tests to Protocol P231, which aligns with the EPA Guide Standard.

If you decide to purchase a water purifier, find one that is certified to the EPA Guide Standard. If the retailer does not clearly advertise this qualification, rest assured the product is not certified. Certification is like winning an Olympic gold medal; you are going to display it for everyone to see.

CHANGING FILTERS

Filter elements must be changed periodically, or the filter will lose its effectiveness. The element can become clogged with debris (usually indicated by reduced water flow), or the surface of the filter can become saturated, leading to reduced performance.[87] The time between filter changes is dependent on the condition of the water being filtered. If you are pumping water directly from streams or lakes, you will need more frequent filter changes than if you are filtering tap water.

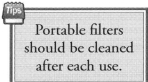

Portable filters should be cleaned after each use.

Follow the manufacturer's recommendations whenever possible. Beyond that, change the filter anytime you notice the flow rate has decreased or the water begins to have an unpleasant taste or odor. Regardless of performance, change the water filter element at least annually for daily use systems.

Portable filters that are used infrequently should be cleaned after each use. This will prevent bacterial growth. A simple way to do this is cycle a weak bleach solution (i.e., 1 tbsp bleach into 1 gallon of water) through the unit. Then allow the filter to air dry.

WATER SOURCES

The fundamental goal of your water plan should be to have access to enough potable and non-potable water to see your family through a two-week disaster. Given the resources necessary to store hundreds

Alternative Water Sources

➢ Hot water heater
➢ Water pipes
➢ Toilet
➢ Waterbed
➢ Swimming pool
➢ River, lake, or spring

of gallons of water, it is likely that in some situations you may find yourself in need of more water than you have stored.

In the best case, you can simply turn on the tap. The water may require purification before drinking, but at least you won't have to forage for water. In the worst case, however, your local water service may have been cut off due to excessive contamination, shortages, or sabotage.

When tap water is no longer available, you will need to locate alternate water sources. There are likely many such sources around you (see tip box).

HOT WATER HEATER

Your hot water heater is an excellent source of potable water. Many units are 75 gallons or larger in size, giving you a sizable emergency stockpile. Water can be drained out the bottom of the tank through the built-in spigot. The five-step process to drain the tank is straightforward (shown below). Like all extraction methods, however, you need to practice on your particular water heater before a crisis hits. Don't assume that you can do it and then find yourself in an "oh, crap" moment later.

Draining Hot Water Heater

1. Turn off power or gas to the water heater.
2. Turn off the incoming water supply.
3. Attach one end of a hose to the spigot, and put the other end into a bucket below the level of the spigot.
4. Open the pressure relief valve near the top of the tank, or turn on a hot water faucet in the home.
5. Open the spigot and collect the water in the bucket. Careful, it's hot!

WATER PIPES

When water is shut off by your local water authority, it may still be possible to drain the water in your home's pipes for potable water needs. However, to have access to the water in your pipes, you must first prevent it from draining back out of your home. This is done by installing an anti-siphon water valve on your incoming main water line. The anti-siphon valve automatically closes when the water pressure gets too low, thereby keeping the water safely stored in your home's pipes.

Personal aside: I was able to recover just over two gallons of water from the pipes in my two-story home. Cut off the water main, and try this water recovery process in your own house to have a better idea of the expected yield.

> **Draining Water Pipes**
> 1. Turn off the incoming water main—simulates losing water with an anti-siphon valve installed on your water line.
> 2. One by one, turn on all of the taps at the highest elevation, collecting the small amount of water that comes out of each. Leave the taps open.
> 3. Turn on the taps at the lowest elevation, collecting the water that comes out. This might take several minutes.
> 4. Once all of the taps stop outputting water, close them.

TOILET

The water from the toilet tanks (not bowls) can be used for drinking, but you should purify it first. Purification is needed because bacteria and rust can collect in the tank. Use a portable filter to retrieve the water, or simply scoop the water out with a cup and soak up any remaining water at the bottom with a rag or sponge. While the thought of drinking water from the back of a toilet may not be particularly appealing, it does represent a valuable secondary water recovery method. Depending on your particular model of toilet, this could yield anywhere from 2 to 7 gallons of water per toilet.

WATERBED, SWIMMING POOL

If you have a waterbed or swimming pool, it can act as an excellent secondary source of water. However, according to the NSF, the water from a waterbed or swimming pool should only be used for non-potable needs.[88] High levels of chemicals and organic contaminants may be present, and conventional methods have not proven effective at making this water completely safe.

RIVER, LAKE, OR SPRING

The water from rivers, lakes, or other fresh water sources can be used directly for non-potable needs, such as toilet flushing, or for potable needs, such as drinking and cooking, if purified first. If the water contains visible particulates, such as dirt or twigs, use a coffee filter or clean cloth to do a cursory filter before purification. Regardless of how clear the water may appear to be, assume all natural water sources are contaminated. If the water is very shallow, use a rag or sponge to soak it up.

Sea water should not be used for potable needs unless purified with a desalination device. Other purification methods will not remove the salt. Never drink brackish water since it will dehydrate you and can lead to death.

NATURAL COLLECTION

Natural Collection Methods
➤ Rain collection
➤ Snow melting
➤ Dew collection
➤ Transpiration
➤ Solar still

> Don't rely on natural collection methods. Treat them as a last resort.

When you have exhausted all known water sources, it may be necessary to extract water from the environment around you. Extraction techniques are referred to as *natural collection*. Collection of this sort should be considered a last resort. Despite what you may have read in various survival manuals, natural collection is very difficult and requires skill, materials, and patience.

Don't make the mistake of relying on natural collection without having spent the necessary time and energy learning how to do the extraction effectively. There are very few people, for example, who can build a solar still and get more water from it than they put out in sweat building it. Before relying on any natural water collection method, practice it ahead of time! Don't wait until you are dying of dehydration to figure out that you don't know what you are doing.

RAIN COLLECTION

> A kid's inflatable swimming pool makes a great rain collector.

Great American Rain Barrel

Rain water can be an excellent source of natural water. The obvious (and significant) disadvantage of collecting rain water is that rainfall is unpredictable. Also, given that in most places it usually rains one inch or less with each rainfall, you will need a large surface area to collect enough water. A child's inflatable swimming pool works well. A six-foot diameter pool collects about 18 gallons of water if it rains one inch.[89]

An alternative is to use a clean, waterproof tarp tied up into a mild "V" shape, sloping downward into a large container. With a 10 ft. × 12 ft. tarp, you can collect up to about 70 gallons of water from a one-inch rainfall.[89] Be sure to secure your tarp so that it won't be blown down by heavy winds. The idea is to face the tarp into the wind, allowing rain to blow onto the surface, down the channel, and into your water storage container.

Another good method of collecting rainwater is to place buckets (or rain barrels) under your home's gutter downspouts. The large surface area of your roof will yield significant water. People have been doing this for years to collect water for use in their gardens, as well as prevent flooding of their yards.

A final, albeit less effective, option for collecting rain water is to hang bed sheets outside your windows. Let them get drenched in the rain, and then bring the sheets inside and ring the water out into a container. Repeat for as long as the rain continues.

All rain water should be purified by one of the recommended methods before drinking.

SNOW MELTING

If you live in a cold climate, snow can serve as an excellent natural water source. Simply scoop the snow up, packing it into buckets, pots, or plastic bags, then take it indoors or put it near a heat source to melt. The amount of water extracted from snow varies greatly, but a reasonable estimate is to assume that it will be reduced by a factor of ten (i.e., ten inches of snow might yield one inch of water). Though not always necessary, it is safest to purify the resultant water as you would any other natural source.

A particularly useful method of collecting snow is to gather it in a cotton pillowcase. Once full, hang it near (not over) a heat source, such as a campfire or wood-burning stove. Put a pot underneath it to catch the water as it drips through. Once the snow is depleted, put the pot over the fire and bring it to a boil. When cooled, it's ready to drink. The pillowcase acts not only as a useful collection container, but also a coarse filter—removing twigs, rocks, and other debris.

DEW COLLECTION

In heavily vegetated areas, dew can be collected off plants early in the morning or immediately after a rainfall. This can be done by dragging absorbent rags across the surface of plants. Once the rags become saturated, wring them out into buckets. Continue the process until the yield starts to decrease as temperatures rise.

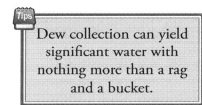

Dew collection can yield significant water with nothing more than a rag and a bucket.

The advantages of dew collection are that it requires only minimal supplies and can yield fairly good results even in the wilderness. The two drawbacks are that the collection process can be mildly arduous, and once again water must be purified before drinking. The method is also not effective in areas with limited vegetation.

Personal aside: Following a rainfall, I was able to collect 12 cups of water in one hour using only a sturdy paper towel and a small bucket—a significant yield.

TRANSPIRATION

Transpiration bags use evaporation and condensation to collect water. Large clear plastic bags are secured over the green foliage of non-poisonous plants. Plants with large root systems work best. The opening of the bag is tied off to make it as airtight as possible. Using a cloth or paper as a gasket in the mouth of the bag will also help. The bag creates a greenhouse effect causing the plant to release water

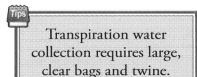

Transpiration water collection requires large, clear bags and twine.

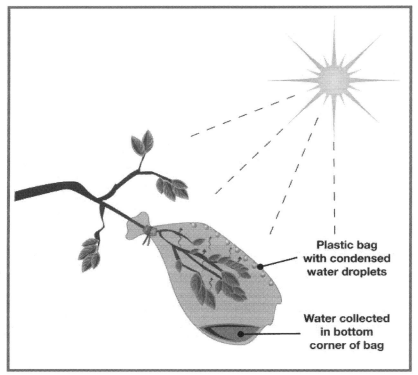

**Plastic bag
with condensed
water droplets**

**Water collected
in bottom
corner of bag**

Transpiration

vapor. The vapor then condenses on the inner surface and pools in the bottom corner of the bag (see illustration above).

The amount of water released through transpiration varies by temperature (as temperature rises, yields increase), relative humidity (as humidity rises, yields drop), plant type, and soil moisture. The biggest advantage of using transpiration bags is that they can be placed with very little energy, allowing for many bags to be used in parallel. However, practical yields are often minimal (perhaps only a cup per bag per day, depending on conditions). Also, be aware that transpiration bags can kill the plants.

A modest secondary benefit can be had by wiping dew from the outside of transpiration bags during the early morning hours.

SOLAR STILL

The solar still is a well known natural water collection method also based on the greenhouse effect. Two simple models are the single-sloped box still and the pit still (see illustration). With the single-sloped still, a sealed box is constructed with a dark insulator material lining the bottom, a sloped clear glass or plastic barrier on top, and a way of introducing and removing the contaminated and distilled water.

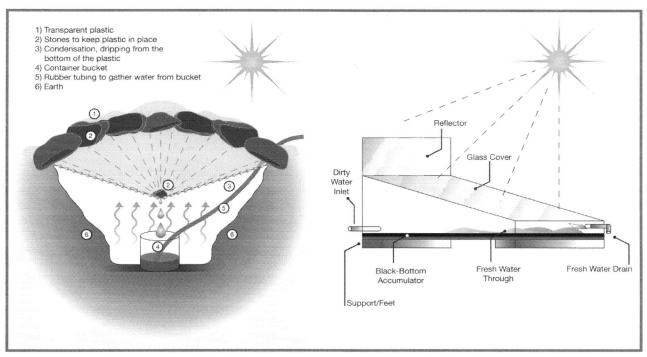

1) Transparent plastic
2) Stones to keep plastic in place
3) Condensation, dripping from the bottom of the plastic
4) Container bucket
5) Rubber tubing to gather water from bucket
6) Earth

Reflector

Glass Cover

Dirty Water Inlet

Black-Bottom Accumulator

Fresh Water Through

Fresh Water Drain

Support/Feet

Solar stills – pit style and single-sloped box

The pit solar still is the type you will find in most survival manuals. It consists of a large hole in the ground, perhaps three feet across, covered with a clear plastic barrier. A collection cup is placed in the center of the pit, and a water source (e.g., shredded vegetation, urine, brackish water) around the cup. A rock is put on top of the plastic to form an inverse apex centered over the cup. Rubber tubing can be used to drink water from the cup so as not to disturb the still.

Both types of stills operate in a very similar manner. Solar energy heats the ground or black background. Moisture inside the greenhouse evaporates, rises, and condenses on the underside of the clear barrier. The solar-distilled water then runs down the slope and drips into a collection channel or cup.

One significant advantage of the solar distillation process over other natural collection methods is that the water recovered *does not* require purification. Plants, sea water, and even urine can be used as the originating source of water. All will ultimately produce clean, drinkable water.

However, there are three notable problems with the solar still. First, it requires materials (plastic, collection cup, and tubing) and a shovel for digging the hole. The still also takes significant energy to set up. Finally, it is very difficult to make work effectively. Experts who have evaluated the effectiveness of the solar still suggest that many things can go wrong— causing it to provide very little if any water. Problems can include failing to get the plastic sufficiently tight, wind disturbance, insufficient transparency of plastic, improper angle to the sun, and a host of other things.[90] Some experts swear by solar stills. Others will warn you away.

> **Tips**
> If you think solar stills are easy to make work, you've never built one.

DP PLAN EXAMPLE

Table 4-6 Sample DP Plan Entry

Need: Water			
Danger	**Goals**	**Needs**	**Implementation**
Shortage	Access to enough water for family of 5 to survive for 14 days	Potable: Minimum of 28 gallons per person = 140 gallons Non-potable: Enough water to flush the commode 5 times per day @ 3 gallons per flush = 210 gallons	Store 250 gallons of treated potable water in a Super Tanker container. Access neighbor's swimming pool for non-potable use. Use 5-gallon buckets for water retrieval. Use two rain barrels on main house gutters.
	A 2-day supply trans-portable by car	Transportable: Minimum of 4 gallons per person = 20 gallons	Transport water using four 6-gallon blue water jerry cans
Contamination	Neutralize or remove all pathogens in water coming from at least one tap in the house	POU water filter purifies tap water A backup plan in case filter becomes inoperable	Install a Seagull IV water purifier on the kitchen tap. Boil water as the backup plan.

Quick Summary - Water

➢ Store, or have access to, enough potable water to maintain your family for 14 days, assuming 2 gallons per day per person.

➢ Have access to enough non-potable water for at least one toilet flush per person per day for 14 days. The amount of water required per flush varies based on toilet model.

➢ Hygiene is critically important during a disaster because falling ill can leave you weak and unable to handle the situation's hardships.

➢ Store potable water in FDA-approved containers, out of light, and away from chemicals or gasoline.

➢ If keeping a permanent stockpile, rotate the water every six months or treat with a water preserver.

➢ Water pre-treatment is usually unnecessary for tap water.

➢ Pathogens found in water include protozoa, bacteria, and viruses. To purify the water, you must remove, kill, or neutralize the pathogens.

➢ Distillers, certified purifiers, or reverse osmosis systems are recommended for home POU or POE use.

➢ Certified purifiers are recommended for portable use. Other options include combining filtering with boiling, UV light, or chemical treatment.

➢ Use secondary sources as necessary, including your home's hot water heater, water in house pipes or commode tanks, swimming pool, waterbed, and rivers or lakes.

➢ Do not try to purify seawater without a desalination device.

➢ Depend on natural collection methods only as a last resort. Natural methods include: rain collection, snow melting, dew collection, transpiration, and solar stills.

➢ Test your water collection methods before an emergency.

Recommended Items - Water

❏ Water Storage
   a. Large water containers to allow for 28 gallons of potable water per person stored in FDA approved containers—e.g., 55-gallon water storage barrels with pump, Super Tankers, Aquatank water bags
   b. Two sets of smaller water containers for retrieving and transporting water (one set for potable water, one set for non-potable water)—e.g., plastic jerry cans, 2-liter bottles, five-gallon buckets, WaterCubes, dromedary bags

❏ Water retrieval
   a. Garden hose and bucket for draining the hot water heater

❏ Chemical disinfectant for preparing containers, treating water, and cleaning surfaces.
   a. Jug of household bleach, dropper

❏ A home POU or POE water purification system
   a. A purifier that meets the EPA Guide Standard, *or*
   b. Water distiller, *or*
   c. Reverse osmosis system

❏ A portable water purifier
   a. A purifier that meets the EPA Guide Standard, *or*
   b. UV light, *or*
   c. MIOX device

❏ Supplies to stay clean
   a. Adequate supply of hand soap, alcohol-based hand sanitizer, and hand/body wipes
   b. Supply of large plastic bags and twist ties

❏ Sanitation needs
   a. Access to enough non-potable water for at least one toilet flush per person per day, *or*
   b. A potty bucket, plastic bags, and treatment chemicals, *or*
   c. A self-contained composting toilet

CHAPTER 5
SHELTER

Challenge

Your microwave oven short-circuits in the middle of the night, igniting a house fire. The fire smolders for several hours before spreading, filling the house with smoke. Do you have the correct smoke alarms to detect the slow-burning fire? How will your family evacuate the house if the primary exit is impassible? Can your children escape without your help? Does everyone know what to do once they get out of the house?

There is a well understood order to wilderness survival: shelter, water, and then food. The idea is simple enough. You must first protect yourself from the environment, whether it is the heat of summer, the frost of winter, or the drenching rains of spring, before worrying about what you are going to eat or drink. This generally holds true for disaster preparedness as well. The only difference is that you are not typically required to build a shelter. Rather, you will need to effectively use and protect the shelters you already have.

> In order of importance: shelter, water, then food.

Your home—a generic expression for a house, apartment, townhouse, or mobile home—is the shelter that your family will most likely depend on when confronting a disaster. Certainly there may be times when you are forced to evacuate, in which case you will have to secure a new shelter. However, for most situations, your family will be much better served by staying put. As shelters go, your home is far superior to anything you can construct ad hoc. If you don't believe this, try spending a cold, rainy night in your yard under the best shelter you can build in half a day. You will quickly come to enjoy the comforts provided by your home, even with the power, water, and gas all turned off.

There are certain threats that require specialized shelters. For example, buried or hardened shelters offer the best protection from radioactive fallout.[91] They can be built from conventional materials (stone block, concrete, etc.), or they can be made from such things as fuel tanks, shipping containers, steel culverts, Quonsets, or pre-made fiberglass inserts. However, despite being well hardened, underground bunkers are not generally the best solution to more commonplace disasters.

Your home as a shelter *(FEMA photo/Dave Gatley)*

Take a look at the list of disasters in *Chapter 1: Introduction,* and you will see that, for most situations, a buried bunker doesn't really offer any advantage over a conventional house. Specialized shelters of this sort are also very expensive for a private individual, not to mention nearly impossible for people living in metropolitan areas. For these reasons, this chapter focuses on issues related to using a conventional residence as an effective shelter. A section in *Chapter 14: Transportation* discusses the special case where you find yourself on the road depending on your automobile to serve as an emergency shelter.

In general, there are three important steps in preparing your home to be an effective shelter:

1. Assess what protection your home provides.
2. Make improvements to increase that level of protection.
3. Safeguard your home to prevent loss, damage, or deterioration.

ASSESSING AND MAKING IMPROVEMENTS

Begin by evaluating the safety that your home provides. Factors that weigh into this evaluation are construction, geography, and distance from likely threats.

- **Construction**—What is your house made from? Brick, wood, vinyl, or sheet metal? Are there any specific threats to which your house's construction will be particularly vulnerable? Are there large exposed windows, and if so, can they be shuttered or boarded? Do you have a basement that might flood? Is the construction solid? Can you identify any obvious weaknesses, such as an exposed carport with insufficient supports or a roof with loose shingles?
- **Geography**—In which part of the country do you live? What weather events are most likely to affect you? Are there any specific geological threats such as mudslides, avalanches, earthquakes, rockslides, or sinkholes? Does the landscape offer any natural protections? Are you on the coast

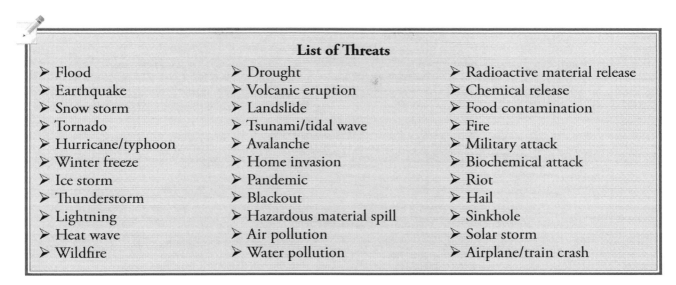

List of Threats

➢ Flood	➢ Drought	➢ Radioactive material release
➢ Earthquake	➢ Volcanic eruption	➢ Chemical release
➢ Snow storm	➢ Landslide	➢ Food contamination
➢ Tornado	➢ Tsunami/tidal wave	➢ Fire
➢ Hurricane/typhoon	➢ Avalanche	➢ Military attack
➢ Winter freeze	➢ Home invasion	➢ Biochemical attack
➢ Ice storm	➢ Pandemic	➢ Riot
➢ Thunderstorm	➢ Blackout	➢ Hail
➢ Lightning	➢ Hazardous material spill	➢ Sinkhole
➢ Heat wave	➢ Air pollution	➢ Solar storm
➢ Wildfire	➢ Water pollution	➢ Airplane/train crash

where ocean-related weather events, such as hurricanes or nor'easters, pose very real concerns? Is your home situated in a valley that might flood from heavy rains? Are you close to a river or lake that could overflow its banks? Are you off the beaten path, making it difficult for repair crews to get to you? Are there large trees that might fall on your home?

- **Distance from likely threats**—Do you live near a nuclear or industrial plant that might release an airborne contaminant? Are you in a large metropolitan area, such as New York or Los Angeles, that might experience a terrorist attack? Do you live near a railway or airport at which a major crash might occur? Are you near a dam that might rupture and cause flooding?

Consider the list of threats in the tip box. How likely is each to affect you? How much protection would your home offer? What steps can you take to improve that level of protection?

To make your home threat assessment a little easier, a blank worksheet is included in the *Appendix*. Table 5-1 serves as an example of what a threat assessment entry might look like. This particular example is assessing my home against floods.[92,93,94]

Assessments are completely subjective—you can rate your home at whatever level you think appropriate. The point of the exercise is to have you think about your home's vulnerabilities, and then take steps to reduce them. Realize that many of the threats listed are very unlikely, requiring little or no preventive action. Again, the key to practical preparation is to identify the likeliest (or most worrisome) threats first, and then make improvements to reduce their impact.

Personal aside: Many years ago, I lived in a mobile home in rural Alabama. I scrutinized the weather regularly, fearing the announcer would utter the single word we all dreaded—tornado. When twisters were spotted anywhere in the vicinity, alarms would sound for miles, and my family would race to a small brick building that the landlord provided for just such occasions. The lesson that I took away from the experience was that it is very important to understand what protections each type of shelter offers. In my case, the small reinforced brick building was much more likely to survive the winds of a tornado than a mobile home constructed of sheet metal and particle board. With that said, I never felt the need to evacuate when thunderstorms, ice storms, blackouts, or winter snow threatened. And in general, I think I had it right. My rather fragile house was perfectly capable of handling certain dangers but ill-equipped for others.

Table 5-1 Example of Shelter Assessment (Flood)

Threat	Likelihood (1-10)	Protection (1-10)	Steps to Improve
Flood	**8** I live on the east coast where flooding is common, usually due to nor'easters and hurricanes.	**5** My home is situated atop a small hill, 42 feet above sea level. It has gutters to route water effectively off the roof. Unfortunately curbside water drains are directly in front of the house and thus prone to backup and flooding.	1. Determine the base flood elevation (BFE) for the area.* 2. Ensure that water-resistant materials have been used in places below BFE. If not, replace them as appropriate. 3. Raise the water heater and other appliances above BFE level. 4. Install backflow valves in sewer pipes to prevent backups. 5. Install flood shields or natural built-up barriers near basement windows and doors. 6. Install sump pump system for basement. 7. Check yard grading to ensure adequate water drainage. 8. Use rain barrels on the two main gutters. 9. Landscape using vegetation that resists soil erosion. 10. Seal walls in basement with waterproofing compounds. 11. Add flood insurance coverage.

*BFE is the height that a flood has a 1% chance of reaching or exceeding for a given year—a.k.a.,the 100-year flood level. Maps showing your base flood elevation and flood zone designation are available online at FEMA's Map Service Center.[95]

STRUCTURAL IMPROVEMENTS

There are several structural improvements that homeowners can make to protect their home from very high winds, such as those experienced with hurricanes. The improvements target the four primary areas of weakness: roof, doors, windows, and garage doors.[284]

Roof: The roof of your home should be designed to transfer wind energy down through the walls to the foundation. Homes with gabled roofs (a.k.a. A-frames) are more likely to suffer damage from hurricane-force winds because the end wall takes a great deal of stress. Several improvements can be made to shore up a roof:

- Inspect your roof for loose or damaged shingles and replace as necessary.
- Install additional truss bracing—2x4s that run the length of the roof and overlap across two center trusses (see illustration on next page). Truss braces should be installed at three levels:

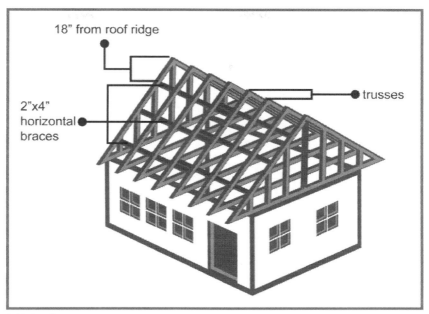

18" from roof ridge

trusses

2"x4" horizontal braces

Gabled roof

eighteen inches from the roof ridge, in the center span, and near the base. Attach them at each truss using 3-inch, 14-gauge wood screws or 16d galvanized nails.

- Install gable end bracing—2x4s placed in an "X" pattern from the top center of the fourth truss to top center of the gable (see illustration below). Attach bracing to the gable and each of the four overlapped trusses using 3-inch, 14-gauge wood screws or 16d galvanized nails.
- Install hurricane straps—galvanized metal straps used to hold the rafters and walls together. These may require professional installation.

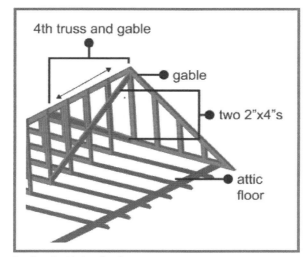

4th truss and gable

gable

two 2"x4"s

attic floor

Bracing of gabled roof end

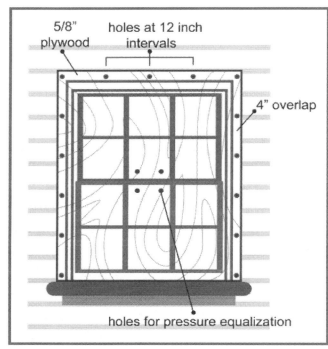

Makeshift shuttering with plywood

Doors: Double-entry doors (a.k.a. French doors) should be secured using reinforced bolts at the top and bottom of the inactive door. In addition to protecting your home from high winds, they also help to make the doors harder for intruders to break in. Doors with windows should have the windows covered with shutters, protective film, or plywood.

Windows: Windows, sky lights, and glass doors should be protected by covering them with storm shutters, protective film, or 5/8-inch plywood. If using plywood, precut it to size by adding 4 inches to each side. Before the threat arrives, attach it around the window frame's periphery every 12 inches using wood or masonry anchors, lag bolts, and large washers (see illustration). Make sure the anchors are securely installed into the wood frame or masonry, not the siding or trim. If replacing your windows, consider installing impact-resistant windows.

Garage Doors: Most garage doors can be strengthened using retrofit kits or braces available at home improvement stores. Garage doors can also be reinforced using 2x4s or light-gauge, metal girds attached

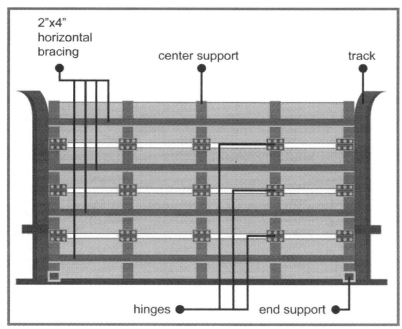

Bracing garage doors

horizontally across the door (see illustration on previous page). Additionally, the springs, end supports, and hinges can be replaced with heavy-duty versions. Check the tracks to be sure that they are firmly attached to 2x4s inside the walls and ceiling. Once strengthened, the door may require rebalancing (done by adjusting the heavy springs), but this can be dangerous and should only be done by a professional. Double-wide garage doors are particularly susceptible to be blown off their tracks and collapse from high winds. Specialty garage doors designed to withstand high winds are available but at a premium cost.

For additional preparations against high winds, such as those associated with tornadoes or hurricanes, see the recommendations given in *Chapter 18: Five Horsemen of Death*.

MAINTENANCE

A home is only as good as the elbow grease put into it. It needs to be maintained with diligence. Major systems include roofing, heating and cooling, appliances, plumbing, electrical, and structural. Each system should be regularly inspected, and any problems discovered should be repaired without unnecessary delay—an empty pocketbook being the most likely cause of a "necessary delay." Think of your house as your personal fortress against the dangers of the world. In doing so, give it the attention that it deserves.

One of the first things you should do is learn how to shut off your utilities (e.g., electricity, water, and gas). Don't just *think* you know how to shut them off. Go through the motions a couple times. Keep the necessary shutoff tools near the cutoff points. This way you won't have to hunt for them when trouble is knocking on your door. One word of caution: don't actually shut off your gas. It may require a professional to come out to turn it back on. That can be a hassle, not to mention expensive.

Even if you do maintain your home properly, damage will undoubtedly occur. Perhaps high winds or a fallen tree are to blame. In the case of a quickly passing disaster, you can simply head to your local home improvement store or call your insurance company to hire appropriate contractors to make repairs.

If the danger is more prolonged, you may need to make impromptu emergency repairs to patch a leaky roof, cover a broken window, or replace a cracked pipe. For these types of home repairs, you will need a few tools and supplies. However, it is not necessary to have a toolkit that would make Bob Vila proud. Your money is better spent elsewhere. For now, a basic set of hand tools that you know how to use (e.g., hammer, screw drivers, wrenches, hand saw, pliers, pry bar) will suffice.

> **Home Repair Supplies**
> ➢ A good set of hand tools
> ➢ Basic home repair materials
> ➢ Home fix-it books
> ➢ Cleanup equipment

Likewise, limit your construction materials to those needed to perform emergency repairs on your home, such as fasteners, plastic sheeting, PVC fittings, and duct tape. You don't need pallets of sheetrock or bags of concrete. Just keep it simple. Pick up a few home fix-it manuals to help you understand the basics of home repair.

> Tips
> Cleanup equipment can be useful for clearing roads, digging out, and putting your property back in order.

It's a good idea to have some cleanup equipment on hand too, such as shovels, an axe, wheelbarrow, gloves, push broom and dustpan, heavy duty garbage bags, and a chain saw if you can afford one. All of these items can be very useful for cleaning up after a major storm blows through.

As for proficiency, that comes with practice and perhaps a bit of one-on-one instruction from a local handyman. Developing some basic know-how will not only help you to make emergency repairs but also to appreciate the details of your home's construction.

Personal aside: When a particularly nasty nor'easter blew through town a few years ago, I had to carefully hang outside my second story window to secure a large tarp over a leaky window. It was an unpleasant (and very wet) job, but I did feel fortunate to have had the right materials available to keep my house from becoming a swimming pool.

HARDENING YOUR HOME

There is one particular area that often doesn't get the attention it deserves, and that is hardening your home from intruders. It is not necessary or even wise to be paranoid, but it is important to be honest enough to accept that bad things can happen to good people. Home invasions and burglaries occur in nearly every community, and during times of crisis, even otherwise law-abiding people may feel compelled to take desperate actions.

Try this simple 30-day challenge. Imagine for a moment that a nasty-looking thug comes up to you and says, "In thirty days, I am going to break into your home and scare the hell out of you and your family."

What would you do? Call the police—for sure. Get a gun—probably. One thing for certain is that you would turn your attention to hardening your home and preparing your family. The point in posing this ridiculous scenario is to motivate you to take some precautionary steps that might one day prevent you from becoming a victim of theft, burglary, or home invasion.

It is possible to transform your house into a "hard target" without turning it into a maximum-security prison. Consider some of the many proven methods used to improve your home security and deter criminals.[96,97,98,99,100]

BE MORE CAUTIOUS

> The first step in being safer at home is becoming more cautious.

Before you even think about hardening your home, teach everyone in your family to be more cautious. Even simple actions like keeping all your doors and windows locked, whether you are at home or away, will go a long way toward keeping your family safe. It is estimated that 90% of all illegal entries occur through doors.[100]

Be careful about how you open your door, and who you open it to. Never blindly open the door—always look first. Ideally use a peephole, chain lock, or floor-mounted retractable door stop to help you better assess the situation. If you see an unexpected repairman, evangelist, or solicitor, simply don't open the door to them. They will get the message and move on to the next house. If an unknown person persistently knocks on your door, call a neighbor to come over to back you up. Finally, never allow your children to open the door unattended.

Personal aside: When I was in college, one night at around 1:00 AM someone began banging on my apartment door. Purely by coincidence, I was up that night cleaning my handgun from a day at the range. I foolishly answered the door without checking the peephole, thinking that someone was in trouble—perhaps needing medical attention. What I found was a scruffy-looking man who smelled of booze and whose intention never became fully known. When he saw the pistol hanging loosely at my side, he apologized and hastily stumbled away. I was fortunate that my mistake didn't cost me my life.

HOME INVASIONS

Home invasions are perhaps the most terrifying physical altercations that anyone could imagine occurring in their home. Intruders are usually seeking to rob their victims, but such encounters often end in terrible violence. Most men assert that they could fight off an intruder. Perhaps under normal circumstances, this might indeed be true. However, people who commit home invasions are not likely to give you a fair fighting chance. Rather, they will outnumber you, use weapons, and act with immediate, decisive violence. This will put you and your family at a critical disadvantage that is difficult to overcome. *I say this as a person who has studied the martial arts my entire life.*

Given that every family situation is different, there is no single best tactic for surviving this type of threat. That said, one action is likely to provide the best chance of a positive and quick resolution for most families. The tactic is simple and straightforward. If one or more people try to push their way into the home, every member of the family should immediately attempt to escape and call for help. Why? Because if even a single person escapes, the situation will almost certainly resolve itself very quickly. The chances that someone will be killed or raped decrease dramatically because the intruder will know that the police (or neighbors) will be arriving shortly. They may strike the person answering the door, perhaps grab a few easily-accessible valuables, and then make a run for it. This outcome is significantly better than one that puts your family at the mercy of violent criminals for countless hours.

> Tips
>
> Every family should have an emergency word.

An emergency code word can be used to signal danger and let family members know that it's time to escape and seek help. An emergency word can also be lifesaving in other situations. For example, if forced to pretend that everything is okay (perhaps on the telephone) while a victim of some crime, incorporating the emergency word into your conversation can let a loved one know that you are under duress and require rescue. The emergency word should be something truly unique so that it cannot be accidentally used in conversation. For example, my family uses the word "jubalo," which a few select readers might recognize. The exact word to be used is not important. What is important is that every family member understand that the word is a very serious cry for help.

Practice the home invasion challenge in *Chapter 19: Trial by Fire* to help your family better prepare for this dangerous threat. Obviously if you live alone or have only a single exit from your home, this strategy does not work well. In that case, you will have to decide whether to fight, attempt escape, or comply with the intruders' demands.

MAKE IT HARDER FOR A PERSON TO SURPRISE YOU

A guard dog can be a wonderful companion and an almost unbeatable security system.

(Wikimedia Commons/Ana Kompan)

Take steps to make it harder for someone to surprise you. Consider installing an electronic home security system. Even if the system just serves as a noise-maker, it will alert your family when someone tries to enter your home while you are asleep. The noise may also be sufficient to scare away an intruder.

An alternative is to have a guard dog act as your early warning system. But don't rely wholly on your pet's natural instincts. Make the effort to teach your dog that security is an important part of his/her role in the family. A point of distinction is in order. A *guard dog* is a dog that will make noise when someone tries to enter your home. It could be as tiny as a Chihuahua. An *attack dog* is a dog that will viciously attack an intruder. Unless you live alone and know how to handle a trained attack dog, you should definitely limit yourself to a guard dog. They are much easier to train and can serve as a loving companion.

Keep the exterior of your home well lit using floodlights and motion-detector lighting under the eaves or in the yard. Keep bushes and tree limbs cut back to prevent anyone from hiding outside your home or using trees to access your upper floors. Put gravel or pebbles under the windows as a noise deterrent. Also, consider planting thorny bushes, such as Barberry or Hawthorne, in front of the windows.

Motion-detector lighting *(courtesy of Heath Zenith)*

DON'T BROADCAST YOUR WEAKNESS

Never let anyone other than family or friends know when you or your spouse will be out of town. If you travel and must leave the house empty, use light timers to make it seem like someone is at home. When particularly vulnerable, such as when home alone, let your closest neighbors know that you'd appreciate them keeping an eye on you. It is also useful to create the illusion of additional security by keeping a large set of boots or a dog bowl with the word "Killer" on the front porch. A home security sign in the yard can act as a similar deterrent.

HARDEN YOUR DOORS AND WINDOWS

Shore up your doors and windows. Install heavy-duty, solid-core external doors and deadbolt locks with 1-inch or longer bolt throw. Replace hinge screws with longer wood screws (2 inches for doors, 3 inches for frames). Replace strike plates with high-security models. At a minimum, replace your existing strike plate screws with 3-inch wood screws. If it isn't possible to install adequate locks, consider using inexpensive security bars that wedge between the doorknob and floor (e.g., Master Lock 265), or keyless security latches that are many times stronger than deadbolts (e.g., Meranto DG01-B).

> **Tips**
> A simple step to improving a door's security is to install longer screws in the hinges and strike plate.

> **Tips**
> Don't use double-cylinder locks in your home because they introduce a fire safety hazard.

If you have a glass window in your door, it may be possible to install shatterproof plastic or security glass—assuming it doesn't already have it. If that's not possible, you can line the glass with security film, such as 3M's Prestige.

For French or double doors, use inset-keyed slide bolts at the top and bottom to lock in place the lesser-used door. For sliding glass doors, you can drill a hole in the door and frame and insert a locking steel pin, use a window bar, or insert a rod (wood or metal) in the bottom track to prevent it from being forced open.

A lock is only secure if the intruder doesn't already have the key. Don't hide a key outside—especially not under the mat, above the door frame, or in nearby flower pot. It is much better to give a spare key to a neighbor or relative. If you move into a new home, immediately rekey the locks. You might also consider using battery-operated electronic locks that can be easily reconfigured with new codes, should they ever be compromised. Don't use a deadbolt requiring a key on both sides (a.k.a. a double cylinder lock), since it might slow your family's escape in the event of a fire.

Courtesy of Schlage

Recent studies have shown that most locks are vulnerable to an intruder drilling out the cylinder, rendering the lock useless. High-security locks, such as those from Medeco, are made with hardened cylinders and offer excellent protection from this very real threat. However, they are quite expensive.

If windows are not going to be used as fire exits, you can nail or screw them closed, install keyed latches, or cover them with metal grillwork. If the window may act as a potential fire escape, consider using a removable window bar (e.g., Master Lock 266), or quick release, hinged interior bars. An alternative for high-risk windows, sliding glass doors, and storm doors is to cover them with protective security film (e.g., 3M's Prestige).

HARDEN YOUR BASEMENT AND GARAGE DOORS

> Garage and basement doors are easy targets for someone wishing to break into your home.

Due to their privacy and weak construction, basements and garages are often easier points of entry for intruders. To help mitigate these vulnerabilities, install solid-core exterior doors to your basement and garage, and consider using motion-sensor lighting outside. Be sure to include your garage and basement in your home security-system.

If you have automatic garage door openers, reprogram them to be different from the factory setting (usually done by changing a couple of internal switches). The same goes for changing the external key-pad access code. Treat your garage remote with the same level of care as you would a house key. Also, if you are guilty of accidentally leaving your garage open at times, consider getting an electronic garage door monitor.

Most garages have manual slide bar latches on each side that can lock the door in place from the inside, making forcing the garage difficult. An alternative to a slide bar is to drill a hole in the track and use a bolt or padlock to keep the door's rollers from moving. Both of these security measures are handiest for use at night or when away on vacation.

Garage door slide latch

SECURE THE COMMUNITY

An important part of home security is being part of a community that looks out for one another. In many neighborhoods this manifests as a formal watch program. If your community doesn't have a neighborhood watch program, at least work to establish an "I'll watch your back if you'll watch mine" mentality with your neighbors. An alternative to a neighborhood watch is a *Neighborhood Ready* program. In this type of community group, neighbors not only commit to watching out for crime, but also to come to one another's aid during a disaster. More information about establishing a Neighborhood Ready program is available in *Chapter 17: Creating a DP Network* and at *http://disasterpreparer.com*.

> **Tips**
> A neighborhood watch program is a great way to find neighbors interested in being part of a DP network.

Neighborhood Ready

SAFE ROOM

During certain events (e.g., terrorist threats, contaminant leaks, tornadoes, hurricanes), your family may be safer confined to a small fortified place within your home.[101] Unless you already live in a small home or apartment, it probably makes sense to create a "safe room." This is a secure area that you and your family can retreat to when a serious disaster threatens.

Select a structurally sound room in your home, perhaps a basement or large closet. Ideally, pick an interior room with easy access to a toilet, running water, and a telephone. Stock the room with a complete set of emergency supplies, including blankets, food, water, medicine, a first aid kit, flashlights, NOAA All Hazards weather radio, batteries, a cell phone, whistle, children's toys, and weapons.

BIOCHEMICAL CONSIDERATION

If because of your location you feel that you are at risk of a chemical or biological threat, you may wish to locate the safe room on the upper floor of your home. Gases tend to settle, so the upper rooms are likely to have less airborne contamination. Also, you will want to consider storing some additional items for protection from biochemical hazards:

- Duct tape and plastic sheeting for sealing around doors, windows, and heating/cooling vents.
- A HEPA air filter. When properly sized to the room, HEPA filters have been shown to be effective in removing vapor contaminants and some poison gases.[102]
- Gas masks or protective hoods, carefully adjusted to fit each family member. Inexpensive disposable respirators, such as N95 masks, can also be used.
- Disposable biochemical protection suits, such as Tyvek F.

For additional details and recommendations regarding sheltering in place and safeguarding against airborne hazards, see *Chapter 9: Air*.

There is no evidence to support the myth that a sealed room poses a suffocation hazard.

As with every other preparation, creating a safe room is about weighing risks versus costs. If you live close to an industrial plant, then a room designed to protect your family from chemical leaks probably ranks high on your action list. If you don't live close to a potential biochemical threat, then you may want to tailor the room to better handle other threats—putting your money and efforts toward the most likely dangers first.

Biochemical protection *(U.S. Navy)*

You may wonder if it is possible to seal the room so well that suffocation becomes a risk. The only documented case found was in 2003 when three Israeli citizens suffocated in a sealed safe room.[103] Their mistake, however, was not in over sealing the room, but in burning a coal-fueled heater in the safe room. The heater consumed oxygen as it burned, eventually creating a deadly deficit. Fuel-burning appliances should *never* be used indoors without proper ventilation. See *Chapter 8: Heating/Cooling* for more heating safety recommendations.

From the evidence available, it appears to be very difficult to seal a room in your home well enough to endanger your family from suffocation. With that said, it would still be prudent to keep an eye out for signs of oxygen shortage and CO_2 buildup. Symptoms include headache, fatigue, shortness of breath, euphoria, and nausea.[104,105] If anyone confined to the safe room experiences any of these symptoms, everyone should immediately seek fresh air.

SAFEGUARDING YOUR SHELTER

Given that your home is likely one of your most important financial investments, as well as your family's primary shelter during most disasters, every effort should be made to ensure that it is protected from damage. There are three key steps to safeguarding your shelter.

Safeguarding your Shelter

1. Minimize hazards in and around your home.
2. Equip your home with safety devices.
3. Adequately insure your home.

MINIMIZING HOME HAZARDS

Set aside an hour every couple of months to inspect your home for hazards (i.e., things that could endanger your family or your home). Make a checklist, and then work your way through the needed improvements over time—starting with the most important first.

Below are sample outdoor and indoor home hazard checklists. A blank home hazards checklist is included in the *Appendix*.

OUTDOOR

	Clear heavy vegetation away from the home.
	Remove any dead trees that pose a threat from falling or catching fire. Ideally, even live trees should be far enough away to prevent them from hitting your home, should they fall.
	Secure items that might act as flying debris, such as a swing set, barbeque grill, bicycles, deck furniture, or trash cans.
	Keep gutters clear to facilitate proper water drainage from the roof.
	Inspect and clean chimneys or wood-burning stove vents.
	Inspect roofing for loose or worn shingles and degraded seals.
	Check gas lines for signs of corrosion or leaks.
	Make sure land is properly graded for water runoff.
	Store flammables (e.g., gas, paint thinner, cleaners, oily rags) in approved containers, away from heat, and in areas with adequate ventilation.
	Check under home for leaks, water-damage, termites, or mold.
	Insulate or drain any external pipes (e.g., sprinklers) that might freeze.
	If flood prone, raise heating/cooling units above the Base Flood Elevation (BFE) level.
	Make sure your house number is visible from the street so emergency vehicles can easily locate the home.

	In hurricane prone areas:
	  - Install storm shutters, or have precut 5/8-inch-thick marine plywood ready to install over windows and sliding glass doors.
	  - Install truss bracing to secure your roof against high winds.

	In flood prone areas:
	  - Install backflow valve on main sewage line.
	  - Ensure adequate water drainage from around home.
	  - Install sump pump in basement, if applicable.
	  - Elevate your hot water heater and furnace.

INDOOR

	Check that electrical outlets and extension cords aren't overloaded.
	Check for frayed electrical cords.
	Check that all electrical outlets near sinks are equipped with properly wired ground fault protection.
	Connect all sensitive electronics to surge protectors (certified to UL1449 330V).
	Use correct wattage light bulbs in lamps and lighting fixtures.
	Check all safety devices at least twice a year (e.g., smoke alarms, fire extinguishers, CO alarms).
	Inspect underneath sinks and around toilets for leaks.
	Ensure that poisonous substances, such as cleaners, pesticides, medicines, and alcohol are locked up or high enough that young children can't reach them.
	Eliminate clutter in your home, clearing emergency escape routes.

	In earthquake-prone areas:
	  - Place large, heavy items on lower shelves.
	  - Hang pictures and mirrors away from beds.
	  - Secure water heater by strapping it to wall studs.

SAFETY DEVICES

> **Tips**
> Equipping your home with the proper safety devices is *the* single biggest action you can take to increase your family's chance of disaster survival.

Home safety is about being responsible and protective of what is most precious to you—your life and those of your family. This is one of those rare times when there is a clear right thing to do; and that is to take every reasonable precaution when it comes to keeping your family safe.

Fortunately, there are a small set of safety devices that can significantly reduce your chances of dying from an accident in your home. Safety devices are like insurance; most of the time they cost you money and give nothing more than peace of mind in return. But if tragedy strikes, nothing could be more valuable. Don't skimp on properly equipping your home.

SMOKE, HEAT ALARMS

According to the U.S. Fire Administration, fires kill more Americans than all other natural disasters combined.[106] Did you miss that? **Fires kill more Americans than all natural disasters combined!** Forget hurricanes, storms, earth-destroying asteroids, or volcanic eruptions. If you want to make a big impact in reducing your chances of dying from a disaster, take steps to reduce your risk of dying in a fire.

Fire—a very real threat! *(photo by Adam Alberti, NJFirepictures.com)*

The National Fire Protection Association (NFPA) reports that an average household will have five fires through the course of a person's lifetime; roughly one every 15 years.[107] Most will be small fires, perhaps a greasy pan in the kitchen, or a candle igniting a tablecloth—easily managed by the homeowner. The odds of having a fire large enough to be reported to the fire department are 1 in 4, and the chances of someone in your household being injured in a fire are 1 in 10.

According to the NFPA, residential fires kill an average of 3,000 Americans each year and injure another 13,000.[108] Perhaps the most tragic part is that 63% of residential fire deaths occurred in homes with no working smoke alarms. For your family's sake, take a stand against fire by being thoroughly prepared.

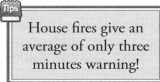

House fires give an average of only three minutes warning!

Early detection is the key to surviving a house fire. Install smoke detectors in the hallways outside sleeping areas as well as in the bedrooms. Keep a minimum of one on every level of your home, even if there are no bedrooms. Put one in the kitchen, as well as the attic, and at the top of the basement stairs. The idea is to put smoke detectors between hazard areas and people areas—providing you with the earliest possible warning. Think about where a fire could start, and where you might be sitting, sleeping, or working; then put a detector between the two locations.

There are three types of smoke alarms available today:

- Ionization alarms—detect flaming, fast-moving fires quickly
- Photoelectric alarms—detect smoldering, smoky fires quickly
- Dual sensor alarms—combined ionization and photoelectric sensors in one unit

Strobe and voice alarms (courtesy of Kidde)

Since both types of fires (i.e., fast-moving and smoldering) are possible, you should equip your home with both ionization and photoelectric alarms, or better yet, use dual sensor alarms. Note that alarms using strobe lights rather than sound are also available for people with hearing disabilities.

Smoke detectors can be powered from batteries or from your home's electrical system. House-powered units with battery backup are preferred. If your house was built after 1993, the alarms installed during construction are *interconnected* alarms—meaning that when one sounds off, it should trigger the others to sound also. This is a great advantage over individual alarms because it ensures that you will receive the greatest warning possible. Did you know that house fires give an average of only three minutes warning for occupants to escape?[109]

If you are installing additional smoke alarms after construction, you will likely have to settle for battery-only models since they are much easier to install. Once again, it is best to select units that are interconnectable—now readily available from First Alert, Kidde, and other manufacturers.

The U.S. Consumer Product Safety Commission recommends replacing your smoke alarm batteries at least once a year.[110] A simple way to remember this is to change the batteries when the time changes in the spring or fall. Always test a smoke alarm after you change the batteries. This is done by pressing the test button on the unit. When you purchase a new smoke detector, you may wish to test the sensitivity of the unit. This can easily be done by lighting a few matches together, blowing them out, and holding them up to the smoke detector. Once the alarm sounds, quickly blow the smoke away, or spray a fine mist of water to clear the air.

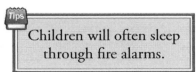

Tips

Children will often sleep through fire alarms.

You might be surprised to learn that children often don't wake up when a smoke alarm sounds. Studies have shown that even at ear-piercing levels, children often remain asleep.[110] To overcome this, there are special "voice" smoke alarms available, some even allowing you to pre-record your own voice as the alarm. Voice alarms of this type have been shown to be more effective at waking sleeping children.

Hint: If you accidentally set off the smoke alarm in the kitchen, spray a fine mist of water underneath it rather than turning off the alarm. Alternatively, kitchen smoke detectors can be replaced with heat detectors to make them less susceptible to sounding from burning food.

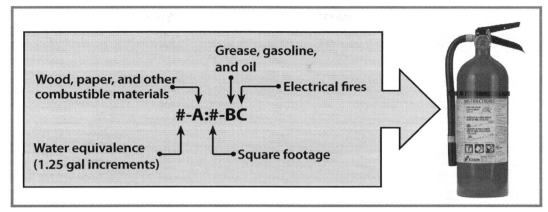

Fire extinguisher labeling

FIRE EXTINGUISHERS

It is important to equip your home with fire extinguishers. The NFPA recommends that you keep at least one primary extinguisher (size 2-A:10-BC or larger) on every level of your home. Supplement these with smaller extinguishers in the kitchen, garage, and car to further reduce your family's changes of being injured or killed in a fire.

> **Tips**
> For general home protection, use ABC extinguishers.

Fire extinguishers spray water or chemicals that either cool burning fuel, displace or remove oxygen, or stop chemical reactions. They must be approved by nationally recognized testing labs, such as Underwriters Laboratory (UL). Fire extinguishers are labeled with an alpha-numeric classification, based on the type and size of fire they can effectively extinguish. It is important to understand the labeling conventions used on fire extinguishers.

The letters A, B, and C represent the type of fire for which the extinguisher is approved: type A is used for standard wood, paper, and combustible material fires; type B is for grease, gasoline, and oil fires; and type C is for electrical fires. A multiclass extinguisher (e.g., BC, ABC) is effective on more than one fire type. For general home protection, use ABC multiclass fire extinguishers. This type of extinguisher is good for nearly any kind of fire except very hot grease fires, chemical fires, and those that burn metals.[111]

> **Tips**
> Caution: Class A air-pressurized water extinguishers (APW) should never be used on grease, electrical, or chemical fires!

The numbers convey information about the size of fire that the extinguisher can put out. The number in front of the A indicates the number of 1.25 gallon units of water that the extinguisher is equivalent to when fighting standard combustible fires. The number in front of the B rating indicates the square footage of a grease, fuel, or oil fire that the extinguisher can put out. There is no number associated with the C rating.

For example, an extinguisher labeled 1-A:10-BC is rated to be equivalent to $1 \times 1.25 = 1.25$ gallons of water for standard combustible fires, and is capable of putting out a grease fire measuring 10 square feet in size.

It is tempting to equip your home with the largest fire extinguishers on the market, but realize that family members must also be able to handle them effectively. A good compromise might be to keep a combination of smaller and larger units distributed throughout the house.

For recommendations regarding the latest fire extinguisher models, consult online reviews, such as the GALT website.[112] If the recommended models are not available in your area, simply find a suitable replacement.

Using an Extinguisher

A fire extinguisher is only effective in the hands of someone who has experience using it. Unfortunately, most people don't know how to properly operate a fire extinguisher—understandable since the average person doesn't have an opportunity to practice with them. To gain this much-needed and valuable experience, set up a well-controlled practice session. It is worth the cost of an extinguisher or two for you and your family to learn how to put out a fire.

Practicing *PASS (photo by U.S. Navy)*

Start by picking a suitable location— ideally, a sandbox in the back yard away from everything else. Be absolutely certain that the location is safe. Please don't burn down your house trying to learn how to use a fire extinguisher! Keep a garden hose ready in case you need to put the fire out quickly. Once you have the site ready, build a small controlled fire that is within the capability of your extinguisher. Give each family member a chance to practice putting the fire out.

Have them follow OSHA's *PASS* method (in tip box below).[113]

PASS Method

➤ *PULL*—Pull the pin. This will also break the tamper seal.

➤ *AIM*—Aim low, pointing the extinguisher nozzle or horn at the base of the fire. Don't point at the flames but rather at the material that is actually burning.

➤ *SQUEEZE*—Squeeze the handle to release the extinguishing agent.

➤ *SWEEP*—Sweep from side to side, aiming at the base of the fire until extinguished.

When fighting a house fire, try to keep your back facing a clear escape path. If the room becomes filled with smoke, leave immediately. Realize that many fires can't be put out with a single fire extinguisher. If you don't get to the fire early, it is better to evacuate your family and let things burn. The primary purpose of having fire extinguishers is to save lives. Property comes second.

ESCAPE LADDERS

If you have upstairs bedrooms, you need a way to escape from them without relying on the main corridors of your home. Some rooms may have direct access to the roof, which can serve as a fire escape by allowing you to go out onto the roof, and hang and drop down to the ground. If a bedroom doesn't have roof access, or if the roof is too high to hang and drop from, you will need to equip the room with an escape ladder. These ladders must be easily accessible. Keeping them buried under fifty pounds of clutter in the bottom of a closet isn't going to help anyone.

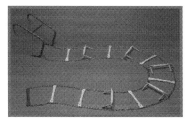

Courtesy of Bold Industries

Escape ladders must be long enough to reach the ground, able to simultaneously support multiple people, and easy to use by those sleeping in the bedrooms.

Check your windows periodically to make sure they can be easily opened. Windows can become stuck or "painted shut" when not frequently used. Escape ladders (or roof access) are of little value if the windows can't be opened. Breaking out the window glass and trying to climb through it is both difficult and incredibly dangerous.

Ideally, the windows should be able to be opened by the people who sleep in the bedrooms. If it is a child's room, then he or she should be able to open the window and deploy the escape ladder. This obviously introduces the risk of a child opening the window and falling out. You as a parent have to weigh that risk against the risk of your child

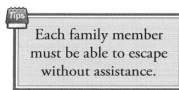

Each family member must be able to escape without assistance.

perishing in a fire. Many families conclude that emphasis should be placed on falling safety when the child is very young, and then on independent evacuation when the child gets a little older. Whatever methods you choose for escape, practice them so that everyone is clear and confident about their respective actions.

CARBON MONOXIDE ALARMS

Carbon monoxide (CO) is a deadly, colorless, odorless gas produced by the incomplete burning of fuels such as liquid petroleum, oil, kerosene, coal, and wood.[114] Grills, fireplaces, furnaces, hot water heaters, fuel-burning engines, and automobiles all produce CO. If equipment is used correctly (e.g., grill used outdoors), and working properly (e.g., fireplace properly vented), then CO won't build up in the home. However, leaks can occur and judgment can lapse, especially during times of crisis. Having one or more CO detectors in your home is a wise precaution.

Carbon monoxide poisoning is deadly, killing about 200 people in the United States each year.[114] Symptoms of CO poisoning include headache, fatigue, lightheadedness, shortness of breath, nausea, and dizziness. If you experience any of these symptoms when operating fuel-burning appliances, such as a range, fireplace, or gas dryer, get fresh air immediately! Then call your fire department and report the situation.

Below are a few rules to follow to minimize your chances of CO poisoning:

Never . . .

- Leave your car running in the garage.
- Burn charcoal inside your home, garage, vehicle, or tent.
- Use gas appliances such as a range, oven, or clothes dryer to heat your home.
- Use portable, fuel-burning camping equipment inside your home, tent, or vehicle.

CO detectors are cheap, small, and very low maintenance—only requiring the occasional battery change. It is especially important to have a unit with battery backup because people tend to use equipment that generates carbon monoxide (e.g., space heaters, generators, cooking grills) when power is interrupted. The best places to put CO detectors are in the locations where the risks reside, such as the garage, near the furnace, and in your safe room.

Most units have a small display that indicates the current CO level, as well as the highest level recorded since the last reset. When a threshold level is exceeded, an alarm will sound. Independent reviews of several units can be found online through Consumer Search.[115] Don't worry if you can't find the particular models recommended. Buying name brand units with the features you want will usually serve you well.

RADON TESTING

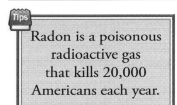

Radon is a poisonous radioactive gas that kills 20,000 Americans each year.

Radon is a colorless, odorless, radioactive gas released by the natural decay of uranium in rock, soil, and water. Once produced, radon rises from the ground and into the air that your family breathes. The gas decays into radioactive particles that become trapped in your lungs. These particles eventually release small bursts of energy that cause lung tissue damage.[116]

Radon is the leading cause of lung cancer among non-smokers. The gas is attributed with causing about 20,000 deaths annually in the United States. That is more deaths than can be attributed to drunk driving, falls in the home, or drowning.[117] It affects about the same number of people as house fires, but with a much higher fatality rate. There are no symptoms from overexposure, and there are no treatments either. Once the damage is done, it's done.

Fortunately, detecting radon is easy and inexpensive. Based on recent testing by Consumer Reports, it is recommended that you use the Accustar Alpha Track Test Kit AT 100.[118] This is also the model currently sold through the National Safety Council (NSC). These units are cheap and simple to use—just a small plastic sensor that you place on a shelf in the lowest level of your home. Leave the device out for at least 90 days, and then return it to the laboratory for analysis.

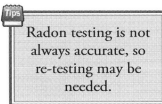

Radon testing is not always accurate, so re-testing may be needed.

In the event that the home test results indicate levels higher than 2 pCi/L, you should repeat the test. Home tests are not always accurate. If the second test disagrees with the first (i.e., first indicated high, second indicated low), then perform a third test before making a conclusion. If two of the three tests indicate elevated levels (i.e., greater than 2 pCi/L), you should to hire a professional radon test company to confirm the findings and take appropriate action.

Remedying high levels of radon requires installing a radon-removal system consisting of a fan-driven, vented system to suck radon away from your home. Correcting the problem can be rather expensive, but given the potential consequences to everyone in the house, there is no question about the necessity of the repair. More information on radon testing and correction is found on the NSC's website.[119]

Personal aside: My own experience is a good example of the confusion that can come with radon testing. My first radon test result indicated that my home had levels of 5.6 pCi/L, well above the 2 pCi/L maximum. Of course, I became concerned. I placed the second unit at a different point in the same room, and those levels were less than 1 pCi/L. Finally, I repeated the test a third time in yet another location in the room, and it also came back as less than 1 pCi/L—no action required. The question burning in my mind was why did the first unit show such a high level? I ultimately concluded that since I had placed the first unit adjacent to several granite book ends, they must have outgassed some radon into the sensor.

MISCELLANEOUS ITEMS

The most important safety items were discussed first: smoke alarms, fire extinguishers, CO detectors, radon detectors, and escape ladders. Consider these safety devices to be on your must-have list. Depending on your particular situation, there are a few additional safety items that you may want to consider.

Radiation Threat

Radioactive materials can be released into the environment either unintentionally (e.g., nuclear power plant accident), or intentionally (e.g., atomic bomb explosion, act of terrorism). If you feel that your family is in danger of a radioactive threat, then you may wish to equip your home with a radiation detector and dosimeter.

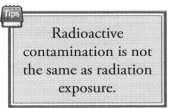

Radioactive contamination is not the same as radiation exposure.

There are two types of radiation poisoning: radioactive contamination and radiation exposure. Radioactive contamination occurs when you come into contact with radioactive materials. Contamination is usually the result of radioactive particulates being inhaled, consumed, or coming into direct contact with your body. Contact with radioactive materials causes tissue and organ damage, and can ultimately lead to death. This is distinctly different from radiation exposure, which is being exposed to the energy that radioactive materials emit—high-frequency rays penetrating the body, such as when you receive an x-ray. Excessive radiation exposure can cause sickness, burns, cancer, and death.

Nuclear power—a potential danger

There are pills, such as potassium iodide, that can be taken to help reduce the amount of radioactive contaminant that your body will absorb. However, they will not protect you from radiation exposure. These preventive measures are discussed in greater detail in *Chapter 9: Air.*

On average, Americans are exposed to about 300 to 400 millirems of radiation per year (where a millirem is 1/1000 of a rem). Most of this is a result of radon, background radiation, and medical imaging.[120] The recommended annual maximum dose is only 100 millirems above the background level for average citizens and 5,000 millirems for radiation workers.[121] A detailed discussion of radiation exposure levels from a wide varity of sources is provided in *Chapter 9: Air.*

A relatively inexpensive radiation detector that fits on your keychain is offered from NukAlert. It claims to reliably detect radiation levels from 100 millirems per hour to more than 50 rems per hour, and can operate for 10 full years without a battery replacement. The detector remains active 24 hours a day, and sounds different alarms depending on the exposure level.[122]

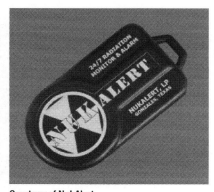

Courtesy of NukAlert

If you detect increased levels of radiation, you should immediately contact your local emergency management services. The best way to protect yourself from radiation exposure or radioactive contamination is to get away from the source. If evacuation is not possible, then put as much solid material (e.g., dirt, concrete, water) as possible between you and the outside world. Even sheltering in your home can significantly reduce your exposure level from contamination. In the event of becoming contaminated with radioactive materials, remove your clothing, store it in a plastic bag away from other people, wash yourself thoroughly, and contact local authorities.[123]

Courtesy of Water-Jel

Additional Fire Safety

For additional fire safety, you may want to consider stocking emergency escape hoods and gel-soaked blankets. Escape hoods can help you get through a smoky building without being overcome by toxic fumes. Gel-soaked blankets are designed to help protect from heat and flame as well as treat burns.

In general, both hoods and gel blankets are good products. Their only drawback is that they both require time to equip. And when it comes to a fire, time is everything. Do you really want to take the time to locate and put on your escape hood or blanket? Escape hoods in particular are not easy to put on properly. More often than not, you will be better served by getting out of the house as quickly as possible.

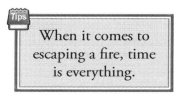

When it comes to escaping a fire, time is everything.

With that said, if you do find yourself in a room with the only exit clouded with smoke or blocked by heat, the use of an escape hood and/or gel-soaked blanket could very well save your life. You need to assess the likelihood of this situation and prepare accordingly. If you know that evacuating would be time-consuming or require you to pass close to where a fire might originate, then one or both of these products might be prudent investments.

If you are trapped in a room with the exit blocked, consider knocking or kicking out the wallboard to get to an adjacent room.

Escape hoods, for example, are popular choices for people who work in high-rise buildings. More information about respirators and escape hoods is given in *Chapter 9: Air.*

Pet Fire Safety

If you have pets in your home, consider getting pet alert stickers. These decals are placed in conspicuous places on your home, usually near the front door, to inform firefighters of how many and what type pets

Courtesy of Pet Safety Alert

you have in the home. If a fire breaks out while you are away, the stickers will let firefighters know that your home is not empty, and that you wish for them to make every effort to rescue your pets.

ADEQUATELY INSURE YOUR HOME

Many people mistakenly assume that the government will step in and help out those who lose property in a disaster. While it is true that mitigating actions are sometimes taken, such as offering low-interest building loans or temporary housing, it is generally up to the individual to rebuild and replace what is lost. This is where insurance comes in.

Property insurance is like every other aspect of disaster preparedness. Start by assessing the most likely risks to your home, and then target your insurance to protect from those risks.

Test your knowledge of homeowner's insurance by considering two scenarios involving the same thunderstorm:

Case 1: High winds tear your roof off. Rains pour in through the exposed roof and flood your home.

Case 2: Heavy rains flood your yard. The water rises to the point that it floods the ground floor of your home.

Did you recognize that these are considered two very different events by an insurance company? Unless you have specific flood insurance, you are likely not covered against Case 2.

Review your current homeowner's (or renter's) insurance policy to answer a few basic questions.

> ### Basic Home Insurance Questions
>
> 1. Are you adequately insured? Are any supplemental structures, such as a carport, fence, or swimming pool, also covered?
> 2. Do you have replacement cost or actual cash value (ACV) insurance?
> 3. What are your deductibles? Are they reasonable?
> 4. Are you protected against earthquake or flood? Should you be?
> 5. Does your insurance provide for temporary housing? What about debris removal?
> 6. Do you have (or should you have) any special riders covering jewelry, guns, or collectibles?

Based on your answers to these questions, contact your insurance agent to make adjustments. Be sure to ask appropriate questions of your agent, and select the coverage that adequately protects your investment. Remember, as a general rule, it is wiser to have a higher deductible and better coverage than a lower deductible and limited coverage.

Insist on replacement cost insurance on both your home and contents. If you don't, you could be out many thousands of dollars in the event of a disaster. With actual cash value (ACV) insurance, the insurer could give you only a fraction of the replacement cost due to their assessment of depreciation. Also, remember to adjust your insurance when property values increase or improvements are made.

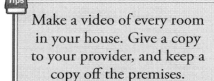

> Replacement cost insurance will fully replace your belongings. Actual cash value (ACV) insurance will pay you only a prorated amount.

If you do suffer a loss, your insurance company will require you to provide a detailed description of those items damaged. There are many horror stories of insurance companies failing to pay for items by claiming that the owner never had them and is therefore committing fraud. For your protection, take a video recording of your home and its contents. Walk through every room slowly,

> Make a video of every room in your house. Give a copy to your provider, and keep a copy off the premises.

verbally discussing all of its contents—noting specific brands, where it was purchased, how long you've had it, etc. Don't forget to include the garage, yard, basement, and attic.

Make at least two copies of the recording. Leave one with your insurance agent, and another one with a friend, family member, or somewhere else outside of the home. Each year when you renew your policy, distribute an updated video. Keeping a video record or your belongings not only protects you, but also makes it much easier for a claim to be processed.

DOCUMENT AND RECORD

Should your property ever become damaged due to an event covered by insurance, you should immediately document the damage. Once again, the best way to do this is with a video camera. Don't discard anything that was damaged, no matter how unusable, since the remnants could serve to better substantiate your claim.

Keep a record of all conversations that you have with your insurance company, appraisers, contractors, and relief organizations. Request that all agreements, repair estimates, and property appraisals be provided in writing. If you purchase emergency relief supplies, such as clothing or food, or if you have to pay for temporary lodging, be certain to keep all receipts. Provide copies of these receipts to your insurance agent as quickly as possible.

EVACUATION

There are two types of evacuations. The first is forced by an immediate threat, such as a house fire, and the second by an imminent, but not yet present, threat such as an approaching hurricane.

In the case of an immediate threat, the rules are simple. Get your family to safety as quickly as possible. Don't stop to grab anything that is not absolutely necessary. It is a good idea to identify exits from each room in your home—the primary escape route likely being the main door, and the secondary route through a window. You should have your family practice evacuating, with everyone managing to get out the house without assistance (very young children excluded). Also, agree on a family gathering point outside the home.

You might think that exercises like these will frighten your children. On the contrary, children like to have a clear understanding of the world around them. Many times our sheltering of children to protect

Escaping danger *(photo by USGS/Don Becker)*

them can in fact cause confusion and lead to dangerous misunderstandings. Rather than shoulder all the responsibilities yourself, explain the dangers to everyone, as well as what actions each person is expected to take. Not only does this empower each individual, but it also makes it more likely that your family will successfully escape, even when you are unable to lead them.

When a threat is approaching but hasn't yet arrived, perhaps giving you hours or even days notice, you will need to make a calculated decision regarding evacuation. The decision hinges on whether you weigh the danger of staying to be greater than that of leaving. The window of time allowing safe evacuation may be short, so understand that you may be stuck with the consequences of your decision.

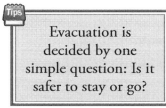

Evacuation is decided by one simple question: Is it safer to stay or go?

As you consider evacuation, take some basic steps to prepare for your possible departure:

- Fully fuel your vehicle and any spare gas cans. Have enough fuel to travel at least 500 miles.
- Identify multiple escape routes from the area being affected. Also, pick at least one alternate retreat location in case traffic flow prevents you from traveling to your preferred one.
- Listen to TV or radio broadcasts to determine when is the best time to evacuate and what are the recommended escape routes.
- Pack your vehicle with supplies, including those you might need for roadside emergencies (see the auto disaster kits in *Chapter 14: Transportation*).

If you decide to evacuate, take additional steps to prepare your home:

- Unplug all electronics except for refrigerators and freezers.
- Turn off unused utilities.
- Lock all doors and windows.
- Brace windows, doors, and garage doors as best you can (if appropriate to the threat).
- Write down the GPS coordinates of your home's location. Some disasters are so destructive that even locating your home can be nearly impossible without a set of absolute coordinates.
- Pack your most valuable items in the car. These could be gold coins or old photo albums—you be the judge.
- Let family and friends know when you are leaving and where you will be heading. If you can't convey this information, then consider leaving a note attached to the front door that indicates who you are, where you have gone, as well as contact information of those who might know your whereabouts. A blank "leave behind note" is given in the *Appendix* for this purpose. Realize that leaving a note like this may make your home susceptible to looting, but that is a price worth paying to give rescuers vital information should you go missing.
- Put on shoes and clothes that will suit you well should your car break down. Be prepared to spend the night in your car.
- Be watchful for washed-out roads, downed power lines, or other roadside hazards. Don't drive into flooded areas! It is very easy to underestimate the depth of water on roadways. If you absolutely must cross a flooded area of still water (not flowing), one person can carefully walk ahead of the vehicle to check for water depth. But this should be done only as a last resort.

Tips
> Waterproof PVC tubes can be used to hide valuables in the soil under your house.

There are times when you may be forced to evacuate but feel uncomfortable taking valuable items with you (e.g., jewelry, coins, guns). Leaving them behind in an empty house is also undesirable because of possible looting. Obviously, if time and money allows, you can store your valuables in a safety deposit box or other secure location. If no other options exist, you can bury them in the yard or under the house in small sections of PVC pipe with threaded or pressure-fitted plastic end caps. They don't have to be buried particularly deep, just far enough beneath the soil that they aren't noticeable. This is a reasonable option for locations that might experience looting. Be sure to check the tubes for water tightness before burying them.

WHEN YOU HAVE NO PLACE TO GO

Some disasters will leave you without a shelter. Hurricane Katrina, for example, displaced tens of thousands of people from New Orleans. Many of those sought temporary shelter in the Superdome. There were countless subsequent stories telling of crimes of opportunity including robbery, rape, battery, and murder. If even a small portion of them are true, the anarchy described is certainly not something you would wish upon your family.

There are many lessons to be learned from Katrina, both by individuals and by our government. Certainly one lesson is that government-provided emergency shelters are less than ideal. They are not the comforting

Hurricane Katrina, 2005 *(FEMA photo/Andrea Booher)*

environments that you want to put your family in during times of great stress.

It is almost always better to seek out family and friends first. If you are already part of a preparedness network, someone will likely have the means (and inclination) to put you and your family up for a short time. If you have nowhere else to go, look for a private shelter—perhaps a neighborhood church or civic organization (e.g., Masons, Elks). Your chances of being treated humanely and meeting likeminded individuals are arguably better than they would be at a government-run shelter.

If you are displaced, consider taking some of the following actions:

- Set up a post office box for mail delivery.
- Write to request duplicates of important documents and identification cards that may have been lost, such as birth certificates, social security cards, and driver's licenses.
- Use the public library or other free services to access email and the internet. This can be a good way to keep family and friends informed of your situation.

> Tips
> Homeless Assistance Programs
> www.hud.gov/homeless/index.cfm

Longer term housing needs can be addressed by checking with local homeless assistance programs. The Department of Housing and Urban Development maintains a website with information on homeless programs for various communities across the United States. The website also contains information about receiving counseling and applying for transitional or Section 8 housing assistance.

TENTS, TARPS, AND TEEPEES

Some of you may wonder why this book doesn't contain details on constructing ad-hoc shelters, such as lean-to's, teepees, and dome shelters. Remember, this guide is targeted toward disaster preparedness, not toward wilderness survival. As such, the decision was made not to fill the pages with detailed plans for building shelters from tarps and twine.

The truth is that, without a great deal of practice, you probably wouldn't be able to build a very good shelter anyway. Also, consider what the chances are of being stranded outdoors without any form of shelter. Keep in mind that your car is a great shelter, much better than anything you can build with makeshift supplies. In the event that you are stranded outdoors without a shelter or a vehicle, it seems inconceivable that you will have the necessary supplies, tools, and blueprints with you to build a protective shelter. From a preparedness standpoint, the best thing you can do is to not put yourself in that very precarious situation.

DP PLAN EXAMPLE

Table 5-2 Sample DP Plan Entry

Need: Shelter			
Danger	**Goals**	**Needs**	**Implementation**
Damage to structure or contents	Mitigate extent of damage to house	Threat assessment	Complete threat assessments and make improvements for likeliest threats.
		Hazards checklist	Complete home hazards checklists.
		Ability to turn off utilities	Practice turning off utilities; keep cutoff tools nearby.
	Safeguard against losses to structure or contents	A safety net to recover losses	Insure home with replacement cost insurance and extra riders for firearms, etc.
Burglary, home invasion	Prevent intruders from easily entering home	Cautious behavior	Keep all doors and windows locked.
			Teach everyone not to open doors blindly.
			Prevent children from opening doors.
			Teach "Rover" to bark when he detects anyone.
			Establish a neighborhood watch.
		Hardened points of entry	Install high-security deadbolts.
			Replace hinge and strike plate screws.
			Block sliding glass door with window bar.
			Use garage slide bar latches when out of town.
	Survive a home invasion	Family escape plan	Teach everyone an emergency word.
			Practice escape plan.
		Means of protecting the family	Secure a firearm in a location accessible by adults.
Fire	Detect fires	Smoke alarms	Install dual sensor smoke alarms in garage and attic.
			Check all smoke detectors in March and November.
	Extinguish a fire	Fire extinguishers	Put a 2-A:10-BC extinguisher in upstairs bedroom, and another in downstairs hallway.
	Escape from a fire	Way to escape from every room	Identify escape routes from every room.
			Equip upstairs rooms with escape ladders.
Carbon monoxide	Detect high levels of CO	CO detectors	Install CO detectors in living room and garage.
Radon	Detect high levels of radon	Radon test kit	Test lowest level of house for radon using long-term test kit.

Quick Summary - Shelter

➤ Make seeking adequate shelter your family's first priority.

➤ Avoid the urge to "run for the hills." Your home is likely to be the most effective shelter during a crisis.

➤ Assess the level of protection that your shelter offers to specific threats, and then make improvements to increase that protection.

➤ Complete a hazards checklist, removing potential threats to your home.

➤ If high winds pose a threat, make improvements to your home's roof, windows, and doors.

➤ Home security begins with teaching your family to be more cautious. A guard dog or electronic security system can help to minimize your chances of being surprised by an intruder.

➤ Your house can be hardened by securing the doors and windows as well as fortifying garage and basement doors.

➤ House fires give an average of only three minutes warning and kill more Americans than all natural disasters combined.

➤ Equip your home with as many safeguards as possible, including smoke detectors, carbon monoxide sensors, fire extinguishers, and escape ladders.

➤ Know how to cut off your utilities, and keep the necessary tools handy.

➤ Set up a safe room in your house that is easy to heat, protect, and seal from chemical and biological threats.

➤ Adequately insure your home and car. Elect to have replacement value insurance for your home and contents. Purchase special riders for guns, jewelry, and collectibles.

➤ Follow OSHA's *PASS* method to extinguish a fire.

➤ When evacuating, seek family and friends first, but also consider church, civic, and government services.

➤ If evacuating, stock your vehicle with emergency supplies (see *Chapter 14: Transportation*), and be prepared to spend the night in your car.

Recommended Items - Shelter

❑ Basic repair and cleanup supplies
  a. A good set of hand tools
  b. Basic home repair materials (e.g., nails, plastic sheeting, duct tape, tarps, PVC fittings, glue, wire, twine)
  c. A few home fix-it books
  d. Clean up supplies: shovel, axe, gloves, wheelbarrow, push broom and dustpan, plastic bags, and a chain saw

❑ Home safety equipment
  a. Smoke alarms
  b. Carbon monoxide alarms
  c. Radon test kit
  d. ABC fire extinguishers
  e. Pet fire safety decals
  f. (Optional) Smoke hoods, Water-Jel blankets

Challenge

Electrical power has been lost due to a major solar storm, and it is expected to remain off for the next two weeks. How will your family function at night?

Although not the first thing to come to mind when considering disaster preparedness, light can play a critical role in not only surviving, but also thriving in dangerous situations. Certainly most people would recognize that a good flashlight is a core component of their supplies.

Light serves four basic DP functions:

1. Allows operation in darkness
2. Deters animals and intruders
3. Provides a sense of comfort
4. Interacts with our bodies to maintain good health; helping to create vitamin D, preventing infection, and combating depression

Without light, you are limited to operating only during daylight hours as it is simply too dangerous to work in the dark. How many times have you tried to get up in the night to perform a simple task, only to jam your toe against an unexpected chair or walk nose-first into a door jamb? Imagine how dangerous it would be if you were forced to hike outside into the dead of night to relieve yourself.

Light has another very real benefit. Having a steady reliable light source allows people to better use their senses, and that helps alleviate the fear that darkness brings. With light, comes a feeling of being in control. Much of survival is keeping the right frame of mind—not melting down. Having this sense of control can help prevent you from making irrational, and perhaps life-threatening, mistakes.

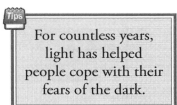

Tips

For countless years, light has helped people cope with their fears of the dark.

Consider for a moment that about one quarter of the world's population (over 1.6 billion people) lives without electricity and is limited to performing most activities during daylight hours.[124] These people know the challenges of trying to function in darkness, and if put in our shoes, could compensate well during an extended blackout. However, the same cannot necessarily be said for the rest of us.

COMPARING LIGHT SOURCES

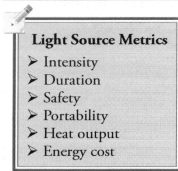

Light Source Metrics

➢ Intensity
➢ Duration
➢ Safety
➢ Portability
➢ Heat output
➢ Energy cost

Light comes from a variety of sources, including natural, electrical, fuel-burning, and chemical. Each source has inherent advantages and disadvantages when comparing them by six metrics: intensity, duration, safety, portability, heat output, and energy cost.

The six metrics give you a way to qualitatively compare the suitability of different light sources. There is no one best light source. Rather, you must select a source that best meets the needs specific to your situation. For example, a very bright flashlight that only burns for an hour without a change of batteries would be a logical choice for hiking to safety from a stranded car. Likewise, a fragile gas lantern that emits both heat and light for an entire evening would be a better choice for extended home use.

NATURAL LIGHT

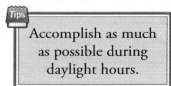

Accomplish as much as possible during daylight hours.

Natural light is light generated by the world around us. This includes sunlight, starlight, and moonlight (i.e., sunlight reflecting off the moon).

Make the most of natural light since it is free in terms of energy cost. Do most of your work during the daylight hours. If you are experiencing a prolonged disaster, use the daytime to prepare for the arrival of nightfall—shoring up shelters, gathering supplies, and seeking out any needed assistance. Whenever possible, evacuation should also be done during the day.

ELECTRICAL LIGHT

Electrical light is light created by electrical current passing through filaments or solid-state devices. For increased brightness, efficiency, and bulb life, the filaments are sometimes surrounded by special gases, such as krypton or xenon.

If you have electrical power available, then light is probably not much of a concern. Modern incandescent light bulbs burn for a couple thousand hours each, so even without a single spare bulb in the house, it is unlikely that you will be faced with a situation in which you are without enough light to function.

Unfortunately, electrical power is often the first utility to fail when a disaster strikes. Without an electrical power source, you are left to use portable light sources that run off of batteries (e.g., lanterns, flashlights) or consume fuel (e.g., candles, lamps, lanterns).

FLASHLIGHTS

Over the past several years, flashlight technology has advanced significantly with the arrival of high-efficiency, white light-emitting diode (LED) bulbs. LED flashlights operate for much longer on a set of batteries than conventional incandescent units, and the LED bulbs never need replacing. The downside is that quality, high-output LED flashlights are rather expensive.

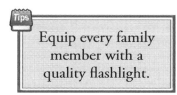

Equip every family member with a quality flashlight.

There are many good flashlight manufacturers, including Surefire, Fenix, Pelican, Streamlight, and Maglite. Flashlight intensity can vary anywhere from 5 to over 500 lumens, with most products outputting between 50-115 lumens.[125] You should consider a quality flashlight to be one of the fundamental pieces of your preparedness gear. Everyone in your family should have his or her own dedicated flashlight. This not only increases their utility, but it also helps to alleviate fear.

Select flashlights that are highly visible, such as those decorated with reflective tape and bright colors. This makes them much easier to find when you accidentally misplace them.

You will also need an adequate supply of batteries for flashlights and electric lanterns. Before stocking up, however, see *Chapter 7: Electrical Power* for a discussion of battery types.

Courtesy of Surefire

SHAKE LIGHTS

Recently, so-called "shake lights" were introduced to the flashlight-loving public. These flashlights work on the Faraday principal of generating electrical energy by moving a magnet through an inductive coil. In theory, having a battery-less flashlight sounds fantastic. Simply give the unit a shake, and bright light pours out the end.

Unfortunately, there are several problems with shake lights. First and foremost, many of the flashlights are not very durable; being closer to toys than tools. The second drawback is that the light they generate is neither bright nor long-lasting. The output from shake lights range from 7 to 13 lumens. Compare this to the 140 lumens that a more compact Surefire 9AN rechargeable LED flashlight emits.

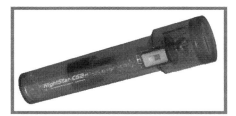

Courtesy of Applied Innovative Technologies

Shake lights require that you physically shake them to store up electrical energy. For example, you might have to shake the flashlight for a minute to get 5 to 10 minutes of light. Also, since they contain a fairly powerful magnet, you must keep them away from sensitive electronics, such as your laptop computer, cell phone, or iPod. They are not good choices for children or the elderly because of the effort required to recharge them. If you plan to purchase a shake light, read reviews to find the better quality choices.[126] Applied Innovative Technologies' Nightstar products are recognized as some of the higher-quality shake lights on the market.

Table 6-1 Lantern Bulb Comparison

Bulb Type	Light Output	Energy Use	Bulb Life (hours)
Krypton	Brightest	Highest	1,000's
Fluorescent	Bright	Lower	10,000's
LED	Dimmest	Lowest	100,000's

ELECTRIC LANTERNS

Courtesy of Coleman

Electric lanterns are used to illuminate wider areas. Nearly all electric lanterns use krypton, fluorescent, or LED bulbs. Each bulb type presents a tradeoff between light output, energy use, and bulb life (see Table 6-1).[127]

Electric lanterns are powered from batteries (typically D-cells), and some are also equipped with hand cranks—you turn the handle every few minutes to keep it going.

Lanterns are particularly useful in lighting your shelter because they use reflectors to spread the light out rather than focus it as a beam. Electric lanterns don't put out much heat, so they are not good choices to warm up an area. They are, however, safe (i.e., not a fire hazard), and most models are fairly durable.

FUEL-BURNING LIGHT

Since the discovery of fire, people have been burning fuel of one kind or another to produce light and heat. Fuel-based lighting includes bonfires, torches, candles, and oil lamps/lanterns. The most significant advantage of fuel-burning light is that heat is also produced, thus serving two important needs. Of course, heat may not always be desirable, such as when you are confined to a small shelter in the middle of summer.

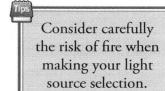
Consider carefully the risk of fire when making your light source selection.

There are also several disadvantages to using fuel-burning light sources. They require you to stockpile and handle combustible fuel. This can be dangerous, messy, and smelly. Also, when fuel is burned, fumes and smoke may be emitted. However, the most significant drawback to any fuel-burning source is the very real risk of fire that it introduces. Consider this risk carefully when making your light source selection. Introducing additional danger into what is already a hazardous situation means extra care must be taken.

FIRE

If you have the supplies, a fire can be a great way to generate both light and heat. To safely burn a fire indoors, you must use a fireplace or wood-burning stove with an adequate ventilation system.

The burning of wood or any other combustible material releases carbon monoxide, which is deadly if allowed to build up in a room. See *Chapter 8: Heating/Cooling* for additional details on indoor heating.

Fires are important when you are stuck outdoors and trying to stay warm or cook food. Knowing how to start a fire is one of life's necessities, like learning to swim. For that reason, you should do a bit of research on fire starting. There are countless books, articles, and websites that offer techniques for starting fires. But be warned, actually starting a fire is not the same as reading about it. Make it your mission to learn how to select tinder, kindle, and fuel wood; stack it properly to ensure adequate air flow; and light it with a single match. Someone telling you how to do it won't be enough (which is why I haven't done so here). This is a skill that can only be learned by doing. Once you are ready to test your fire-building skills, hold a class for your family, teaching everyone the basics. It is convenient to combine this class with the "how to use a fire extinguisher" class recommended in *Chapter 5: Shelter*.

CANDLES

Candles are the staple of many emergency supply kits. They are inexpensive and burn for many hours. They don't put out much light, but the placement of a few candles in a small room will allow you to function. Some candles even put out enough heat to do very minor cooking, such as Nuwick's 120-hour candle.

UCO's Candlelier

There are also lanterns that use candles as their light source, such as Uco's Candlelier and Eastern Mountain Sports' candle lantern. Candles are simple to use, reliable, and easily stored. However, they introduce a fire hazard, so once again caution is needed.

FUEL-BURNING LAMPS / LANTERNS

Fuel-burning lamps and lanterns are good alternatives to electric lanterns. Their merits are compared in Table 6-2. The distinction between *lamps* and *lanterns* is subtle but important. Lamps are kerosene-filled vials topped with a cotton or felt wick inside a glass globe. To adjust the lamp's brightness, you adjust the wick up or down, leaving more or less exposed. Once lit, lamps must remain stationary because the oil can spill, and the wick is very fragile while burning.

Courtesy of BriteLyt

Lanterns are designed to be portable hand-held lights and are usually equipped with a handle. Fuel-burning lanterns emit both light and heat by expressing flaming gas onto a cloth mantle saturated with a rare earth oxide. As the oxide heats, it emits visible light. Lighting a lantern requires manual pumping to build up the gas pressure; this takes practice to get right. Lanterns that use pressurized propane tanks eliminate the need for pumping.

Nearly all lamps burn kerosene, but lanterns may burn different fuels, including lamp oil, white gas, or propane. Generally, kerosene or lamp oil should be used indoors in a lantern. Other fuels are more likely to emit dangerous levels of carbon monoxide.

Table 6-2 Comparing Electric to Fuel-burning Lanterns

Metric	Electric	Fuel-burning
Brightness	Poor to Fair	Good to Excellent
Fire hazard	No	Yes
Emit odors	No	Yes
Durable	Good	Poor to Fair

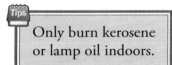

Only burn kerosene or lamp oil indoors.

Fuel-burning lamps and lanterns are very good sources of light and modest secondary sources of heat. For convenience, it is a good idea to use a lamp or lantern that burns the same type of fuel as your heating and cooking equipment. Both lamps and lanterns require frequent maintenance, so be sure to keep an ample supply of mantles, chimneys, and wicks.

There are many good choices for lamps and lanterns, including Aladdin, BriteLyt, Coleman, and others. Each has its particular strengths and weaknesses. For example, Aladdin lamps emit a good deal of light but are expensive and very easy to break. BriteLyt lanterns run on multiple types of fuel and are more durable but are expensive and can be tricky to operate.

Lamps and lanterns share the same disadvantages as other fuel-burning methods. They introduce a risk of fire, require you to store and handle dangerous fuel, and emit an odor when burning that can be offensive to some people. Because of the fire hazard, you may prefer to use electric over fuel-burning lanterns when children are present.

CHEMICAL LIGHT

Light sticks are a relatively new invention and are useful for a variety of military as well as commercial purposes. They operate by a chemical process that causes chemiluminescence. The sticks generally contain hydrogen peroxide and color dye inside a plastic sleeve, as well as a glass vial filled with phenyl oxalate ester. When you break the glass vial, a chemical reaction occurs, causing light to be emitted.[128]

This chemiluminescent process can be accelerated by heating up the light stick. For example, if you put the light stick in boiling water, it will glow very bright. Likewise it can be slowed down by cooling the stick. If you want a light stick to last until the next day, store it in the freezer.

Since light sticks are waterproof and don't require electricity, you can use them under water as well as in environments where explosives are present. For general DP purposes, light sticks are primarily used for night safety. Each person can wear a light stick, making it easier to keep track of everyone's location without consuming flashlight batteries. Light sticks are not a replacement for flashlights or lanterns. They output only a few lumens and provide merely a soft glow. However, if no other means

Light sticks are great for night safety.

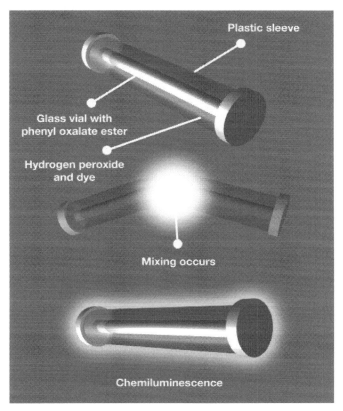

Light stick chemiluminescence process

exists, they can be used as a low-intensity flashlight by cutting open one end of the foil package and allowing light to shine out through the open end.

LIGHT INTENSITY

Light intensity is an important metric when comparing light sources. Obviously, you want your light source to be as bright as possible; unfortunately, there is no single measurement unit to indicate this brightness. Rather, light intensity is reported using several different units, making pre-purchase comparisons difficult. However, with a bit of patience and a dose of math, different light sources can be fairly compared.

Consider this section to be a mathematic aside that is not essential for you to understand when selecting the most appropriate light source. If you feel your eyes glazing over at the sight of equations, simply skip over them and find yourself the brightest, most durable flashlight or lantern you can. For those curious about the details of light intensity, read on.

Light is measured in two different ways. One method describes the amount of light actually emitted at the source—defined as *radiance*. The other method defines the amount of light that illuminates an area some distance away—defined as *illuminance*.

Radiance is the amount of light emitted from a source and is usually measured in candelas (English units) or lumens (metric units). One candela is roughly the amount of light emitted by a single birthday candle. If the light source emits light in every direction, then you simply multiply the number of candelas by 12.6 to get the number of lumens. More generally, you must know the radiation angle (i.e., the vertex angle of the cone of the flashlight) to perform the conversion. Below is the equation used to convert between candelas and lumens. If you are not a mathematician or engineer, you may need to break out your scientific calculator to apply it.

$$\# \text{ } lumens = \# \text{ } candelas \cdot \left[2\pi \cdot \left(1 - \cos\left(\frac{angle}{2}\right) \right) \right]$$

For example, assume you have a flashlight specified as emitting 100 candelas and having a 10 degree angle of radiance. Converting from candelas to lumens would be done as shown.

$$\# \text{ } lumens = 100 \cdot \left[2\pi \cdot \left[1 - \cos\left(\left(\frac{10°}{2}\right)\right) \right] \right] = 2.39$$

Fortunately, lumens are becoming the norm by retailers of flashlights and lanterns, so it is rare that you will need to perform this conversion from candelas.

Electric lanterns that use krypton or fluorescent bulbs are usually specified in watts rather than lumens. There is no direct conversion between watts and lumens because they measure different quantities. Some approximate conversions are given in Table 6-3, which can be used to provide an estimate of equivalent light intensities.[129,130,131]

For example, a 100-watt incandescent light bulb is roughly equivalent to 1,700 lumens. Likewise, a 13-watt compact fluorescent bulb is equivalent to about 900 lumens.

Illuminance is a measure of how much light falls on a given surface some distance away. It is usually measured in foot-candles (English units) or lux (metric units). The foot-candle is the amount of light a birthday candle shines on a surrounding sphere of one-foot radius. To convert between the two units, multiply the number of foot-candles by 10.8 to get the number of lux. No special equations are needed for this one. Thank goodness.

Think of illuminance as how bright an area is. For example, a bright office is lit to about 400 lux, moonlight represents about 1 lux, and midday sunlight might reach as high as 100,000 lux.[132]

Table 6-3 Converting Watts to Lumens

Bulb type	Approx. Conversion (Watts to Lumens)
Tungsten incandescent	17
Halogen/krypton	20
Compact fluorescent	70

A collection of emergency light sources

Most of you will probably never need to be numerically rigorous in comparing light sources. The proof is in the pudding—turn the flashlight or lantern on in a dark room and see how well it works. For the few who are curious about this sort of thing, you can draw upon these simple terms and conversions.

DP PLAN EXAMPLE

Table 6-4 Sample DP Plan Entry

Need: Light			
Danger	**Goals**	**Needs**	**Implementation**
Blackout	Function safely at night for a minimum of two weeks	Portable lighting	Equip each family member with a Surefire G2L flashlight. Stock three changes of spare batteries.
			Keep a supply of light sticks on hand for outdoor night safety.
		Area lighting	Light main living area and bedrooms with six Garrity hand-crank rechargeable LED lanterns.
	Safely evacuate from office building	Portable lighting at workplace	Keep a Fenix TA21 rechargeable flashlight in desk and spare batteries in charger.

Quick Summary - Light

➢ Light provides both comfort and a measure of safety to your family. It also allows you to function safely and effectively in the dark.

➢ Lighting sources include natural, electrical, fuel-burning, and chemical.

➢ Take advantage of sunlight by accomplishing as much as possible during the day.

➢ When comparing different light sources, consider six metrics: intensity, duration, safety, portability, heat output, and energy cost.

➢ Recent advances in LED technology enable electrical light sources to be very energy efficient.

➢ Fuel-burning light sources are very bright and radiate heat. However, they also introduce a fire hazard.

➢ Fires should never be built indoors without proper ventilation due to the risk of CO poisoning.

➢ Assign a flashlight to every person in your family. Use lanterns (electric or fuel-burning) for wider-area illumination.

➢ Most shake flashlights have questionable durability, limited light output, and short storage duration.

➢ Light sticks can increase night safety but are not a replacement for flashlights or lanterns.

Recommended Items - Light

❏ Portable illumination
    a. A quality flashlight for each family member
    b. Spare batteries

❏ Area illumination
    a. Electric lanterns and spare batteries, *or*
    b. Fuel-burning lanterns and fuel in approved containers, *or*
    c. Emergency candles

❏ Night safety
    a. Light sticks

CHAPTER 7
ELECTRICAL POWER

Challenge

Your city has suffered a direct hit from a powerful hurricane. Thousands of power poles are taken down by high winds, leaving your community without electricity. Authorities indicate that it will be anywhere from two to four weeks before power is fully restored. How will you perform basic functions like keeping warm at night, washing clothes, cooking food, and lighting your home?

Our modern society depends on electrical power for its needs and conveniences; from providing heat, to cooking, to powering our televisions and telephones, we need electricity. Unfortunately, electrical power is often the first service to fail in a crisis. Loss of power can be the result of many different events, including downed trees, power grid failures, vehicle collisions, ice storms, and high winds, to name just a few.

Being without electricity for a few hours is usually little more than annoying, perhaps missing your favorite football game, or having to eat dinner by candlelight. However, as those who have experienced extended blackouts can testify, being without electrical power for more than a few days can introduce real hardships.

TERMINOLOGY

Let's begin with an explanation of a few electrical terms. A useful analogy compares electricity to water. Just as water flows through pipes, electricity flows through wires. Water pressure forces water to move, and electrical pressure, known as *voltage*, forces electrons to move. The higher the pressure, the greater the flow of water (or electricity). This flow of electricity—quantified by counting the number of electrons that pass in a given amount of time—is called *current*, and is measured in amps. If you multiply the voltage and current together, the result is *power*, which is measured in watts. More generally, the resultant is in volt-amps,

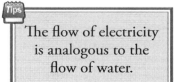

Tips

The flow of electricity is analogous to the flow of water.

but for our purposes, we can neglect this detail. Power can be thought of as the amount of usable energy at any given time that is being delivered to a device.

There are two types of current: *direct current* (DC), and *alternating current* (AC). With DC, current flows in one direction—from point A to B, perhaps from the battery to your cell phone. Alternating current is a bit more complicated. With AC, current flows from point A to B, and then reverses direction, flowing from B back to A. For your house power, this reversing action occurs 60 times in a single second, with the number of cycles being measured in the unit of hertz (Hz). If you live outside the United States, the AC frequency may be different (perhaps operating at 50 Hz), but the principles of alternating current flow remain the same.

HOUSE POWER

Your electric power company generates AC electricity and distributes it to your home using overhead or buried wires. For efficiency reasons, this electricity is transmitted at very high voltages (perhaps hundreds of thousands of volts). When the electricity gets close to your neighborhood, it is stepped down to a more usable voltage using transformers. The electricity is then routed into your home over three wires: two 120-volt "hot" lines, and one "neutral" line. The three wires can be configured for use as 120 volts and 240 volts (see figure below).

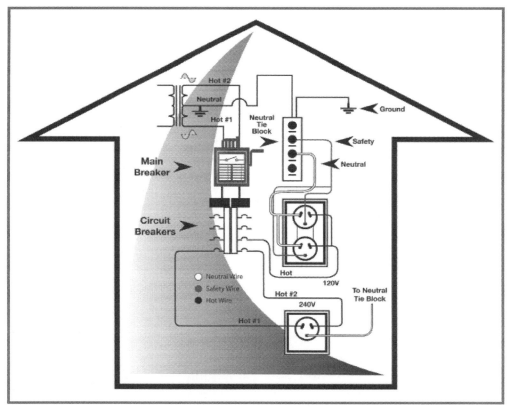

House wiring

Most of the electronic items in your home are powered by 120 volts AC. Your dishwasher, house lights, televisions, and stereos all require standard 120-volt AC power. A few high-power appliances, such as clothes dryers, air conditioners, and stoves, may require 240 volts. Portable devices that plug into an electrical socket for recharging (e.g., laptop, handheld gaming device, cell phone) convert the 120-volt AC voltage into a DC voltage either internally or through a small converter module in the cable or plug.

To develop a DP plan for electricity, start by determining your electrical needs—both AC and DC, and then select and size a method to provide this level of electricity.

AC POWER SOURCES

Providing your own AC power is generally done in one of two ways: using fuel-burning generators or battery-powered inverters. Other more exotic options, such as steam, hydroelectric, and wind-powered AC generators, are generally not practical for individuals due to high costs and low-power output.

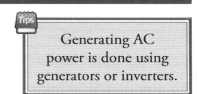

> **Tips**
>
> Generating AC power is done using generators or inverters.

GENERATORS

Generators are fuel-burning engines that generate AC electricity. Most run on standard gasoline, diesel, propane, or natural gas. They are rated according to the maximum power they can produce—typically measured in 1,000's of watts (a.k.a. kilowatts). Generators must be placed outdoors because, like all gas-powered engines, they vent poisonous exhausts. They also require regular maintenance, including changing the oil, replacing spark plugs, and cleaning filters.

Generators are classified as either *standby* or *portable*. Standby generators are permanent units mounted outside your home or office to provide backup power. They are usually configured to automatically turn on when power drops out. Most standby generators are fueled by propane or natural gas. They are rather expensive but offer an excellent solution to power outages. If money is no obstacle, then

Standby and portable generators *(courtesy of Generac and Briggs and Stratton)*

having your home configured with a large standby generator is the optimal solution because they use a clean, reliable fuel source and require little operator interaction.

Portable generators are the type seen frequently at home repair stores. As their name implies, these units are wheeled outdoors and put into service only when the need arises. Portable generators typically run on either gasoline or diesel fuel but can also be converted to run on propane or natural gas.

The biggest shortcoming of portable generators is that they burn a lot of fuel. Using them as a direct replacement for your utility power is not practical. The best way to illustrate this is through an example.

Example: A 10-kilowatt gas-powered generator burns roughly 2 gallons of gas per hour. To run it for a full 24-hour day would require 48 gallons of gas. To run it constantly for a week would require over 300 gallons of fuel!

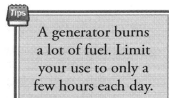

> A generator burns a lot of fuel. Limit your use to only a few hours each day.

It is not practical to store this amount of fuel since it would present a tremendous fire hazard, and almost certainly void your homeowner's insurance policy. For this reason, portable generators are best used to provide short-term electrical power needs, sometimes spread across a long period of time. (See *Draw up a Schedule* for more details on effectively using a generator.)

If you are planning to use a generator for emergency power needs, you will need to make five important preparations (see tip box).

Preparing for Generator Use

1. Select the right size and type of generator.
2. Have the necessary hardware installed in your home.
3. Draft a schedule of how you will use it in an emergency.
4. Stockpile the necessary fuel.
5. Learn how to use the generator safely.

SELECT A GENERATOR

Begin by selecting the right type and size of generator. First, decide whether you need a standby or portable generator. Standby units are permanent systems professionally set up outside the home or business. Portable generators are the type you roll out of the garage when needed. Next, select the type of fuel you wish to use. Most portable generators burn either gas or diesel. Generators that you find at your local hardware store will usually be gas-powered, portable units. Diesel generators are more expensive (perhaps double in price) but also more durable and require less fuel. Diesel fuel is also less combustible than gasoline. For even greater utility, portable generators can be modified using carburetor conversion kits to allow them to run on multiple fuel types, including propane and natural gas. Products that enable generators to run on gas, propane, or natural gas are referred to as "tri-fuel"

> Tri-fuel conversion kits enable generators to run on gasoline, propane, or natural gas.

conversion kits. Propane and natural gas are clean, easy to use, and do not have the risks of fouling up the carburetor like gasoline. Propane can be stored in various sized tanks for several years without the need for any special treatment. Natural gas must be provided by a utility company but is considered one of the most dependable services.

Whatever type of generator you select, it should be large enough to meet your electrical needs. However, resist the temptation to drastically oversize the unit. Generators are typically most efficient when the load is about 70% of the rated generator capacity.[133] To properly size a generator, make a list of those items that will require electrical power. Pay special attention to the voltage that your heating/air unit requires. If it runs on 240 volts, and you plan to power it during an emergency, you will need to select a generator that outputs both 120 and 240 volts.

Generators are sized with both "constant power" and "surge power" ratings. Make certain that the generator you select provides enough constant *and* surge power for your particular loads. Anything with a motor requires additional surge power during startup (approximately two to three times the constant level). Total constant power is determined by adding up all the constant power loads that will be running concurrently. To calculate

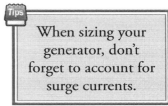

When sizing your generator, don't forget to account for surge currents.

total surge power, determine the extra surge power needed for your single largest appliance, and then add that difference to your total constant power value. You don't need to add up all the surge values because operationally you will ensure that you only start up one appliance at a time.

To make an accurate list of power usage, check appliance labels and product paperwork for power requirements. The amount of electricity listed on an appliance is the maximum it uses, such as when a refrigerator's compressor is running. The normal standby level may be much less. Some devices also consume a small amount of power even when turned off. To get the most out of your power source, it is best to unplug anything that isn't being used.

Table 7-1 lists example power requirements for common electrical appliances. Your loads may be different, but this table gives you some points of reference. You can also directly measure an appliance's power consumption using inexpensive electricity usage meters, such as the *Kill A Watt*.

Table 7-1 Wattage Reference Chart[133,134]

Tool or Appliance	Rated Watts (Running)	Surge Watts (Starting)	Required Voltage
Essentials			
Light bulbs	40-100	-	120
Deep freezer	500-600	1,200	120
Sump pump	800	2,000	120 or 240
Refrigerator	500-800	2,000	120
Water well pump	1,000-2,500	5,600-7,500	120 or 240
Electric water heater	4,000	-	120 or 240

Heating			
Heat pump	4,000-15,000	2 × Rated	120 or 240
Electric furnace	8,000-26,000	2 × Rated	120 or 240
Furnace fan (gas heater)	875	2,300	120 or 240
Space heater	600-1,800	-	120
Electric blanket	200-400	-	120
Cooling			
Central A/C (2.5 ton)	1,500-6,000	4 × Rated	120 or 240
Window A/C	1,200	3,000-4,800	120
Window fan	300-800	600-1200	120
Kitchen			
Microwave oven	800-1,000	-	120
Coffee maker	1,500	-	120
Electric range (one element on)	1,500	-	120 or 240
Toaster	800-1,000	-	120
Dishwasher	1,500	3,000	120
Oven	3,400	-	120 or 240
Family Room			
Color TV	300	-	120
DVD/CD player	100	-	120
Stereo receiver	450	-	120
Computer system	300-800	-	120
Laptop	100	-	120
Other			
Security system	180-500	-	120
AM/FM clock radio	300	-	120
Garage door opener	480-750	-	120
Washing machine	1,150	3,400	120 or 240
Clothes dryer, electric	4,000-5,400	6,750	120 or 240
Clothes dryer, gas	700	2,500	120 or 240

INSTALL THE NECESSARY HARDWARE

Once you have selected your generator, decide how it is to be used. There are two different methods of using a generator: direct and whole house.

Direct Method

With the direct method, you plug lamps and appliances directly into the generator sockets. Every generator is equipped with a different set of sockets, so before purchasing, check that the generator has the right plugs for your appliances. All generators offer numerous standard 120-volt AC sockets, but some also provide 240-volt AC, or even 12-volt DC sockets—useful for recharging lead-acid batteries

Since generators can only be operated outdoors, heavy-duty extension cords are needed to connect your appliances to the generator. The direct method is inherently safe to your home's electrical system. It also requires very minimal setup since there is no hardware to install. A serious drawback of direct generator use, however, is that you can only power equipment that has a standard AC plug. This means that you can't easily power your water heater, house lights, or heat pump—all of which can be very important.

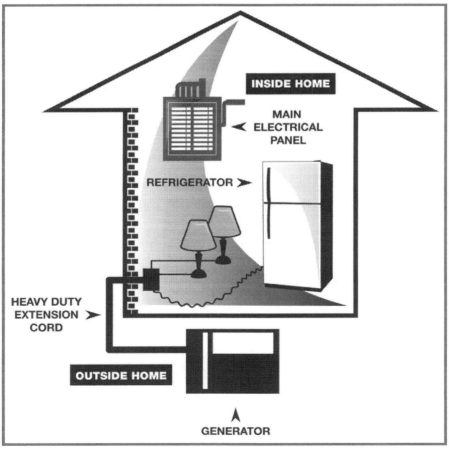

Direct generator connection

Direct generator use is also inconvenient for powering several appliances located in different rooms of your house because each one requires its own extension cord.

Whole House Method

With the whole house method, the generator is connected to your home's electrical system. This allows you to use your wall sockets and lights as you normally do today. Using a generator to power your whole house requires that you install a mechanical interlock or transfer switch. This work should only be done by a licensed electrician. Mechanical interlocks and transfer switches are hardware devices that transfer the source from which your home draws electricity, choosing between the power company's lines or your generator. Interlocks and transfer switches both perform the same general function but in very different ways.

> **Tips**
> To power your home using a generator, you must use an interlock or transfer switch.

The simpler of the two mechanisms is the mechanical interlock. An interlock manually switches your home's main breaker off anytime you switch your generator's breaker on. The interlock is a custom-fitted piece of metal that mounts inside your existing breaker box. This is the cheapest method to connect

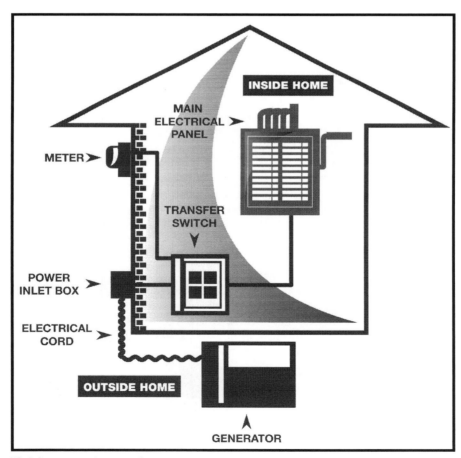

Whole house generator connection

Mechanical Interlock Use

1. Turn off the main breaker.
2. Shut off all the home's circuit breakers.
3. Make sure the interlock is in the *home* position.
4. Connect and start the generator.
5. Slide the interlock into the *generator* position.
6. Turn on desired breakers one at a time, giving each load a moment to stabilize before throwing the next breaker.

your generator to your home's electrical system, requiring only the mechanical linkage (installed inside your panel), a generator breaker (also installed inside your panel), and a power inlet box (installed outside where you wish to plug in the generator). Mechanical interlocks have the advantage of allowing you to power any of your home's appliances using the existing secondary breakers.

Expect to pay a few hundred dollars to have a licensed electrician perform the installations. This is not a do it yourself job since wiring your home without a license can void your homeowner's insurance, not to mention kill you.

If wired correctly, power will flow from the generator, through your existing breaker box, and out to your selected lights and appliances. This flexibility to power anything in your home is a major advantage. No additional secondary breakers are required. The only disadvantage is that mechanical linkages don't allow you to see how much power is being drawn from your generator. However, given its cost and convenience, the mechanical interlock is the preferred method of connecting a generator to your home's electrical system.

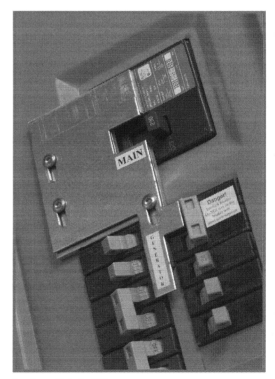

Courtesy of Interlockkit.com

A second method of using a generator is to connect it through a transfer switch that contains its own set of breakers. For this type of installation, the electrician moves wires from your main breaker panel and connects them to the transfer switch breakers—usually mounted next to your existing panel. The transfer switch has only a limited number of breakers, so you will likely have to make some sacrifices as to what systems you power from the generator.

This limitation forces you to allocate transfer switch breakers, starting with the most important loads first (e.g., water heater, cooking range, lights). If you use this method, select a transfer switch that has at least twelve breakers. Some transfer switches have large sockets that allow you to plug the generator directly

Transfer switch *(courtesy of Reliance Controls)*

into them using a heavy-duty cord. If the transfer switch doesn't have a socket, you must install a power inlet box outside the house near the generator. One advantage of transfer switches is that they often have power meters to show the level of power going out to the house loads. This feature helps to prevent overloading of the generator.

Having an interlock or transfer switch is absolutely necessary to power your home's wall sockets and over-head lights. If you wire a generator directly into your home's electrical system without one of these isolation devices, you risk backfeeding the power lines, which can electrocute linemen working to restore electricity. There are stories of people turning off their main breaker and wiring a generator to their dryer socket in an emergency, but it is simply not worth the risk to your home, your family, and the hard-working linemen. Do it right or don't do it at all.

In theory, wiring a generator to your existing breaker panel using an interlock or transfer switch should enable you to use your home appliances just as you do today. In reality, however, you will have to take into account the capabilities of the generator. It is very likely that the generator will be unable to power all your appliances simultaneously. For that reason, you will need to limit the electrical load to only what is needed at the time. This prioritizing is accomplished by keeping off some of the breakers in your main power or transfer switch panel. You may have to cycle some appliances in and out, perhaps waiting for the washing machine to finish before turning on your space heater. If you overload your generator, it will choke out the engine and may even damage the unit over time.

DRAW UP A SCHEDULE

Once you have compiled the list of equipment you want to power and have all the hardware installed, draw up a schedule detailing how you will operate the generator. For most situations, it is best to run the generator for a couple of hours in the morning and a couple more at night. The goal is to take care of all your day's electrical needs during these few hours. Activities could include:

- Recharging your batteries and handheld devices (cell phones, laptops, radios)
- Heating or cooling the house
- Cooking the day's meals
- Cooling the refrigerator and freezer
- Heating the water in the hot water heater
- Tuning in to TV, radio stations, or shortwave broadcasts
- Washing and drying clothes
- Operating your water pump (if using an underground well) to flush commodes and replenish your water supply

Adopting a usage plan like this will still leave you without AC power for most of the day, so life is certainly not going to be normal. However, with brief periods of electrical power, your family will be able to better prepare for the coming day or night.

STORE FUEL

Another important preparation is to stockpile sufficient fuel to handle your generator needs. Consult your homeowner's insurance provider and local fire department to determine how much fuel may be stored without voiding your policy or violating local law. The nation's Uniform Fire Code limits storage to 25 gallons, but local regulations may supersede this rule. If you want to prepare for a weeklong power outage and are operating on a four-hour daily schedule, you will likely need between 30 and 60 gallons of gas—some of which you may wish to pick up just before the emergency. The exact level depends on the size of your generator and how heavily you load it. If you have a diesel generator, you may only need about 20 to 40 gallons since they typically burn less fuel per kilowatt-hour.

Consult your insurance provider regarding fuel storage to prevent voiding your policy.

Store fuel in UL-approved containers in a well-ventilated area (most likely your garage or a nearby shed). Never store fuel inside your home. Gas and diesel are both environmental hazards, and storing them underground is strictly regulated, making it very expensive for an individual to do so legally. Given these restrictions, you will probably end up using jerry cans, or slightly higher-capacity fuel caddies. Note that 60-gallon metal drums are the only large fuel containers approved by the Uniform Fire Code. Regardless of how you decide to store it, fuel must either be in a container light enough to lift or one that is able to be pumped or siphoned. Remember that siphoning only works when transferring to a level below the fuel level. If the fuel level in your gas container is at two feet off the ground, and your generator fuel tank is at three feet, you will not be able to siphon fuel into the generator.

Moeller Marine gas caddy

Over time, gasoline deteriorates, losing octane and turning to a gummy lacquer-like substance. Under optimal conditions, gasoline can be stored up to one year.[135] Optimal conditions are defined as out of sunlight, at a constant, moderate temperature ($< 80°F$), and in a tightly closed container. A more conservative recommendation is to consume or replace your fuel

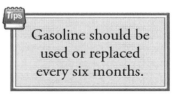

Gasoline should be used or replaced every six months.

within six months. This requirement presents a dilemma. Fuel has a relatively modest shelf life, and yet you only need it when power fails—something you certainly can't schedule.

Fortunately, fuel can be treated with special additives that help extend its shelf life. Two very popular products for this are PRI-D/G and STA-BIL. By adding a stabilizer, fuel may be safely stored for several years. These products mix at small ratios, so very little is needed. If you are going to store fuel, stabilizers are a necessity.

Fuel stabilizers *(courtesy of STA-BIL and PRI-G)*

OPERATE THE GENERATOR

To learn how to use a generator correctly, begin by consulting the operator's manual that comes with your specific unit. A few things are common regardless of the particular model. You will need to add oil and fuel, as well as install one or more filters. To start the generator, either use the pull string or electronic ignition button. Below are some additional operating tips:[136]

- Always operate the generator outdoors. Never use it in a garage, crawlspace, or shed. Generators exhaust a great deal of pollution that must be safely vented outdoors.
- Maintain a safety zone of at least four feet around the generator. Generators situated too close to a home can melt siding or cause a fire.
- Identify the exhaust port and ensure that it faces away from your home. Close all nearby windows or doors to prevent the exhaust from entering your house.
- Before refueling, shut down the generator and allow it to cool for a few minutes. Fuel spilled on a hot generator can ignite.
- Don't overfill the gas tank. Leave a couple of inches of space at the top to allow for the fuel to expand as it heats up.
- Wipe up any fuel spills before starting the generator.
- Keep the generator dry. Don't operate it in the rain unless it is covered with a overhead canopy.
- Be mindful of the generator's power rating, and try not to exceed it. A generator will choke out if overloaded. Continued overloading may cause damage to some units.

Xantrax inverter

INVERTERS

Inverters are a less expensive option for generating AC power. They convert DC battery voltage into AC voltage. Similar to generators, they are rated by the maximum power they output. Small inverters provide up to a few hundred watts, while larger ones can output several thousand watts. There is little need for inverters larger than this because they would quickly drain any reasonable-sized battery source.

Most inverters accept 12 or 24 volts DC as input, and output 120 or 240 volts AC through standard three-prong outlets. The 12-volt input makes them convenient to use since they can be connected directly to automotive or marine batteries.

Inverters are designed for direct plug applications only, routing power to your devices through extension cords. Don't connect an inverter to your home wiring. There is a danger of making the battery terminals live with the full 120-volt AC level.[137]

Most inverters are classified as *modified sine wave* devices. This means their output is not a true sine wave but a distorted representation. Normally this imperfection is not a problem since most electronic equipment will run fine without perfect sine wave power. However, some sensitive electronics, such as computers and televisions, may not operate properly with modified sine wave inverters. For those devices, you may need a more expensive, *true sine wave* inverter.

Inverters are not designed to power devices with high start-up currents, such as washers, refrigerators, or heat pumps (see Table 7-1). Some inverters offer "soft start" circuitry to slowly bring up the current level. This may help power up some equipment with high surge requirements.

> **Tips**
> Caution: Many inverters don't have reverse polarity protection. To avoid damage, pay attention when connecting the battery input terminals.

Features you may wish to consider when selecting an inverter include:

- Overload and high temperature shutdown
- Low-voltage alarm and shutdown
- Soft starting to better handle surge currents
- Low idle current
- Overload protection
- Battery monitoring

Inverters are not a direct replacement for generators for three reasons: they don't output as much power; they are not as capable of starting up equipment with high surge currents; and they require large battery banks to output significant amounts of sustained power. An example will make this last point clear.

Large Load Example:

Assume you wish to use an inverter rated at 2,000 watts to power a space heater and a computer that require a total of 1,900 watts. You connect the inverter to a 12-volt, 100 amp-hour car battery (pulled out and sitting on the floor in this case). Although not precisely accurate, the 100 amp-hour rating will be taken to mean that the battery can output 100 amps for 1 hour, or 1 amp for 100 hours. The total available power is calculated as 12 volts × 100 amp-hours = 1,200 watt-hours.

The inverter is rated as 90% efficient, meaning that 90% of the power that goes into it is usable, and the remaining 10% is lost in heat. To output the 1,900 watts of AC power from the inverter requires 2,111 watts of DC power from the battery. This converts to only about half an hour of run time before the large battery is completely drained (1,200 watt-hours capability ÷ 2,111 watts needed = 34 minutes).

This example illustrates that running a modest load of 1,900 watts for any reasonable length of time will require a significant bank of batteries. You might think that this limitation can be circumvented by keeping the battery in your vehicle and using it while the engine is running. The problem is that an alternator typically only outputs 600 to 800 watts, so it can't keep up with your load. The battery will still deplete rapidly. You may also experience car problems, such as overheating the alternator.

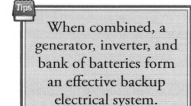

When combined, a generator, inverter, and bank of batteries form an effective backup electrical system.

A particularly effective backup electrical system consists of a generator, battery bank, and an inverter. Teamed together, the three allow you to function with limited capacity throughout the entire day. The generator is used to provide high-current needs, such as cooking and clothes washing for a few hours each day. It is also used to recharge the battery bank. The batteries and inverter are used to provide for the day's lower power needs, such as radios, lamps, computers, and small appliances.

DC POWER SOURCES

Batteries provide DC power to portable devices, such as flashlights, cameras, cell phones, weather or pocket radios, and remote controls. There are many different types of batteries available, each with its respective advantages and shortcomings.

Batteries should be stored in a cool, dry location. Some people store batteries in the refrigerator thinking that the cool temperatures will help the shelf life. Studies have shown that the cool temperatures

Table 7-2 Single-use Battery Comparison[140,141,142,143,144,145]

Type	Energy Density	Shelf Life	Cost	Risk of Leaking	Comments
Heavy duty	Poor	2+ years	Low	Higher	Not recommended
Alkaline	Good	5+ years	Low	Higher	Acceptable for most uses
Nickel Oxyhydroxide (NiOOH)	Very Good	6+ years	Moderate	Lower	Better than alkaline for high drain devices
Lithium (Li)	Excellent	10+ years	High	Lower	Best performing single-use battery; limited sizes

do help to slow the chemical reaction and increase battery shelf life, but only by a few percent—hardly worth the effort. Refrigerated batteries should always be warmed up prior to use. Not only do batteries have higher output levels when warmer, but if they are not allowed to warm, condensation can form inside the electronic device being powered.[138,139]

You will need to decide between single-use or rechargeable batteries. The advantages of single-use batteries are their low cost and good shelf life. Rechargeable batteries offer many more hours of use (assuming you can recharge them repeatedly) but generally suffer from poor shelf life and self-discharge. Most rechargeable batteries also require charging before using them the first time. The decision to use either single-use or rechargeable batteries will likely depend on whether you have a method of recharging them, such as an inverter, generator, or solar charger.

SINGLE-USE BATTERIES

There are four common types of single-use batteries: heavy duty, alkaline, nickel oxyhydroxide, and lithium. They can be compared using four metrics: cost, likelihood of leaking, energy density, and shelf life (how long they can sit on a shelf and still be usable, holding at least 70% of charge). See Table 7-2 for a comparison across these metrics.

Given their low cost and reasonable performance, alkaline batteries are the preferred choice for general purpose uses. For improved performance, such as in cameras, consider using nickel oxyhydroxide batteries. Finally, if cost is not an issue and you want the very best, select lithium batteries for their outstanding performance and shelf life.

An assortment of single-use batteries

RECHARGEABLE BATTERIES

> **Tips**
> To reduce the chances of batteries leaking, only pair up batteries that come from the same package at the same time.

Rechargeable batteries are compared using slightly different metrics as shown in Table 7-3. The good news is that they aren't prone to leaking, and their energy densities are comparable to single-use batteries. Unfortunately, rechargeable batteries suffer from *self-discharge*. This means that even when just sitting on the shelf, they will discharge—some very rapidly. Their initial costs are also higher than single-use batteries, but they offer the advantage of numerous recharges, making them much cheaper to use over the long run.

Rechargeable batteries are great for daily-use devices because they can be recharged hundreds of times. However, because of their self-discharge and shorter shelf life, most are not ideal for disaster preparedness. To keep several sets of rechargeable batteries ready for use would require cycling them in and out of the charger every month—both inconvenient and easy to neglect. Most rechargeable batteries are also not suited for devices that are expected to last for many months, such as clocks or remote controls because once again, the self-discharge makes it necessary to do frequent battery cycling.

> **Tips**
> Low self-discharge NiMH batteries allow hundreds of recharges and have reasonable shelf life.

Low self-discharge NiMH batteries are a new addition to the battery world. They still allow hundreds of recharge cycles but have a much lower self-discharge rate, meaning that once charged, they will remain ready to use for a year or more. Low self-discharge NiMH batteries are also marketed as "ready to use," "pre-charged," or "hybrid" batteries, and generally come charged right out of the package. Although more expensive initially, pre-charged NiMH batteries are an excellent choice for nearly any need—including disaster preparedness.

Lithium-ion batteries offer excellent initial energy capacity but suffer from a unique problem. They degrade in capacity about 20% per year due to rising internal resistance. After a couple of years, they

Table 7-3 Rechargeable Battery Comparison[140,141,142,143,144,145]

Type	Energy Density	Self-discharge	Shelf Life	Cost	Comments
Nickel Metal Hydride (NiMH)	Good	30% per month	1-2 months	High	Acceptable for daily use
Low Self-discharge NiMH	Good	1.5-3% per month	1-2 years	High	Best rechargeable
Nickel Cadmium (NiCd)	Low	10-15% per month	3 months	Moderate	May have limited life due to memory effect
Lithium Ion (Li Ion)	Very good	5-10% per month	6 months	High	Suffers from declining capacity

Pre-charged NiMH batteries in charger

are no longer usable for most applications. For this reason, lithium-ion batteries are not recommended. Additionally, they are typically only available in sizes suitable for camera use.

LEAD-ACID BATTERIES

Lead-acid batteries are large rechargeable batteries typically used in your automobile or boat. They have low energy-to-volume ratios, meaning they are large in size for the power they output. However, lead-acid batteries can supply the high surge currents needed to start motors. Most batteries have an output of a little over 12 volts when fully charged.

Lead-acid batteries are rated by several metrics, the most important of which is their sustained output capability (measured in amp-hours). The amp-hour rating indicates how much current over time the battery can deliver. A battery rated at 100 amp-hours would ideally be able to provide 1 amp for 100 hours, or 100 amps for 1 hour. In reality, batteries perform better with lighter loads, so a 100 amp-hour battery might actually be able to provide 1 amp for 120 hours, but only 80 amps for 1 hour.[146]

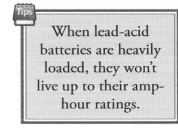

When lead-acid batteries are heavily loaded, they won't live up to their amp-hour ratings.

Excessive charging of lead-acid batteries can emit hydrogen and oxygen, which if not released, can explode. Fortunately most modern batteries are equipped with vents to release these gases. These self-venting batteries are known as valve-regulated lead-acid (VRLA) batteries. The two major types of VRLA batteries are gel cell (widely used) and absorbent glass mat (used primarily with high performance electric vehicles and motorcycles). Unlike previous-generation lead-acid batteries that required periodic inspection and adding water to account for electrolysis, modern sealed lead-acid batteries are relatively maintenance-free.

There are a several types of lead-acid batteries, but two are particularly important: starting batteries and deep cycle batteries.

Starting Batteries

Some lead-acid batteries are designed with numerous thin plates for maximum surge current, making them ideal for starting internal combustible engines. They are, however, not suited to deep discharge (i.e., draining them), which can cause permanent damage and lead to their premature failure. These types of batteries are often referred to as starting, lighting, ignition (SLI) batteries.

Deep Cycle Batteries

Deep cycle batteries are designed to be fully discharged without damage. They are useful for such things as electric vehicles, uninterruptable power supplies, boats, and photovoltaic systems. Their design also makes deep cycle batteries ideal for emergency power needs since you will want batteries that can be fully discharged and recharged repeatedly—perhaps recharging them each morning with a generator.

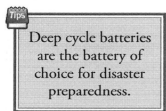

Deep cycle batteries are the battery of choice for disaster preparedness.

There are two ways to use lead-acid batteries when power fails. The first is to use them directly as DC power sources, providing power to your portable electronics or hand tools. The easiest way to do this is to use a power socket with battery clips. One end clips onto the battery; the other end is a conventional car cigarette lighter socket. This configuration enables you to plug your device into the socket using a standard car adapter (see illustration).

The second method is to connect the lead-acid battery to an inverter, which then provides AC power. As illustrated in the *Large Load Example,* even a hefty car battery won't last very long (perhaps only minutes) when used to power home appliances. However, it can last for many hours if used for powering small loads. Again, an example will make this clear.

Small Load Example:

Assume you connect a 12-volt, 100 amp-hour, deep cycle, lead-acid battery to a modest 500-watt inverter with 90% efficiency. You use it to power a 10-watt emergency radio, and recharge a cell phone (2.5 watts/hour) and three NiMH flashlight batteries (5 watts/hour). Assume it takes four hours to recharge your batteries and that you wish to listen to the radio during this time. The power needs are calculated as shown:

Emergency Radio: 10 watts per hour × 4 hours = 40 watt-hours

Cell phone (recharge): 2.5 watts per hour × 4 hours = 10 watt-hours

Flashlight batteries (recharge): 5 watts per hour × 4 hours × 3 batteries = 60 watt-hours

Total power needs = 110 watt-hours

With the inverter's 90% efficiency, the battery must provide 110 ÷ 0.9 = 121 watt-hours. This is only about 10% of the battery's 1,200 watt-hour capacity.

For additional capacity, lead-acid batteries can be wired in parallel: positive terminals tied together; negative terminals tied together. Banks of batteries can power heavier loads such as lights, heaters, or a

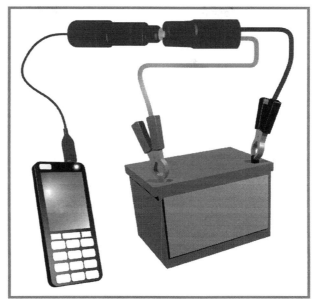

Recharging portable devices with a large battery

microwave oven. You should only connect batteries together like this if they are of the same charge level (i.e., fully discharged or fully charged).

RECHARGING LEAD-ACID BATTERIES

Given that lead-acid batteries will likely be your primary source of DC power, you will need an effective method to recharge them. If electrical power is available, such as a generator, then using a conventional battery charger is the quickest method. Note that some battery chargers can charge parallel batteries concurrently. If your charger can't do this, then disconnect and charge each battery separately.

If electrical power is not available, you will need an alternative method to recharge your lead-acid batteries. From a practical standpoint, there are three ways to do this: solar arrays, wind turbines, and human-powered machines.

SOLAR POWER

Solar chargers exist for both NiMH and lead-acid batteries. Some chargers fit in the palm of your hand, while others may cover the roof of your home. At first glance, solar power would seem to be the perfect solution to your long-term energy needs. Simply hook your batteries up to the solar charger, and a short time later, your batteries are as good as new, right? Unfortunately, it is not quite that simple.

The problem with most solar chargers is the very small amount of current they provide. Consider that one small commercial solar charger, designed to recharge AA and AAA batteries, outputs only 90 mA

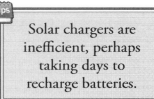

Solar chargers are inefficient, perhaps taking days to recharge batteries.

of current under full-sun conditions. Assuming six hours of bright sunlight each day, batteries would charge about 540 mA-hours per day. Typical high-capacity AAA and AA batteries are 900 and 2,500 mA-hours respectively. To fully recharge a set of four AAA batteries would require 3,600 mA-hours, taking about 7 days. Likewise, recharging four AA batteries would require 10,000 mA-hours and take about 18 days!

The solution to this slow recharge rate is to use larger solar panels. However, most large solar panels are only designed to recharge 12-volt or larger batteries. To charge small NiMH batteries requires you to convert the 12 volts coming off the solar panel to the 1.5-volt level needed for AA or AAA batteries. One method of doing this is to use a recharger, such as the Duracell Mobile Charger, that runs directly off 12 volts. A more general solution is to use an inverter, which allows you to use any recharger with an AC plug.

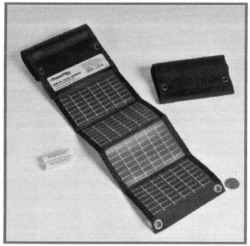

PowerFilm AA solar charger

The problem of minimal solar-panel-energy-yield applies to larger lead-acid batteries as well. A modest solar panel charger measuring 17 in. × 13 in. might provide 400 mA of current in full sunlight. Considering that a heavy duty, lead-acid battery requires about 100 amp-hours, recharging will take 250 hours of full sunlight. If the sun shines with full intensity on the panel for 6 hours a day, it will take about 42 days to recharge the battery. If you increase the size of the solar panel to about 50 in. × 40 in., you might get 5 amps of charging current, reducing the recharge time down to a more manageable 20 hours (or three days of bright sunlight).

Fortunately, solar charger technologies are improving. For example, a modern Powerfilm AA battery charger outputs over 400 mA into two parallel batteries for each hour of full sunlight (a factor of about nine times better than the previous generation of compact solar chargers). At this peak rate, a set of high-capacity AA batteries could be recharged in about 6 hours. Likewise, larger Powerfilm chargers can generate up to 4 amps per hour into a 12-volt lead-acid battery, leading to a recharge time of about 4 days. Modern chargers like these certainly reduce the recharge time, but be forewarned, the improved performance comes with a steep price tag.

Backup electrical system

The point of the examples is to illustrate that using solar chargers to recharge batteries can be a slow process, especially if older generation chargers are used. Despite this limitation, solar chargers can be an viable technology for recharging batteries when conventional electrical power is unavailable. Just do the math ahead of time so you can be realistic with your expectations.

WIND POWER

Wind is used to produce about 1.5% of the world's energy. In 2009, over 52 terawatt-hours of wind power were produced in the United States.[147] Large turbines can now be seen dotting the landscapes of Texas, Colorado, and Oregon. Wind power has become a practical supplement to coal and nuclear energy.

How viable a personal energy source it is depends largely on where you live. If steady, strong winds are prevalent in your area and you have unobstructed access to the wind, then wind-driven systems might yield tangible power. Turbines must be mounted high above the ground, such as atop a roof or tall pole, to gain unfettered access to the wind stream. Also, getting any significant power output requires a large turbine.

It is important to realize that a turbine's specified power output assumes ideal conditions (perhaps 35-mph winds). Practical yields can easily be a mere 10% of the values specified by the manufacturer. A turbine rated with a 400-watt output might only produce an average of 40 watts per hour under normal wind conditions.

Wind turbine

The notable drawbacks to using wind turbines for personal use are cost, availability, and the unpredictable nature of wind. The hardware and installation costs of a small turbine with 26-inch blades and a mounting tower might cost you several thousand dollars. Larger wind generators are significantly more expensive. If a small turbine averages 3 amps of current 24 hours a day, it will yield a respectable 72 amp-hours per day—almost enough to recharge a deep cycle, lead-acid battery.

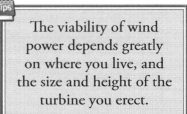

The viability of wind power depends greatly on where you live, and the size and height of the turbine you erect.

MECHANICAL MOTION

It is possible to convert mechanical energy to electrical energy by moving a permanent magnet through a coil of wire. When doing this, you are mimicking a generator. Three common methods to create this movement are to pedal a specially-equipped bicycle, use a hand pedal, or climb a stair-step machine.

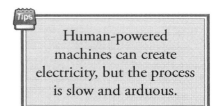

Human-powered machines can create electricity, but the process is slow and arduous.

Windstream human-powered generator

Bicycles give the greatest energy yield, with units from Windstream Power claiming to produce 10 to 16 amps. Recharging a 100-amp-hour lead-acid battery would take about 6 to 10 hours of steady pedaling—a *very* hard day's workout. Hand-pedaled units and stair steppers provide 3 to 4 amps of output power, driving the recharge time up to a whopping 25 to 33 hours of heavy exertion.[148]

There are, however, some advantages to human-powered recharge systems. They are available day and night, and are independent of sun and wind. Also, family members can take turns sharing the work load. The two significant disadvantages are the heavy exertion required and the limited power output.

DP PLAN EXAMPLE

Table 7-4 Sample DP Plan Entry

Need: Electricity			
Danger	**Goals**	**Needs**	**Implementation**
Power outage	Operate lights and appliances for a few hours each day for up to 10 days	AC power for house lights, primary heating unit, and appliances	Use a 10-kilowatt gas generator to power the whole house with select loads switched in and out.
			Have electrician install a mechanical interlock.
			Draw up a schedule that uses the generator 2 hours in the morning and 2 hours at night.
			Stock 25 gallons of stabilized fuel in 5-gallon jerry cans.
			Convert generator to tri-fuel use.
			Stock four 20-lb propane tanks as alternate fuel.
		DC power for recharging batteries, operating small loads (radios, cell phones, electric lanterns)	Use a 200 amp-hour deep discharge battery and 2,000-watt inverter.
			Stock low self-discharge NiMH batteries and chargers.
		DC power to recharge 12-volt lead acid batteries using alternative energy source	Recharge batteries with a Powerfilm 60W foldable solar panel placed along the side of the house with optimal sun exposure.

Quick Summary - Electrical Power

➤ Having a backup power source can make handling a disaster much easier.

➤ Power can be either direct current (DC) or alternating current (AC). Household appliances and lights run on AC, while most small portable devices use DC power.

➤ Generators and inverters are used to provide AC power.

➤ Determine your electrical needs in order to select the right-sized generator.

➤ Either plug directly into your generator using extension cords, or wire the generator to your house through a mechanical interlock or transfer switch.

➤ Limit fuel use by running your generator only a few hours each day. Develop a schedule of tasks to carry out during each power cycle.

➤ Check with your insurance company and fire department to determine how much fuel you are allowed to store at your home.

➤ Add stabilizers to freshly-stored fuel to extend its shelf life.

➤ Inverters convert battery power into AC power but are not suited to powering equipment with high startup currents.

➤ Lithium batteries have the highest capacity and longest shelf life of all single-use or rechargeable batteries. Alkaline are much less expensive and good general purpose batteries.

➤ Low self-discharge NiMH batteries are the only small, rechargeable batteries recommended for disaster preparedness. They are especially useful for powering radios and flashlights.

➤ When teamed with an inverter and recharger, lead-acid batteries are an excellent emergency power source.

➤ Rechargers that use natural energy include solar panels, wind turbines, and human-powered machines. All are viable methods of recharging batteries, but they suffer from limited output.

Recommended Items - Electrical Power

❑ Backup power source
   a. Electric generator system including: a gas or diesel generator, interlock or transfer switch, power inlet box, heavy-duty power cord, and maintenance items (e.g., filters, spark plugs), *and/or*
   b. Bank of deep cycle batteries, conventional charger, high-wattage inverter, and extension cords

❑ Backup fuel source
   a. Gas or diesel fuel in approved storage containers, treated with a stabilizer, *and/or*
   b. Solar, wind, or mechanical battery recharger

❑ Spare batteries
   a. Alkaline, *or*
   b. Low self-discharge rechargeable NiMH batteries with recharger

CHAPTER 8
HEATING / COOLING

Challenge

An arctic freeze has enveloped your community. Snow and ice have made roads impassable. Your heating unit unexpectedly shuts down, and it will be days before repairmen can service your home. Temperatures fall to dangerous levels each night. How will you keep your family warm?

There are numerous situations that can leave you without heating or cooling. In moderate climates, this may not be a serious concern, but most people are reminded annually of the bitter sting that winter brings. Staying warm becomes not merely a matter of comfort but one of survival. Conversely, there are places in the country where summer heat waves can bring suffering and even death if your home's temperature is not adequately controlled.

BODY TEMPERATURE

Simply put, getting too cold or too hot can kill you. At one end of the spectrum is hypothermia—literally freezing to death. At the other end is hyperthermia—overheating to the point of heatstroke. Both conditions are equally deadly and must be prevented.

HYPOTHERMIA

If more heat escapes your body than it can produce, you will cool and eventually develop hypothermia. Normal body temperature is 98.6°F. If your internal core temperature drops below 95°F, it signals hypothermia. Symptoms may include gradual loss of motor skills and mental acuity, fatigue, mumbling, slowed breathing, slurred speech, and cold, pale skin. Since symptoms usually develop gradually, victims are often unaware that they are succumbing to the cold. If the condition is allowed to progress, it can lead to death. In the United States alone, nearly 700 people die from hypothermia each year.[150]

Being in extreme cold without adequate clothing, wearing wet clothes (especially in windy conditions), or being exposed to very cold water can all cause hypothermia. Treatment for hypothermia includes moving to a warmer environment, removing wet clothing, insulating from any cold surface, sharing body heat, and drinking warm beverages. Never massage the skin, apply direct heat using a heating pad or heat lamp, or drink alcoholic beverages.[150]

Prevention is best done by wearing clothing appropriate to the environment, including gloves, hat, and footwear. Avoid overexertion that can cause you to sweat. If you do become wet for any reason, change out of the wet clothes as soon as possible.

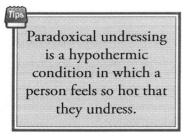

Paradoxical undressing is a hypothermic condition in which a person feels so hot that they undress.

Twenty to fifty percent of all deaths relating to hypothermia are associated with what is known as *paradoxical undressing*.[151] Though the causes are not entirely understood, it is a condition in which a person feels extremely hot even though his core temperature is falling to deadly levels. The victim begins to shed clothing in an effort to cool off, which only further speeds his demise. Unfortunately, the confusion and impaired judgment caused by hypothermia keep the person from recognizing the cause of their self-destructive actions.

HYPERTHERMIA

Hyperthermia is the condition in which your body absorbs or produces more heat than it can dissipate. The term is often used to describe several heat-related illnesses, including heat cramps, heat exhaustion, and heatstroke.

Heat cramps are the first stage of hyperthermia. Symptoms include profuse sweating, fatigue, thirst, and muscle cramps. Strenuous physical activity in hot weather is usually the cause, although exertion is not necessary if the environment is hot enough. Heat cramps can be treated by drinking fluids, especially those containing electrolytes, and resting in a cool location.

If you do not take the steps necessary to cool down when experiencing heat cramps, they will develop into heat exhaustion. You will become thirsty, giddy, weak, uncoordinated, nauseated, and continue to sweat profusely. Your body temperature will likely remain normal, but your pulse may become elevated, and your skin can feel cold and clammy. Heat exhaustion is a serious warning that you are getting too hot and need to immediately rest in a cool location and drink cold liquids.[152]

Continued exposure to heat or exertion could lead to heatstroke—a life-threatening condition. This occurs when your body temperature rises to 104°F or higher. Other symptoms may include cessation of sweating, hyperventilation (rapid, shallow breathing), muscle cramps, weakness, confusion, combativeness, irrational behavior, strong rapid pulse, and delirium. The high body temperature can cause irreversible brain damage, organ failure, and death.[153]

If you suspect you are suffering from heatstroke, get to a cool location and rest. Drink cold liquids, especially water and fruit juice, but avoid caffeinated or alcoholic drinks. If possible, shower, bathe, or sponge off with cool water. Additional medical attention may be necessary—see *Chapter 11: Medical/First Aid* for more treatment information.

COOLING

It may be hard to believe, but humankind survived for a very long time without air conditioning (A/C). You'll find that you can too simply applying a little common sense. If you find yourself without A/C on a very hot day, you can prevent heat-related illnesses by limiting your activities, drinking cold beverages, and establishing some basic air circulation in your home.

KEEPING YOURSELF COOL

Consider the following suggestions to keep *yourself* cool:

- Take it easy. Give yourself permission to be lazy.
- Drink plenty of cold, non-alcoholic liquids.
- Wear loose-fitting, warm-weather clothing, such as shorts and a t-shirt.
- Enjoy a Popsicle, frozen fruit, or other icy treat.
- Soak your t-shirt in water and sit in front of a fan or breeze.
- Take a cool bath or shower.
- Stay in the lowest level of your home. Heat tends to rise to the upper levels. Better yet, sit under a large shade tree outside.
- Fan yourself—improvise as necessary with a magazine, piece of cardboard, or other lightweight flat object.
- Soak your feet in a bucket of cool water.

KEEPING YOUR HOME COOL

Consider the following suggestions to keep *your home* cool:

- Close your blinds on windows facing the sun.
- Open windows on opposite sides of your house to create air circulation. If possible, use fans to draw off heat from the hottest parts of your house, and pull in cooler air from outside.
- Turn off anything creating unnecessary heat (e.g., lights, computer, dishwasher, television).
- Run appliances in the evening when it is cooler.
- Blow air across a bowl full of ice cubes or water frozen in 2-liter bottles to create a mini air conditioner.
- As a longer-term solution, plant shade trees around your home.

Beyond these basic preventive measures, no other special preparations are usually necessary to stave off heat-related illnesses. For most families, there isn't a pressing need to maintain a secondary cooling system for the home. With that said, if someone in your family is particularly susceptible to heat (perhaps an elderly parent, a young child, or even a pet), then you should plan accordingly. This usually means having backup electrical power to run fans or a small A/C system—see *Chapter 7: Electrical Power* for details on electrical preparations.

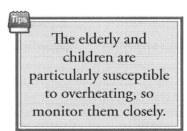

The elderly and children are particularly susceptible to overheating, so monitor them closely.

Emergency Blankets

Emergency blankets (a.k.a. space blankets) were originally developed by NASA for the space program back in 1964. They are small and light-weight, making them easy to toss into a glove box or grab-and-go bag, and are generally con-structed of a thin sheet of PET plastic coated with a metallic reflecting agent (typically aluminum). If the packaging is to be believed, this feather-weight miracle will keep you warm regardless of the temperature or weather conditions. In reality, if you actually use an emergency blanket in very cold weather, you will almost certainly freeze!

Wrap up tight to prevent heat loss

The problem is not with the blanket but with your method of using it. Emergency blankets will only keep you warm if they are large enough to completely envelope your entire body, including your head. This prevents body heat from escaping and cold air from getting in. Blankets that are fabricated as bags (a.k.a. bivouac sacks) are better than conventional tarp-like emergency blankets because they make wrapping up much easier.

Here are a few suggestions regarding the use of emergency blankets:

1. Don't lie directly on the cold ground. If you do, your body heat will efficiently conduct through the bag to the underlying ground. It is much better to rest on top of an insula-ting material or structure—anything from old clothes to a thick bed of pine needles.
2. If your clothing is damp, remove it before getting in the bag. Some people prefer to wrap up nude (or in their underwear), but you may find that a layer of clothing is preferable since it acts to insulate you from the cold surface of the bag.
3. If possible, use a sleeping bag or regular blankets in conjunction with the emergency blanket. This will make a big difference because once again, if blankets are placed over and under the emergency blanket, they will act as insulation.
4. Secure the emergency blanket or bivouac bag so that it covers your head and most of your face—see figure. Ideally, you should have only a small hole at the top of the bag through which to breathe. The larger the hole, the more body heat that will escape. If you leave your head completely exposed, a great deal of body heat will be lost. Better emergency blanket products are designed with drawstrings or zippers, making them easy to seal up. Less expensive models must be taped up by the user.

Like all lifesaving preparations, you should never rely on emergency blankets until you have tested them thoroughly. Learn how to effectively use these technological marvels, and you will find them to be a valuable emergency product.

HEATING

Keeping warm starts with having the right kinds of clothing and bedding. Your family should already have clothing appropriate to the climate in which you live, so additional garments are likely unnecessary. Bedding may be another matter. You are accustomed to maintaining a comfortable living environment by simply adjusting the thermostat. But what if that heat fails, and your home is left no warmer than the frosty outdoors? Your family's usual bedding probably won't be enough to stay warm at night.

The easiest way to prepare is to consider what your family would need to spend the night outdoors. Winter sleeping bags, down comforters, and heavy wool blankets should all come to mind. How much and what kind depends on where you live. Whatever blankets you select, plan as if Jack Frost is maliciously working against you—worst winter on record, heating system failure, a couple of windows broken out. Unrealistic yes, but keep-

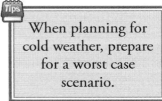

When planning for cold weather, prepare for a worst case scenario.

ing warm is critically important and not overly difficult to do with a few basic preparations. You can't go wrong by having extra blankets. If you end up not needing them, loan them to neighbors or friends who might be less prepared. Also, remember to stash blankets in the car for roadside emergencies—see the auto kits described in *Chapter 14: Transportation* for more information.

If you want to do much more than huddle under blankets, you will need some kind of area heating. Not only will heating help your family to function, but it will also help them to feel more comfortable—and that leads to a happier family.

LOSS OF HEAT

Heat can be lost in two very different ways, and each requires different preparations. The first is a disruption in your fuel source. For most people, this is manifested through a loss of electrical power or the shutting off of the gas supply. The second is when your heating unit fails, perhaps due to cold weather or just a case of bad luck. Regardless of the cause, a reasonable goal is to have sufficient heating supplies to support your family for a minimum of two weeks.

LOSS OF FUEL

Many situations can result in your heating system's fuel source becoming unavailable, whether that fuel is electricity, propane, or natural gas. For these occasions, you need a supplementary source of fuel and a method to use it in your heater. Depending on the type of heater, this provision is handled in different ways.

If you have an electric heating system, you can probably power it using a large generator. Take care to size the generator to handle both constant and surge current requirements—see *Chapter 7: Electrical Power*. If your heater burns propane, or if you've converted your natural gas heater to run on propane in an emergency, you will need a backup fuel tank and a valve that allows you to switch over to that secondary propane source. You will also likely need a small backup AC electrical source, such as an inverter or generator, to power the blower motor, thermostat, ignition system, and backup strip heating.

If you have a wood-burning stove, kerosene heater, or other local heat source, you will probably not be directly affected by the centralized disruption of fuel. But once again, you may still require modest levels of electrical power for the blower or pellet auger. Investigate your particular system to fully understand its operational needs.

LOSS OF HEATING UNIT

If you have ever witnessed the numerous repair trucks servicing neighborhoods during a cold spell, you likely realize how frequently heating systems fail. This is especially true when the weather is unusually frigid, and the heater is worked harder than usual.

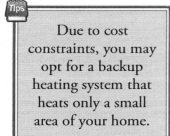

Due to cost constraints, you may opt for a backup heating system that heats only a small area of your home.

To be prepared for a heating system failure, you will need a backup heat source. Due to costs, having a fully redundant secondary heating system is probably not feasible. Instead, you may opt to have a smaller backup heater that is capable of heating only a portion of your home—enough to help you get by until your system can be serviced. There are many options for backup heating, including kerosene heaters, electric space heaters, and wood-burning stoves, all of which are discussed later in this chapter.

Personal aside: On three separate occasions, and at three different residences, my family's heating system failed during harsh winter storms. In all three cases, I was unable to get a service person out to my house in less than 48 hours—leaving us to fend for ourselves. The first time, my family and I were ill prepared and had little choice but to abandon our house and seek a warm hotel. When the second incident occurred, we had enough blankets and knowhow to weather the event, albeit with a few shivers. By the third time, my family was equipped with a kerosene heater, a wood-burning fireplace, and stacks of blankets—making our home a welcome neighborhood refuge.

SAFETY FIRST

Before considering backup heat sources, realize that the dangers of improper use far outweigh the dangers of freezing to death. Two significant risks are worth noting: carbon monoxide (CO) poisoning and fire.

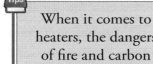

When it comes to heaters, the dangers of fire and carbon monoxide poisoning far outweigh the dangers of freezing.

Any type of heater that burns fuel, such as wood, coal, propane, and kerosene, also produces CO. Carbon monoxide is a colorless, odorless gas, so you won't see or smell it. Symptoms of CO poisoning include headache, fatigue, lightheadedness, shortness of breath, nausea, and dizziness. If you experience any of these symptoms while operating fuel-burning heaters, get fresh air immediately! Carbon monoxide poisoning is deadly, killing about 200 people in the United States each year.[114] The surest way to prevent CO poisoning is never to use fuel-burning appliances indoors without proper ventilation. It is also a good idea to put a CO detector in any room where a fuel-burning appliance is used (see *Chapter 5: Shelter* for a discussion of safety devices).

Using a fuel-burning heater requires either direct venting to the outdoors using a pipe that routes through the wall or ceiling, or keeping the area well ventilated with open windows or doors. Most heaters come with manufacturer's recommendations regarding ventilation. Read and follow them. If recommendations are not furnished, a useful rule of thumb is to provide one square inch of window opening for each 1,000 BTUs of heater rating.[154] For example, a heater with a 20,000 BTUs/hour rating would require that a 10-inch wide window be open 2 inches (i.e., 20 square inches of opening). This recommendation should be treated as the minimum ventilation for all fuel-burning heaters that are not directly vented outdoors.

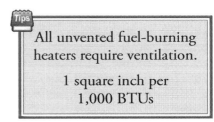

All unvented fuel-burning heaters require ventilation.

1 square inch per 1,000 BTUs

The second risk of using a fuel-burning heater is that you might catch your house or shelter on fire. Heaters must be kept away from flammable materials, and that means basically everything in your home except stone, metal, or brick. Normal construction materials, such as sheetrock, wood flooring, and carpet, as well as furniture, clothing, and bedding can all catch fire from backup heaters. The hotter the heater, the more likely it is to cause a fire. For this reason, never leave heaters unattended, and solicit help from everyone involved to keep a watchful eye.

Preventing Heater Fires

1. Keep the heater away from anything flammable.
2. Never leave the heater unattended.
3. Explain the dangers to everyone involved, and ask them to help keep a watchful eye. Be especially mindful of very young children who might tip the heater or burn themselves.
4. If the heater plugs into the wall, avoid using extension cords. If you must use one, select a heavy duty cord with at least 14 AWG wire.

AMOUNT OF HEAT NEEDED

The amount of backup heat you need is largely determined by the climate in which you live. It is safe to say that if you live in Florida, you will need less backup heating than someone who lives in North Dakota. However, in all but the most temperate climates, some form of backup heat is advisable.

Start by assessing your current heat source. Does your house or apartment use a heat pump, gas furnace, or electric heater? If your heater burns fuel, such as natural gas or propane, does it also require electricity to power a blower? Perhaps you are out in the country and use a wood-burning stove or kerosene heater. If your primary fuel source was interrupted in the dead of the winter and remained out for two weeks, how would your family stay warm? The time to address this question is now while comfortably sipping iced tea, not when you can see your breath in the air.

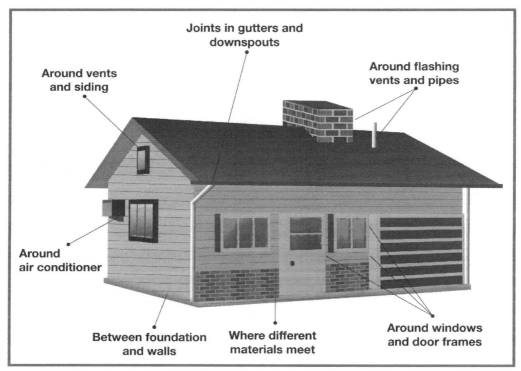

Weatherproofing your home

Conserve heat by weatherproofing your home: seal gaps, insulate windows, and weatherstrip around doors.

Regardless of what type of secondary heating system you select, the first step is to conserve heat. Keep your house well sealed unless you are specifically using an unvented, fuel-burning heater. If you add up the gaps in an average home, they total about 14 square inches, and those gaps result in significant energy loss.[155] Obvious corrective actions include applying weatherstripping around leaky doors and window sashes, caulking gaps and joints, and using double-paned windows.

SELECTING A HEATER

When considering the type of backup heater to use, many factors will come into play, including cost, safety, heating area, and the climate in which you live.

Seven steps for selecting and readying a backup heating system are given below:

1. If your backup heater cannot heat your entire home, identify a well-insulated space in your house to where your family will retreat. It should have limited windows or doors to the outside but easy access to a bathroom. If you decide to consolidate in the main living area of your home, seal off adjoining rooms.

2. Determine how much heat you will need (see *Sizing Your Heater*).

3. Select the most appropriate type of heater for your situation (see *Types of Heaters*).

4. Learn how to safely use your heater.

5. Install any necessary venting or duct work.

6. Stockpile the necessary fuel.

7. Test your heater on a very cold winter day to ensure it will meet your family's needs. Supplement or resize the heater as necessary.

SIZING YOUR HEATER

There are heaters available in nearly every size; some are big enough to heat a barn, while others are barely able to warm your feet. Properly sizing your backup heater can mean the difference between being cozy and being miserable.

Sizing a heater requires you to consider three things:

- The volume of air to be heated
- The heat that escapes through the walls, floor, and ceiling
- The minimum allowable temperature difference between indoors and outdoors on the worst possible winter day

If you're installing a large, fully-redundant heating system, such as a heat pump or furnace, have the retailer inspect your home to determine the correct heater size.

If you opt for a smaller, localized heat source—perhaps a wood-burning stove, kerosene heater, or electric space heater—you will need to determine the necessary heater size by making a few measurements and performing some simple calculations.

There are many online tools and resources to help you perform heater sizing.[156,157,158,159] Unfortunately, their methods and results frequently disagree. That is because precise heater sizing requires detailed knowledge of your home's exposure, orientation, insulation, sealing, and air flow. Each tool makes different assumptions, some more conservative than others.

The following guidelines are considered to be reasonable methods of estimating the required size of a heater. The three methods discussed are in order of increasing accuracy.

METHOD 1

The simplest method to size a heater is to multiply the square footage to be heated by 25. The result is the number of BTUs/hour needed to stay warm.

For example, if you wish to heat 200 square feet, you would need a heater capable of outputting a minimum of:

$$200 \times 25 = 5{,}000 \text{ BTUs/hour}$$

METHOD 2

A second, more accurate method requires measuring the volume of air to be heated and specifying the minimum temperature difference desired between your home and the outdoors. Once again, a straight-forward equation can be used to estimate the heater size.

$$Volume \times \Delta T \times 0.133$$

For example, to heat a room measuring 200 square feet and having 8-foot ceilings, would require heating 1,600 cubic feet of air. If the outside temperature can fall as low as 10°F, and you wish to maintain the inside temperature at 60°F, the temperature difference is 50°F. The minimum heater size then would be determined to be:

$$1,600 \times 50 \times 0.133 = 10,640 \text{ BTUs/hour.}$$

METHOD 3

An even more accurate method takes into account your home's level of insulation but no longer considers the specific temperature difference you are trying to achieve. This method has you multiply the size of the area to be heated by an insulation factor (given in Table 8-1).

$$Area \times Insulation\ Factor$$

For a house with average insulation, heating a 200-square-foot area would require between:

$$\text{Best case:} \quad 200 \times 50 = 10,000 \text{ BTUs/hour}$$

$$\text{Worst case:} \quad 200 \times 70 = 14,000 \text{ BTUs/hour}$$

Table 8-1 Insulation Factors[160]

Category	Insulation	Factor
Poor Insulation	No insulation in walls, ceilings, or floors; no storm windows; windows and doors not well sealed	90-110
Average Insulation	R-11 insulation in walls and ceilings; no insulation in floors; no storm windows; doors and windows fairly tight	50-70
Good Insulation	R-19 insulation in walls, R-30 in ceilings, R-11 in floors; tight-fitting storm windows or double-pane windows	29-35
Super Insulation	R-24 wall insulation, R-40 in ceilings, R-19 in floor; tight-fitting storm windows or double-pane windows; vapor barrier sealed carefully during construction	21-25
Earth-sheltered	Earth-sheltered house with little exposure; well insulated	10-13

SUMMARY

As you can see, the three estimates range anywhere from 5,000 to 14,000 BTUs/hour. With these methods you can get a first-order estimate of your heating needs (perhaps by taking the average of the three). Once you have that estimate, you will need to select a heater that functions well in your house for the temperature extremes you are likely to experience. This can often be a process of trial and error—buying a heater that you think will work, and then supplementing or replacing it as needed.

Most heaters are specified by British Thermal Units per hour (BTUs/hour). Sometimes retailers will get sloppy and simply state the value as BTUs, but they are always referring to BTUs/hour. A BTU is defined as the amount of heat required to raise one pound of water one degree Fahrenheit. It makes sense that a heater should be rated by how much heat per hour that it outputs.

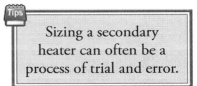

Sizing a secondary heater can often be a process of trial and error.

Electric heaters are an exception to this method of specification. They are sized in watts rather than BTUs/hour. For comparison purposes, you can convert from watts to BTUs/hour by multiplying the number of watts by 3.4. For example, a standard 1,500-watt electric space heater will output heat comparable to a 5,100-BTUs/hour propane heater. There are other considerations to make this conversion more accurate, such as energy efficiency, but the factor of 3.4 is a reasonable estimate.

Another method to size an electric heater is to assume that every square foot requires about 10 watts of power.[161] This is not precise and does not consider cubic feet or temperature drop, but it gives a preliminary estimate. Using this method, a 200-square-foot room would require 2,000 watts of electric heating (or roughly 6,800 BTUs/hour).

TYPES OF HEATERS

There are many possible sources of backup heat, each with its respective advantages and disadvantages. A list of likely candidates is given in the tip box.

Regardless of the type of heater you select, you should fully understand the dangers associated with it. If you select an electric, propane, or kerosene space heater, pick a unit that has an automatic shutoff feature in case it is accidentally tipped over. It is worth repeating that any heater that burns fuel, whether it is wood, propane, kerosene, or gas, must either be vented to the outdoors or reside in a well-ventilated room to prevent carbon monoxide buildup.

Backup Heaters
> Fireplaces
> Masonry heaters
> Wood/Coal/Pellet stoves
> Electric space heaters
> Kerosene space heaters
> Oil-filled radiators
> Propane or natural gas heaters

FIREPLACES

Modern fireplaces typically come with artificial logs and glass covers and are designed more for style than heat. They offer a wonderful instant fire ambiance but unfortunately don't output much heat to

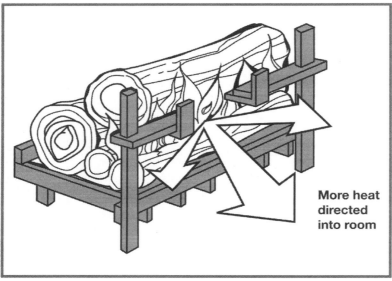

More heat directed into room

Texas Fire Frame

the room. To improve their utility, many instant-on fireplaces can be equipped with internal blowers that help to circulate the heat. If using blowers, don't forget to address this requirement in your electrical power preparations.

> Fireplaces are terribly inefficient, but using a Texas Fire Frame can help.

If you have a traditional stone or brick wood-burning fireplace, then you can get a little more heat from it—but not much. Wood-burning fireplaces are very inefficient, losing about 90% of their heat up the chimney.[162] This can be improved through the use of a "Texas Fire Frame" or other fireplace radiant grate that focuses heat back into the room, perhaps raising the efficiency up to 30%. Another option is to install a wood- or pellet-burning stove insert into the fireplace. Stoves inserts significantly improve the efficiency as well as reduce smoke emissions.

MASONRY HEATERS

Modern masonry heaters are based on traditional designs dating back to the 14th century. They are known by many names, including Russian stoves, Kachelofen, Tulikivi, Grudofen, and Contraflow heaters. Each heater consists of a large stone or brick burn box, a chimney, and a labyrinth of channels to trap the heat and smoke. This trapped heat is transferred to the masonry hearth and surrounding walls. The fire itself only burns for a brief period of time, but the heat is radiated out from the masonry for 12 to 24 hours, meaning that only a single fire may be needed for an entire day's heat. This slow, steady release of heat is the primary advantage of masonry heaters over other heating methods.

> Even with a small morning fire, masonry heaters radiate heat throughout the entire day.

Masonry heaters typically burn wood, but they have also been used to burn straw, dry vegetable matter, and even garbage. The fire burns very

Courtesy of Bob Weaver Masonry

hot (over 1,400°F), leading to complete and rapid fuel consumption that releases very little air pollution. Despite the high temperature, the surrounding stone is massive enough not to be hot to the touch. The heaters offer exceptional (80-90%) efficiency because most of the energy is trapped in the masonry rather than vented up through the chimney.

However, there are several drawbacks to masonry heaters. They require a couple of hours to heat up, which can be a long wait on a cold winter morning. They also do not have centralized heat distribution, leaving most of the radiant heat to remain in the room with the heater. Finally, anyone considering a masonry heater must take into account their weight and cost. The stonework can easily weigh two thousand pounds or more, often requiring that the floor be reinforced. Masonry heaters are custom built by expert craftsmen and can cost many thousands of dollars. Additional information and a listing of licensed builders can be found online at the Masonry Heater Association of North America.[163]

COAL/PELLET/WOOD STOVES

Coal, pellet, and wood stoves are in many ways the epitome of disaster preparedness. They can often serve as both heater and stove—meeting two very important needs simultaneously. Some stoves fit into existing fireplaces as inserts, while others are free-standing. Newer models control the oxygen level so that fuel lasts longer and burns more completely, making them more efficient and less polluting than stoves of the past. Stoves are excellent backup heaters, but they are fairly expensive, require periodic maintenance, and professional installation is definitely recommended. There are many vendors for fuel-burning stoves, including Woodstock Soapstone Company, Lopi, Harman, Napoleon, Reading Stove Company, Lehman's, and Vermont Castings.

Coal Stoves

Coal stoves emit more pollution than EPA-certified wood stoves. The primary advantage of a coal stove is its higher fuel efficiency and lower fuel costs. Table 8-2 gives a useful comparison of fuel efficiencies and costs for many types of vented room heaters. This table makes clear the advantage of coal over other fuels since costs can be half those of a pellet stove or one-third those of electric power.

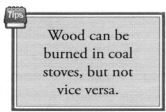

Wood can be burned in coal stoves, but not vice versa.

Much like masonry heaters, coal and wood stoves allow burning of improvised fuels, such as lumber scraps, furniture, newspapers, and books. However, when burning makeshift fuels, pay attention that the chimney doesn't become clogged and catch fire. Burning wood in a coal stove is generally acceptable, but burning coal in wood or pellet stoves will quickly deteriorate the stove floor due to the higher burn temperatures.

Pellet Stoves

Pellet stoves are designed to burn cleaner and more efficiently than conventional wood stoves. They burn short, thin rods of compressed sawdust or other renewable materials. These "pellets" are poured into a hopper that feeds automatically into the stove, typically using an internal electric auger. Note that these stoves can be picky about the types of pellets they will handle without jamming, so be sure to experiment with products before stocking up.[165] Pellet stoves emit very little pollution and therefore do not require EPA certification.

Wood Stoves

There are two types of wood stoves: catalytic and non-catalytic. Most stoves are non-catalytic, which means they are essentially large metal fireboxes with baffles. Catalytic stoves are a bit more complicated since they pass the smoky exhaust through a coated, ceramic honeycomb inside the stove where the smoke gases and particles burn. The honeycomb elements must be replaced every two to six years. Catalytic stoves also come equipped with a lever-operated bypass damper that is opened when starting or reloading.

Do your homework and select an EPA-certified stove that provides adequate heat output and burns the fuel you have selected. The smoke emissions, efficiency, and heat output ratings are listed on the back of certified stoves. Modern wood-burning stoves emit much less smoke than older stoves. For a stove to be EPA-certified, particulate emissions must be below 7.5 grams per hour for non-catalytic wood stoves and 4.1 grams per hour for catalytic stoves. Some states impose even more restrictive limits. Compare this to previous-generation stoves that emitted between 15 and 30 grams per hour.[166]

Table 8-2 Fuel Comparisons for Vented Room Heaters[164]

Fuel Type	Fuel Unit	Fuel Heat Content per Unit (BTUs)	Fuel Cost per Million BTUs	Efficiency	Fuel Cost per Million BTUs (efficiency included)
Coal (Anthracite)	Ton	25,000,000	$8.00	75%	$10.67
Solid Wood	Cord	22,000,000	$9.09	55%	$16.53
Corn (kernels)	Ton	16,500,000	$12.12	68%	$17.83
Natural Gas	Therm	100,000	$12.27	65%	$18.88
Fuel Oil (#2)	Gallon	138,690	$17.01	78%	$21.81
Pellets	Ton	16,500,000	$15.15	68%	$22.28
Propane	Gallon	91,333	$21.27	65%	$32.73
Electricity	Kilowatt-hour	3,412	$33.85	100%	$33.85

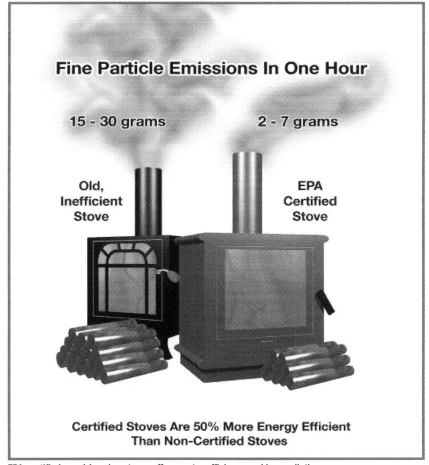

EPA-certified wood-burning stoves offer greater efficiency and less pollution

Table 8-3 Fuel Equivalence[249]

Fuel Type	Equivalent Quantity
Wood	1 cord
No. 2 Fuel Oil	150 gallons
Liquid Propane	230 gallons
Natural Gas	21,000 cubic feet
Electricity	6,158 kilowatt-hours

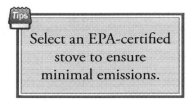

Select an EPA-certified stove to ensure minimal emissions.

When researching stoves, don't just read product descriptions; check user reviews to see what types of problems others have experienced. Also, visit the retailer and personally inspect each stove for durability and quality. This is an investment that your family will benefit from for many years, so take the time to make a well-informed selection.

Calculating Wood Needed

If you are relying on a wood-burning stove, either as your primary heat source or as a backup emergency heater, you will need to determine your fuel needs. Wood is purchased in cords, with each cord measuring 4 feet × 4 feet × 8 feet. A cord should consist of tightly-stacked, cut wood. Most of the time, individual wood pieces are sold in lengths smaller than 4 feet to facilitate easier burning. The amount of energy in a cord of wood can be compared to other fuel sources as shown in Table 8-3.

Using these equivalence values and the fuel efficiencies given in Table 8-2, you can determine the number of cords of wood needed to meet your heating needs. Consider the following equation:

$$Cords = \frac{PF \times E_{PF}}{W \times E_W}$$

PF is the amount of primary fuel used, W is the wood equivalence number from Table 8-3, and E_{PF} and E_W are energy efficiencies of the primary fuel and wood (given in Table 8-2).

Example: Assume your 2,000-sq. foot home normally uses 60,000 cubic feet of natural gas per year. The efficiency of natural gas is about 65%, and the efficiency of wood is about 55%. The number of cords of wood needed to supply the same level of heating for a full year would be given by:

$$Cords = \frac{60,000 \times 0.65}{21,000 \times 0.55} = 3.4$$

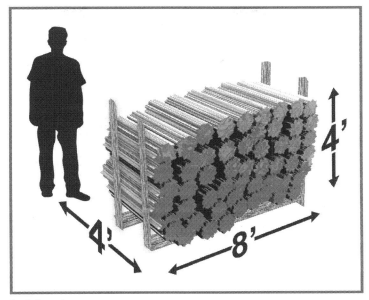

Cord of wood

ELECTRIC SPACE HEATERS

Electric space heaters are inexpensive and easy to use; simply plug them in and set the desired output level or thermostat temperature. They obviously require electricity, provided either by your electric company, a generator, or an inverter and battery combination. The Electrical Code limits the size of residential electric heaters to a maximum of 1,500 watts. If you need more heat than that, you must use multiple heaters. Each heater should be plugged into a different circuit in your home (e.g., one in the living room outlet, one in the kitchen—assuming they are on separate breakers). If you plug two heaters into the same circuit, they will almost certainly overload and trip your breakers.

If you have consolidated your family to a small area, even a single 1,500 watt electric space heater may be adequate to keep warm. This depends on the size of the room, the insulation, and how cold it is outside. Electric heaters do not emit carbon monoxide. However, as with all heaters, ensure that there is adequate space around the heater to prevent a fire.

> **Tips**
> Plug each electric heater into a separate circuit to prevent overloading your breakers.

Features to look for in an electric space heater include:

- Automatic shutoff when tipped over
- Overheat protection
- Long power cord to allow easier placement of the heater
- Thermostat or multiple output settings
- Circulating fan (for convection models)

The two most common types of electric space heaters are *convection* and *radiant*. Convection heaters flow air across a hot thermal conductor (usually ceramic plates or metal coils). Air is circulated into

Vornado convection and Soleus radiant heaters

the room either naturally through convection or by an electric fan. Convection heaters are useful for heating small rooms.

Radiant heaters produce directional heat using an element, such as quartz, and a reflector. They are best suited for heating a small area directly in front of the heater because very little heat is radiated at any other angle. The heat is immediate but dissipates quickly when the unit is turned off.

There are many manufacturers of electric space heaters including Bionaire, Soleus, DeLonghi, Duraheat, Honeywell, Vornado, Presto, and Lasko.

KEROSENE SPACE HEATERS

Kerosene heaters can either be vented or unvented. Larger units generally feed directly from outdoor kerosene tanks and vent back outside through pipes (similar to wood stoves). Smaller, unvented units have modest kerosene tanks attached directly to the unit and are meant to heat one or two rooms. Beware that the term "unvented" is misleading. Like all fuel-burning appliances, unvented kerosene heaters consume oxygen and emit poisonous fumes, meaning that venting of the room is still required. Kerosene heaters also emit an odor that many people find unpleasant—some heaters worse than others.

Portable, unvented kerosene heaters generate significant heat, most outputting from 10,000 to 30,000 BTUs/hour. Surface temperatures can reach up to 500°F, so use caution and keep children a safe distance away from them.

Below are additional safety recommendations for using kerosene heaters:[154]

1. Use heaters that have been tested and listed in accordance with Underwriters' Laboratories (UL) Standard 647. This compliance should be present on the nameplate.

2. Use 1-K kerosene fuel, which has about ⅛ the sulfur content of 2-K kerosene. The use of 2-K kerosene leads to increased sulfur dioxide emissions and frequent wick maintenance. The terms "water clear" or "clear white" are often used to describe 1-K kerosene. Never use diesel fuel, gasoline, jet fuel, or No. 1 fuel oil in a kerosene heater.

3. Keep heaters at least 36 inches from anything combustible.

4. Perform periodic maintenance, including cleaning the unit, trimming the wick, and inspecting for fuel leaks.

5. Ensure adequate ventilation. Follow the manufacturer's guidelines. If recommendations are not given, provide a minimum of one square inch of window opening for each 1,000 BTUs of the heater rating.

6. Don't operate kerosene heaters when everyone is asleep. A malfunction could lead to asphyxiation.

Sengoku kerosene heater

7. Allow the heater to cool a minimum of 15 minutes before refueling. If kerosene is spilled onto a hot heater it can ignite and cause a fire. Also, refuel the heater outside in a well-ventilated area. Don't overfill it; leave a space at the top for fuel expansion.

8. Keep fuel outside of the main living area, stored in approved blue safety containers with "Kerosene" clearly marked on them.

9. Have smoke and CO detectors in the area being heated, as well as fire extinguishers nearby.

Manufacturers of kerosene heaters include Sengoku, KeroSun, Toyostove, and Corona.

OIL-FILLED RADIATORS

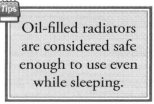

Oil-filled radiators are considered safe enough to use even while sleeping.

Oil-filled radiators are sealed metal canisters filled with heating oil. When the heater is plugged into AC electrical power, the oil heats up and warms the air around the radiator. Oil-filled radiators have many advantages: they are easy to use, provide convective heat in all directions, are equipped with thermostats, and do not emit carbon monoxide. They are the safest of all space heaters and are the only type recommended to leave running while asleep. With outputs ranging from 600 to 1,500 watts, radiators are best suited to heating small rooms. Heat output is much slower than from electric or kerosene heaters, but radiators stay warm long after power is turned off. Manufacturers include Pelonis, Honeywell, Lakewood, and DeLonghi.

PORTABLE PROPANE HEATERS

Portable propane heaters are a relatively new addition to space heaters. They come in many different sizes. There are large, outdoor units that provide up to 400,000 BTUs/hour of heat and run off 100-lb tanks. These units are not meant to be used indoors, being better suited to sporting events, barns, or construction sites.

DeLonghi oil-filled radiator

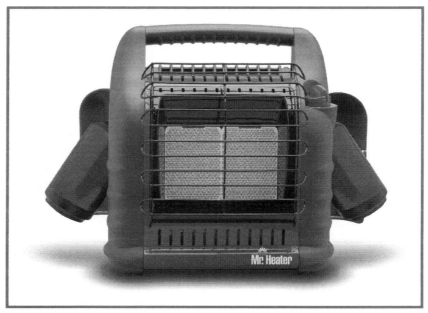

Mr. Heater propane heater

There are also smaller hand-carried heaters that use standard 1-lb or 20-lb propane tanks. They typically output between 4,000 and 18,000 BTUs/hour. Small units can be used indoors as long as ventilation rules are followed—check the owner's manuals. If no specification is given in the owner's manual, allow a minimum of one square inch of ventilation for every 1,000 BTUs of heat.

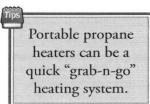

Portable propane heaters can be a quick "grab-n-go" heating system.

The small, portable propane heaters offer the unique advantage of providing heat on the go. You can quickly pack the heater and a few bottles of propane and take them on the road. However, be aware that 1-lb propane tanks may only last a few hours, so be sure to stock enough fuel. If you decide to use portable propane heaters, invest in a disposable tank adapter that allows you to refill 1-lb bottles from a 20-lb tank.

PREVENTING PIPES FROM FREEZING

Water expands when it freezes and, as any homeowner who lives in a cold climate can attest, this expansion can crack pipes. The most susceptible pipes are those directly exposed to the cold, including sprinkler and swimming pool supply lines, and your home's main water line. However, pipes in exterior walls or unheated areas of your house, such as in the basement, attic, garage, or crawlspace, are also prone to freezing.

Depending on the size of your backup heating system, it may or may not keep the temperature of your entire home above freezing. If it doesn't, then some of your water pipes may freeze. This is an important consideration since being without water makes a bad situation far worse.

Below are several recommendations to help prevent pipes from freezing.[167]

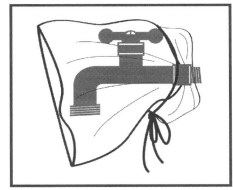

Protect your faucets from freezing

Before the cold arrives

- Drain sprinkler and pool supply lines.
- Remove, drain, and store hoses.
- Cut off the water supply to external hose spigots. If that is not possible, cover hose spigots with insulating covers.
- Insulate all exposed pipes using fiberglass, pipe sleeves, or heat tape.

When the cold arrives

- Keep your garage doors and crawlspace access panels closed to keep in any available heat.
- Open kitchen and bathroom cabinet doors to allow house heat to warm the pipes in the walls more efficiently.
- Let water trickle from several faucets (both hot and cold lines open), especially overnight.
- If you have heat available, keep the thermostat set no lower than 55°F.
- Consider turning off the water entering your house and draining your house pipes until the worst of the cold has passed. To do this, cut off the main line (typically located in your yard somewhere). It may require a special tool that can be purchased at most home improvement and hardware stores. An alternative is to use the interior cutoff (sometimes in the garage or a nearby closet), but the water may still freeze as it comes into your home. Once you cut off the flow of water, turn on all faucets until the water drains out of the interior house pipes. Note that the main water cutoff should be a ball valve. These types of valves keep the operating surfaces out of the water flow, which helps to prevent them from building up scale and becoming inoperable.

After the cold passes

- If your pipes freeze but don't crack, turn on your faucets and leave them open, giving the pipes a way to flush themselves.
- If you can access the frozen pipe, try thawing it with a hair dryer or small electric heater—don't use a propane torch.

COOKING

Cooking and heating are often related preparations because the two can frequently be used interchangeably—stoves can be used to heat the house; fireplaces can be used to cook food. At a minimum, you should try to select cooking and heating equipment that uses the same type of fuel. This helps alleviate the burden of multi-fuel storage and allows you to use your limited fuel for the more pressing of the two needs.

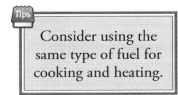

Consider using the same type of fuel for cooking and heating.

There are several ways to do small-scale cooking in an emergency. They include using a fuel-burning stove, fireplace, barbeque grill, microwave oven, camp stove, or solar oven.

FUEL-BURNING STOVE

Courtesy of Reading Stove Company

Many coal, wood, and pellet stoves have flat surfaces that are intended for cooking. This design is a great combination since it takes care of two very important needs with one appliance. These stoves generally don't allow traditional baking, limiting you to stovetop cooking. The disadvantages include high initial setup cost, difficulty in regulating temperature, and the need to store large quantities of fuel. But even with their limitations and drawbacks, fuel-burning stoves are an excellent method of providing both heat and a means of cooking.

FIREPLACE/OPEN FIRE

Texsport campfire tripod

If you are going to cook over a fire, you will need some way to support the food. In the case of a fireplace, it is easiest to install a swinging-arm crane that folds out and allows you to hang a pot directly over the fire. A heavy, fire-safe pot, such as a Dutch oven, is definitely recommended since fire can wreak havoc on regular cooking equipment.

Likewise, when cooking over an open fire, you can either cook using a cast iron support tripod or a folding camp grill that straddles the fire.

BARBEQUE GRILL

Masterbuilt grill

Americans love a good barbeque. When the power fails, breaking out the grill can be an excellent way to cook. One particular benefit of the barbeque grill is that you are probably already proficient with it. A good grill master knows not only how to prepare meats, but also fresh vegetables, foil-wrapped dinners, and food in a heavy pot.

If you are not experienced at cooking over the grill, there are numerous cookbooks available to teach you the necessary skills. Depending on the model of grill, you will need either a supply of propane or natural gas, or charcoal briquettes and starter fuel. Barbeque grills should only be used outdoors because they release deadly carbon monoxide. Never use them inside your home or garage, even with the windows or doors open.

MICROWAVE OVEN

You may have never considered your microwave oven to be an emergency cooking source, but it does offer several advantages. Most units require only 800 to 1,000 watts of electricity, well within the range of a small generator or inverter—see *Chapter 7: Electrical Power*. Microwave ovens cook very quickly and consume far less power than a range or conventional oven. If you rely on a microwave oven to be your backup cooking source, become proficient ahead of time at preparing many different types of foods.

CAMP STOVES

Camp stoves are portable cooking units fueled by small propane canisters or tanks. Some stoves can burn a host of different fuel types, including white gas, kerosene, diesel, and jet fuel. Smaller stoves are little more than a single burner centered over a fuel canister. Larger units resemble a conventional stove top with multiple burners and fit the needs of a family better.

The main advantage of a camp stove is its portability—great for packing up and taking with you on the road. One thing to note about any portable stove that burns fuel is that it is not safe to use indoors. Also, store the compressed fuel canisters or tanks in a safe place that is cool and dry.

Coleman camp stoves

SOLAR OVENS

You might think that cooking with the sun is the stuff of survivalist lore. In reality, solar ovens that can cook a daily meal using nothing more than the sun do exist.

The advantages of solar ovens are easy to identify: they use free energy, don't release any poisonous gases, and can be used to cook just about anything that your conventional oven can. But alas, nothing is perfect. The disadvantages are equally significant. Solar ovens cook significantly slower than other stoves, hours of direct sunlight are required, and winds can cool the food or disturb the reflectors.

Slow cooking can be partially offset by cutting the food into bite-sized pieces, and careful placement of the oven can remedy wind disturbances. The need for sunny days, however, precludes the solar oven from serving as a general purpose backup cooking source.

Sun Ovens International solar cooker

DP PLAN EXAMPLE

Table 8-4 Sample DP Plan Entry

Need: Heating/Cooling			
Danger	**Goals**	**Needs**	**Implementation**
Disruption of heating fuel source, *or* Loss of heating unit *addressing both dangers with the same preparations*	Heat the main living area for a minimum of 14 days without primary heating system	A backup heater capable of heating 800 square feet with a minimum inside-to-outside temperature difference of 50°F	Operate a vented, wood-burning stove in the living room that provides a minimum of 28,000 BTUs/hour. Seal off unused upstairs rooms to consolidate heat to the main living area. Sleep in main living area. Stock 2 cords of wood at rear of property, covered with tarp.
	Cook food without electrical or gas services	A stove capable of cooking two hot meals each day	Use the wood-burning stove as a stovetop. Use the microwave oven with backup electrical power.
	Stay warm if stranded in car	Blankets for two adults and three children	Keep five wool blankets in the car's trunk.
		Blankets that could be taken when evacuating a vehicle	Keep two emergency blanket bags in the glove box.

Quick Summary - Heating/Cooling

➢ Hypothermia is caused by allowing your body to get too cold. Likewise, hyperthermia is a result of your body getting too hot. Both conditions can be deadly.

➢ Keeping cool can be accomplished by limiting your activities, wetting yourself down, drinking cold liquids, and establishing some basic air circulation in your home.

➢ When stockpiling blankets, consider the worst case scenario. Extra blankets can also benefit less-prepared neighbors or family members.

➢ Emergency blankets are good for automobiles and grab-and-go bags but are only effective if used correctly.

➢ Heat can be lost by a disruption in your fuel source or a failure of your heating unit. You must be prepared for either one.

➢ Carbon monoxide poisoning, fire, and burns are very real dangers associated with backup heaters. Take the necessary precautions, including installing CO alarms and fire extinguishers, and training everyone on how to safely operate the heater.

➢ Several methods exist to estimate the size of a secondary heater. Use them to establish a rough estimate of your heating needs. Trial and error is often needed to get the size right.

➢ Many backup heating options exist, including fireplaces, masonry heaters, fuel-burning stoves, and space heaters.

➢ Coal, wood, and pellet stoves are the epitome of disaster preparedness because they reliably serve two vital needs: heating and cooking.

➢ Backup cooking can be accomplished with a fuel-burning stove, fireplace, barbeque grill, microwave oven, camp stove, or solar oven.

Recommended Items - Heating/Cooling

❑ A method to stay warm while inactive or asleep
    a. Heavy blankets and/or sleeping bags
    b. Emergency bags

❑ An emergency area heat source
    a. Fireplace with Texas Fire Frame, *or*
    b. Masonry stove, *or*
    c. Coal, wood, or pellet stove, *or*
    d. Propane, kerosene, oil-filled, or electric space heater

❑ Fuel for emergency heater *and* backup stove
    a. Coal, wood, pellets, *or*
    b. Gas or diesel for a generator, *or*
    c. Kerosene, propane, or natural gas

❑ Backup stove or oven
    a. Coal, wood, or pellet stove, *or*
    b. Fireplace with swinging-arm crane, *or*
    c. Barbeque grill, *or*
    d. Microwave oven, *or*
    e. Camp stove

CHAPTER 9

AIR

Challenge

News outlets issue an urgent warning that an accident has occurred at a nearby nuclear power plant, releasing dangerous levels of radioactive iodine into the air. Authorities are advising that residents immediately shelter in place. How will you protect your family from this potentially deadly airborne threat?

There are numerous disaster scenarios that can threaten your air supply. These threats could come in different forms, from an industrial accident releasing sulfuric acid into the air, to a volcanic eruption spewing ash across the country, to a terrorist attack spraying chemical or biological agents over metropolitan cities. To effectively respond to these types of airborne threats, you must learn to shelter in place as well as understand the benefits and limitations of air purifiers, facemasks, and respirators.

Before even considering disaster response, however, it is important to recognize that the world is already in the midst of a global air calamity.

AIR POLLUTION

On average, each person breathes about 3,000 gallons of air per day.[168] It is fair to say that air is the ocean we breathe. When the air becomes polluted, it affects not only *your* health but also those of plants, animals, and fish. Air pollution doesn't just hinder natural processes; it also damages buildings and monuments, and interferes with aviation. Those with respiratory problems are particularly sensitive to air pollution, including the 30 million people in the United States who suffer from asthma.

Air pollution is particularly dangerous to the 30 million Americans with asthma.

In October 1948, a polluted cloud lingered over Donora, Pennsylvania, for five days, killing 20 people and sickening 6,000 others. During a similar incident in 1952, over 3,000 people died in London

Deadly London smog of 1952 *(Wikimedia Commons/N T Stobbs)*

from a polluted fog so thick that people had to carry lanterns to walk through it. These events and others alerted governments to the dangers of air pollution. In 1963, the United States adopted the Clean Air Act, which now targets reducing 189 different toxic air pollutants.[169]

The Environmental Protection Agency (EPA) has set national air quality standards for six common air pollutants:

- carbon monoxide
- ozone
- lead
- nitrogen dioxide
- particulate matter
- sulfur dioxide

Each year, the EPA tracks levels of these air pollutants across the country. There is no denying the effectiveness of the Clean Air Act. Since 1980, the pollution levels for these contaminants have decreased as much as 92% (in the case of lead).[170] Trends for your particular community can be found at *www.epa.gov/airtrends/where.*

Roughly 60% of the U.S. population lives in areas where air pollution is at unhealthy levels.

Even with emission regulations adopted around the world, serious air pollution hazards remain. Poor air conditions can be found in developing countries, such as China, Vietnam, and India, as well as industrialized nations, such as Russia and Japan. While it is true that, comparatively speaking, the United States is not as polluted as many other countries, its citizens still face very serious pollution-related health risks. Consider that the United States is currently ranked 17th in overall air quality, and roughly 60% of its citizens live in areas where air pollution is at unhealthy levels of either ozone or particulate contamination.[171]

As pointed out by the World Health Organization, even more sobering statistics exist if the entire globe is considered:[172,173]

- Air pollution causes approximately two million premature deaths every year. Most of these deaths are due to respiratory infections and chronic obstructive pulmonary disease (COPD).
- Indoor air pollution (primarily due to the burning of solid fuels) is estimated to be the 8th most important risk factor for disease, and responsible for 2.7% of all global diseases. In developing countries, this number rises to 3.7%, making it the most lethal killer after malnutrition, unsafe sex, and lack of clean water and sanitation.
- Globally, acute respiratory infections represent the single most significant cause of death in children under age five. Exposure to indoor air pollution more than doubles this risk.

Air pollution

Air pollution is both a local and a global problem—one that must be considered an ever present threat to every corner of the world. Just as with other potential threats, you need to better understand the air hazards that your community faces. Real-time air quality assessments and projected forecasts for the entire United States can be found at *www.airnow.gov.* Likewise, a scorecard for your community's air, including a listing of the worst polluters, can be found at *www.stateoftheair.org.*

A COLLECTIVE SOLUTION

There is no short-term solution to air pollution. Remedies will take years, if not decades, to make a significant difference. Any sustainable solution to air pollution begins by recognizing that we all share one planet. What happens across the world ultimately affects you. If air pollution continues to worsen, humankind will collectively bring about a disaster of global proportions—one that may not be fixable. This prediction does not refer to global warming or other atmospheric issues still open to scientific and political debate but is based solely on current, measurable pollution levels and the deaths and disease caused by them.

Regardless of political views, it is impossible to deny that air pollution is a serious problem, one that no individual or single nation can remedy. Much of the solution lies in the hands of world leaders, including reducing the emissions of coal-burning power plants and responsibly managing deforestation. However, individuals must also play a critical role in ensuring cleaner air by adopting conservation-friendly actions (see Table 9-1), none of which should come as a surprise to anyone.[174,175]

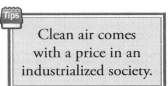

Clean air comes with a price in an industrialized society.

Realize that reducing air pollution (or at least stemming its growth) comes with a price, whether it be lost productivity, the inconvenience

Table 9-1 Individual Actions to Reduce Air Pollution

Conserve Resources	
	Turn off lights and appliances when not in use.
	Install thermostats that shut off when you are away from home.
	Replace incandescent with fluorescent bulbs.
	Run the dishwasher only when full.
	Wash clothes in cold or warm water.
	Turn off the water when brushing your teeth.
	Connect your outdoor lights to a timer or motion detector, or use solar lighting.
	Buy Energy Star appliances (see *www.energystar.gov*).
	Plant deciduous trees around your home to provide summer shade and winter warmth.
	Lower your water heater's thermostat to 120°F.
Reduce the Impact of Your Car	
	Take shared transportation whenever possible, using car pools, buses, trains, and airplanes. Better yet, get some exercise by biking or walking.
	Drive a fuel-efficient or low-emission vehicle. A useful tool to compare the environmental impact of different car models can be found at *www.epa.gov/greenvehicles*.
	Don't leave your vehicle idling for more than 30 seconds (except when in traffic).
	Keep your car tuned up and the tires properly inflated.
Reduce Waste	
	Recycle paper, plastic, glass bottles, cardboard, and cans.
	Use rechargeable batteries.
	Recycle cell phones, laptop computers, CRT monitors, and other electronics.
Minimize Your Pollution	
	Don't burn wood or trash unnecessarily.
	Never smoke indoors.
	If purchasing a wood, coal, or pellet stove, select one with reduced emissions, preferably EPA certified. Once in use, maintain wood stoves (and fireplaces) so that they burn efficiently.
Get Active	
	Influence your public transportation and local school systems to drive modern low-emission buses.
	Get politically involved. Politicians will only take action if their voter base directs them to do so. Make it known that you are concerned about air quality and want steps taken to reduce emissions.

of recycling, sacrificing your gas guzzler, or increased production costs. In an industrial world, clean air isn't free, but the price of doing nothing and simply letting the corporate bottom line drive the rampant polluting of our planet will ultimately prove too costly for everyone. Assume a responsible posture toward the planet and its citizens by doing your part, however small or large that may be.

INCIDENT-DRIVEN EVENTS

Air quality can also be affected by sudden, immediate disaster events, such as a volcanic eruption, wildfire, chemical or nuclear contaminant release, biochemical attack, or airborne pandemic. Protecting your family from these threats requires insulating them from the contaminated air as thoroughly as possible. Tools at your disposal include a well-sealed shelter, masks and respirators, and air filters.

UNDERGROUND SHELTERS WITH AIR FILTRATION

Underground shelters equipped with highly efficient air filtration systems clearly fall on the survivalist end of the preparedness spectrum. With that said, they can be designed to offer excellent protection from dangerous air hazards (as well as other threats). Many factors must be considered before constructing an underground shelter, including water level, air flow, temperature and humidity management, electricity, and access. If you are contemplating building an underground shelter, do your homework, consult professionals, and thoroughly plan *before* breaking ground. This topic is not covered in more detail here because for most people, installing an underground bunker is prohibitively expensive and perhaps physically impossible (e.g., apartment dwellers). For additional information on underground shelters, consult Philip Hoag's *No Such Thing as Doomsday*.

SHELTER IN PLACE

Even if a full-up underground bunker is out of the question, there are some basic measures you can take to create a mini-shelter in your own home. This process is typically referred to as sheltering in place, and is designed to protect against radiological, biological, and chemical threats. This creation of a contaminant-free area is an addendum to the safe room discussed in *Chapter 5: Shelter* with additional consideration given to protecting against airborne threats. Sheltering in place is a short-term strategy, useful to insulate your family from an immediate threat that will diminish or pass quickly (i.e., within minutes, hours, or days). Specific recommendations are given below:[176,177]

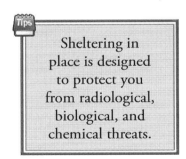

Sheltering in place is designed to protect you from radiological, biological, and chemical threats.

1. Select an interior room in your home (ideally one without windows or exterior doors) to serve as your family's shelter. It should have a minimum of 10 square feet of space for each person, which will provide adequate air volume for at least five hours of sheltering. A room with an adjacent bathroom is often a good choice. You may wish to locate the safe room on the upper floor of your home because gases tend to settle.

2. Pre-stock your room with necessary emergency supplies, such as water, food (for you and your pets), a radio, television, first aid kit, flashlights, hardwired telephone, medications, blankets, and material to effectively seal the room.

3. When directed by emergency management services, or when you feel that something poses an immediate threat to air quality, gather your family in the safe room. You may also wish to bring your pets into the room with you. At a minimum, bring your pets into your home from the outdoors.

4. Cut off all heating, air conditioning, and circulation systems. Close fireplace dampers. Close and lock all exterior doors and windows.

5. Cover all doors, windows, and air vents in the safe room with plastic sheeting (minimum thickness of 4 mils, where a mil is 1/1,000 of an inch). Cut the plastic several inches wider than the openings being covered, and secure the sheeting with heavy-duty duct tape (minimum thickness of 10 mils). For expediency, pre-cut and label the plastic sheets, and have them ready in the room.

6. Seal any water pipes that enter the area using appropriate sealants.

Seal around doors, windows, air vents, and water pipes

Shelter in place

7. Use a portable HEPA air filter to help purify the air in the safe room. When properly sized to the room, HEPA filters have been shown to be effective in removing vapor contaminants and some poisonous gases.[102]

8. If appropriate to the threat, put on protective hoods or respirators. Although not as effective, you can also use inexpensive N95 disposable masks for limited protection. Disposable biochemical protection suits, such as Tyvek F, can also be worn to protect against skin contamination.

9. Monitor the radio or television for updates. Officials may call for evacuation, so be ready to move.

UNDERSTANDING RADIATION

Radiation is broadly classified as either low-energy, non-ionizing waves or high-energy, ionizing waves. Low-energy waves include visible light, radio waves, and microwaves. Non-ionizing waves of this type are generally not harmful to humans except in very high doses. High-energy waves include alpha, beta, gamma, and x-rays. Gamma and x-rays are particularly dangerous because they are able to travel through the air and disrupt the atomic structure of living creatures. This atomic disruption can lead to bleeding, cancer, mutation, and death. Table 9-2 compares the different types of radiation.

Radiation—a deadly threat

Table 9-2 Comparing Radiation Types[178,179]

Radiation Type	Examples	Range	Penetration	Shielding	Danger	Detectable
Low-energy, non-ionizing	Visible light Radio waves Microwaves	Long	Shallow	Easily blocked	None unless in very high doses	Naked eye, test equipment, or radio receivers
High-energy, ionizing	Alpha	Short (inches)	Generally unable to penetrate skin	Easily blocked	Harmful if swallowed or inhaled	Geiger-Mueller counters, scintillation counters, survey meters, dosimeters, film badges
	Beta	Medium (yards)	Can penetrate human skin to the germinal layer	Clothing offers some protection	Skin and eye injury	
	Gamma, X-ray	Long	Can penetrate clothing and the human body	Dense materials (e.g., lead) required to block	Cellular damage, bleeding, cancer, and death	

RADIATION EXPOSURE

Everyone is exposed to low levels of radiation on a daily basis. Even the most mundane actions cause radiation exposure, including sleeping next to someone, using a smoke alarm in your home, flying on an airplane, and simply being on this planet. Fortunately, these levels of radiation are quite harmless. The chart below provides a summary of the radiation exposure from many common sources. On average,

Radiation Exposure for Various Events[247, 248, 249]

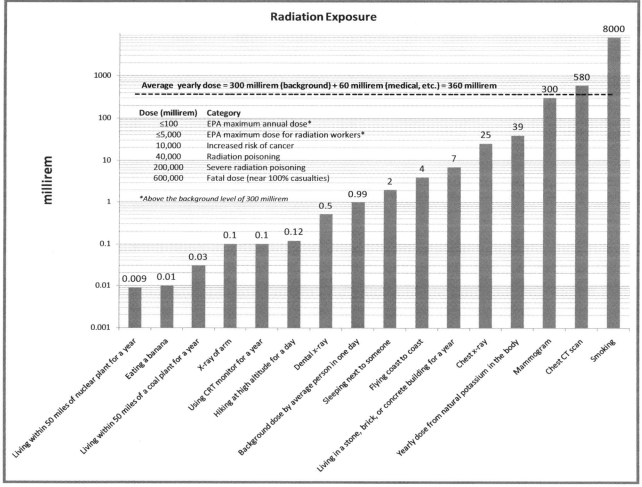

Additional Notes:

1. Smoking 1½ packs of cigarettes a day can result in up to 16,000 millirem of radiation per year.[247] Tobacco has a high concentration of polonium-210, a naturally occurring radioactive element—yet another reason not to smoke.
2. Flying in an airplane reduces your atmospheric shielding from cosmic radiation, resulting in about 1 millirem for each 1,000 miles you fly at high altitude.
3. Surprisingly, living near a nuclear power plant exposes you to less radiation than living near a coal plant. This is due to the thorium and uranium intrinsic to coal.

people are exposed to about 360 millirem of radiation per year, of which 300 millirem is from unavoidable background sources. The remaining 60 millirem is primarily due to medical procedures such as x-rays. The EPA recommends a maximum dose of no more than 100 millirem above the background dose for the average citizen. Those working in occupations that deal with radiation are limited to 5,000 millirem (i.e., 5 rem). The lowest level of radiation recognized to increase the long term risk of cancer is 10 rem. Noticeable effects of radiation poisoning begin at a dose of about 50 rem.

Radiation levels are often discussed using several different units of measure including the roentgen, rad, rem, and sievert. Distinctions between the four units are made below:[250]

> **roentgen (R):** a measure of the ionizing ability of radiation, named after Wilhelm Konrad Röntgen

> **rad (radiation absorbed dose):** a measure of the amount of radiation energy absorbed by a material; equal to 0.01 gray (Gy)

> **rem (roentgen equivalent man):** a measure of the equivalent dose of radiation received by a human that takes into account a quality factor that is larger for more dangerous forms of radiation such as alpha particles; equal to 0.01 sievert (Sv)

> **sievert (Sv):** the International System of Units derived unit for equivalent dose; equal to 100 rem

For purposes of radiation protection in humans, the roentgen, rad, and rem can all be taken to be equivalent. One roentgen of radiation causes one rad (or rem) of absorbed radiation in humans.

RADIOLOGICAL THREATS

As touched upon in *Chapter 5: Shelter,* radiation threats come in two different forms: radioactive contamination and radiation exposure. Radioactive contamination occurs when radioactive materials are released into the environment. This can be the result of a nuclear power plant accident, a medical or industrial device malfunction, or an intentional act of terrorism. When the contaminants are inhaled, ingested, or come into contact with your skin, they can cause tissue and organ damage, and ultimately death.

Radiation exposure occurs when you are exposed to high-energy, ionizing waves. This exposure causes atomic disruption, leading to sickness and death. The severity of illness depends on the exposure level and type of radiation. The immediate and latent biological effects of acute radiation exposure are summarized in Table 9-3. The latent effects are cumulative, meaning that all the lower-level effects are also suffered.

It is important to understand the difference between radioactive contamination and exposure because the preparations and subsequent actions to be taken are quite different. In the case of radiation exposure, put as much distance between you and the source of radiation as possible—distance means safety because the energy of the waves decreases rapidly with space and time. If evacuation is not possible, or if the radiation levels are too high to evacuate safely, retreat to a well-shielded location. This area could be an underground storm shelter, a fallout shelter at the local library, mall or subway, or even a cave in nearby mountains. The idea is to put as much dense material (e.g., dirt, concrete, metal, water) as possible between you and the radiating source. If you suspect that you have been exposed, contact local emergency management services for medical treatment.

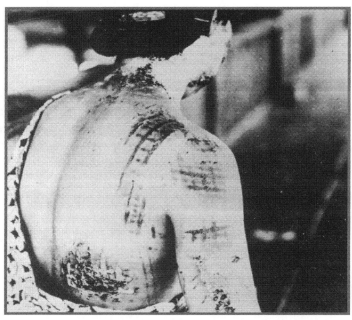

Radiation burns on Japanese woman following atomic bombing, 1945 *(U.S. Army)*

Table 9-3 Biological Effects of Acute, Total Body Irradiation without Treatment[179,180,181,182]

Exposure (rem)	Time to onset, duration	Immediate effects	Time to latent effects	Latent effects (cumulative)	Lethality after 30 days
<10	-	No detectable injury or symptoms	-	None	0%
10-50	-	No detectable injury or symptoms	-	Increased risk of cancer	0%
50-100	3-6 hrs, up to 1 day	Headache, nausea, fatigue, vomiting, diarrhea (intensity and frequency increases with higher exposure levels)	-	Temporary male sterility possible	0%
100-200	3-6 hrs, up to 1 day		10-14 days	Illness, fatigue, premature childbirth possible	0-10%
200-300	1-6 hrs, up to 2 days		7-14 days	Loss of hair, permanent female sterility possible	10-35%
300-400	1-6 hrs, up to 2 days		7-14 days	Bleeding in mouth, under skin, and in kidneys	35-50%
400-600	½-2 hrs, up to 2 days		7-14 days	Same as 300-400, with greater intensity	50-90%
600-1,000	15-30 minutes, up to 2 days		5-10 days	Damage to bone marrow and intestinal tissue	Near 100%
>1,000	5-30 minutes		Hours to days	Same as 600-1000	100%

In the case of radioactive contamination, remove your clothing and put it into a plastic bag away from other people. Next, wash yourself thoroughly with soap and warm water, and contact local authorities for decontamination and medical treatment.[183] The effects of internal radioactive contamination (i.e., contamination that is ingested or inhaled) may be reduced by taking medicines. The two most well-known medicines are potassium iodide and Prussian blue.

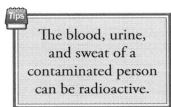

> The blood, urine, and sweat of a contaminated person can be radioactive.

POTASSIUM IODIDE[184]

Potassium iodide (KI) is a salt of stable iodine used by your body to produce thyroid hormones. In the event of a radiological or nuclear event, radioactive iodine may be released into the air, food, or water. If you ingest or inhale this iodine, it will be quickly absorbed by your thyroid gland. This can injure your thyroid, leading to cancer and death. If you take potassium iodide in advance, it essentially fills your thyroid, preventing the absorption of radioactive iodine and potentially saving your life.

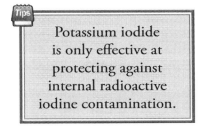

> Potassium iodide is only effective at protecting against internal radioactive iodine contamination.

The effectiveness of potassium iodide depends upon three things: (1) how much time has passed between the contamination of radioactive iodine and the taking of KI, (2) how fast the KI is absorbed into your blood, and (3) the total amount of radioactive iodine to which you are exposed.

It is also important to understand what KI cannot do. Potassium iodide only protects the thyroid against ingested or inhaled radioactive iodine. It will not protect against other radioactive materials such as those released by a "dirty bomb"; neither will it protect any part of the body except the thyroid. It will also not protect against the effects of radiation exposure.

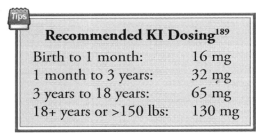

Recommended KI Dosing[189]	
Birth to 1 month:	16 mg
1 month to 3 years:	32 mg
3 years to 18 years:	65 mg
18+ years or >150 lbs:	130 mg

The thyroid glands of unborn fetuses, infants, and young children are particularly susceptible to radioactive iodine, making KI pre-treatment critical. Potassium iodide will protect the thyroid for 24 hours, so depending on the duration of the threat, repeat dosing may be required. Potassium iodide comes in both tablet (65 mg and 130 mg) and liquid (65 mg per mL) forms. These concentrations align well with the FDA's recommended dosages—see tip box.

Potassium iodide is generally considered very safe but can be harmful to people with certain medical conditions, so consult your doctor before taking KI. A prescription is not required to purchase KI; your pharmacist can dispense it directly to you. It is also widely available on the internet. The product iOSAT is recommended since it is the only full-strength, FDA-approved KI tablet for radiation blocking legally sold in the United States. It is best to purchase iOSAT directly from the manufacturer at *www.anbex.com* to guarantee optimal freshness. The only FDA-approved KI liquid is ThyroShield, and is especially useful for administering to young children (see *www.thyroshield.com*). Potassium iodide is inexpensive and has a shelf life of at least five to seven years, making it a low-cost, yet potentially life-saving preparation.

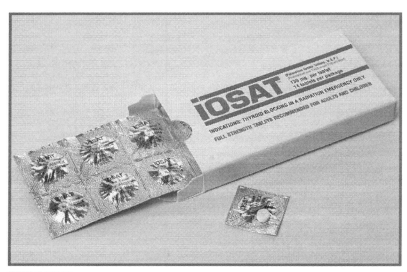

iOSAT KI tablets

A recent alternative to KI is potassium iodate (KIO3). KIO3 tablets are very similar in function and shelf life to the more common KI tablets. According to the World Health Organization, both have been shown to work at roughly the same level of effectiveness.

The Medical Corps makes three claims regarding KIO3:[252]

1. Extended shelf life—KIO3 is non-hygroscopic (meaning that it doesn't degrade in humid air), whereas KI is water soluble. However, KI tablets are almost always sold in airtight packaging, so the water solubility is not likely to be an issue. Both products have a shelf life of five to seven years and are usually safe and effective for years beyond that.
2. No bitter taste—KI tablets have a bitter taste when compared to KIO3 tablets. However, KI tablets are very small and easy to swallow, so it is hard to imagine that taste is an important consideration when evaluating the two products.
3. Cheaper per dose—Both KI and KIO3 are close in price, depending on the brand. Note that you have to take two KIO3 tablets to get roughly the equivalent dose of a 130mg KI tablet. The requirement to take two can actually be beneficial when administering smaller doses to children since you won't have to worry about accurately breaking or cutting the pills.

The bottom line is that KI and KIO3 are both perfectly acceptable products for protection against ingested or inhaled radioactive iodine, and either could save your life if faced with the threat of contamination.

PRUSSIAN BLUE[185]

Prussian blue is an FDA-approved drug used to treat internal contamination caused by radioactive cesium-137 or thallium. It is sold under the name Radiogardase and has been available since 2003. Along with potassium iodide, Prussian blue is currently stockpiled by the United States government for emergency response.

Table 9-4 Comparing KI and Prussian Blue

	Potassium Iodide/Iodate	Prussian Blue
Effective Against	Internal contamination from radioactive iodine	Internal contamination from cesium and thallium
Possible Source	Nuclear power plants	Dirty bomb
Taken	As a preventive before contamination	As a treatment after contamination
Prescription	Not required	Required

Unlike potassium iodide, which blocks the absorption of a specific contaminant, Prussian blue helps to speed your body's excretion of radioactive cesium and thallium. It does not protect against radioactive iodine likely to be released from a nuclear power plant. However, it may offer some protection against a "dirty bomb" from which radioactive cesium has the potential to be released. Prussian blue is available only with a doctor's prescription. It is not taken as a preventive to radioactive contamination but rather as a treatment after poisoning has occurred. Table 9-4 provides a brief comparison between KI and Prussian blue.

SURVEY METERS

Survey meters are portable meters used for detecting and monitoring radiation levels. Two common types are scintillation and Geiger counters. Despite being based on two very different interactions, the devices perform the same general function—to provide portable detection of radiation.

As was previously discussed in *Chapter 5: Shelter*, a relatively inexpensive radiation detector that fits on your keychain is offered from NukAlert. It claims to reliably detect radiation levels from 100 millirem per hour to more than 50 rem per hour and can operate for 10 full years without a battery replacement. The detector remains active 24 hours a day and sounds different alarms depending on the exposure level.[122] If you live in an area where nuclear contamination is a viable threat, the NukAlert or other radiation detector is a reasonable precaution capable of providing your family with the early warning vital to evacuation or other preparatory steps.

Geiger counter *(Wikimedia commons/Horst Frank)*

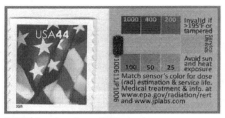

RADSticker *(courtesy of KI4u.com)*

Portable radiation dosimeter *(Wikimedia Commons/Prolineserver)*

DOSIMETERS

In the event of a nearby radiological disaster, such as a dirty bomb or nuclear power facility accident, it is important to monitor the cumulative dose to which you are exposed. This can be accomplished through blood work analysis (i.e., biodosimetry) or by using portable radiation dosimeters. There are many types of dosimeters available, including pen-shaped units that fit in your pocket and pager units that clip to your belt. An inexpensive alternative is the RADSticker™, a postage-stamp-sized dosimeter developed through funding from numerous government organizations. RADStickers™ are a type of self-indicating instant radiation alert dosimeter (SIRAD).[251] They are constructed of a radiation-sensitive material that quickly changes color when exposed to radiation. The greater the dose, the darker the color change. By comparing the color strip to a reference chart, the wearer can quickly determine his level of total radiation exposure (anywhere from 25 to 1,000 rad).

> **Tips**
>
> RADStickers™ can be stored in the freezer to extend their shelf life up to 10 years.

AIR PURIFIERS

You should consider using a portable high-efficiency particulate air (HEPA) purifier in your safe room. HEPA filters were designed as part of the Manhattan Project to prevent the spread of airborne radioactive particles, and as such are capable of removing a significant number of very small airborne contaminants. HEPA filters operate using a combination of interception, impaction, and diffusion. Even for the most penetrating particle sizes (MPPS), they remove with an efficiency of at least 99.97%.[123,186] HEPA purifiers are routinely used in hospitals to remove bacteria and viruses from the air. They have also been shown to be effective in removing biological and radiological contaminants.[187] HEPA filters alone do not remove chemicals, gases, or odors, but are typically combined with carbon filters to remove such things as cigarette smoke and pet odors.

> **Tips**
>
> Select a HEPA purifier with a CADR of at least: square footage ÷ 1.5.

The amount of air flow that a portable HEPA purifier provides is rated by its clean air delivery rate (CADR). The higher the CADR, the more air flow it provides—meaning it cycles the room air more frequently. The CADR value is expressed in cubic-feet-volume per minute. A CADR of 250 indicates that the filter cycles 250 cubic feet of air per minute. A rule of thumb is to size your unit so that the

CADR multiplied by 1.5 is equal to or greater than the square footage of the room. For example, a purifier with a CADR of 250 would be effective filtering rooms up to 375 square feet in size. It is recommended that you purchase a purifier with a lifetime HEPA filter and a replaceable carbon pre-filter to significantly lower your ownership costs.

Larger HEPA filters can also be equipped onto your home's primary air circulation system. However, because your entire home is not well sealed, the protection from sudden high-levels of contamination will be limited. Whole-house filters are, however, excellent at reducing allergens in the home, such as pollen, pet dander, dust mites, and mold.

Honeywell HEPA filter

FACEMASKS/RESPIRATORS

Facemasks and respirators have proven to be effective at preventing respiratory illnesses in dangerous workplaces for many years. The term *facemask* usually refers to disposable cloth masks approved by the Food and Drug Administration for use as medical devices. They typically tie around the head and may be loose cloth or have a pre-molded shape. Facemasks are designed to provide limited protection from blood and body fluids. They are, however, not designed to protect against small airborne particles or gases. Facemasks should be used only once and then discarded.

The term *respirator* refers to more effective protective masks that are certified by the National Institute for Occupational Safety and Health (NIOSH). Respirators are specifically designed to protect from chemical, biological, radiological, or nuclear airborne contaminants. As such, respirators must fit more snugly to the face than facemasks, making prolonged breathing through them more arduous (unless they have forced air systems). Many respirators cannot be used by people with beards or thick mustaches because facial hair prevents adequate sealing. Achieving the greatest protection requires training regarding proper fit and use.[188]

There are numerous types of respirators, each with its own limitations and capabilities. They range from single-use, disposable masks to reusable models with replaceable cartridges. Some masks cover the entire face, while others cover only the nose and mouth. Many masks allow you to speak or drink while wearing them, but some require that you keep a mouthpiece clenched between your teeth. The key points are to: select a respirator that protects from the right contaminants, ensure that it fits properly, and keep it ready and in working order.

Respiratory equipment is generally divided into five categories: particulate filtering facepiece respirators, negative pressure air-purifying respirators (APRs), powered air-purifying respirators (PAPRs), self-contained atmosphere-supplying respirators, and escape respirators.

PARTICULATE FILTERING FACEPIECE RESPIRATORS

Particulate filtering facepiece respirators use wool, plastic, glass, or other materials to capture contaminants from the air that pass through the filter. These budget respirators are worn in a manner similar to

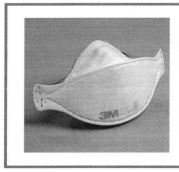

Particulate filtering facepiece respirators *(Copyright 3M)*

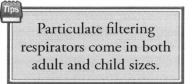
facemasks, although designed to fit more snugly and filter more thoroughly. The filters cannot be cleaned and should be discarded after each exposure. They are the least expensive type of respirator but also the least effective. Common types of particulate filters are the N95 (removes 95% of airborne particles), N99 (removes 99% of airborne particles), and N100 (removes 99.97% of airborne particles). There are also oil-resistant equivalent models (e.g., P95, P100). A listing of NIOSH-approved particulate filtering facepiece respirators can be found online at the CDC website.[189]

Particulate filtering facepiece respirators are low-cost safety devices that provide limited protection from airborne contaminants, such as during a flu pandemic. They are only designed to protect against particles, not gases or vapors. For people who wish to take a modest approach to protecting against airborne threats, these are likely the respirators of choice. Smaller sizes are also available to fit children. Tests conducted by the CDC showed that achieving a tight fit is critical for optimal mask effectiveness.[190]

AIR-PURIFYING RESPIRATORS (APRS), A.K.A. "GAS MASKS"[191]

Air-purifying respirators (APRs) provide a much higher level of protection by removing air contaminants through filtering, adsorbing, absorbing, or chemical reaction. Most are negative pressure devices, requiring the wearer to force air through the filtration system by breathing. Some units are positive pressure systems, circulating forced air with battery-powered electric blowers.

APRs can be certified by the NIOSH to protect against chemical, biological, radiological, and nuclear (CBRN) contaminants. All CBRN-certified respirators provide a specified minimum level of protection against a host of dangerous atmospheres, including biological aerosols and chemical gases. They are rated by their protection duration (e.g., Cap 1 = 15 min, Cap 2 = 30 min, Cap 3 = 45 min). CBRN respirators are targeted primarily to emergency first responders but can be purchased by anyone.

Avon Protection APR

3M PAPR

POWERED AIR-PURIFYING RESPIRATORS

A powered air-purifying respirator (PAPR) uses an electric blower to circulate filtered, ambient air into the facepiece. PAPR systems consist of a motor blower unit with filter cartridge, headpiece, battery, and breathing tube. The positive air pressure not only relieves the user from having to force air in and out of the filter, but it also creates a positive pressure in the face mask, preventing inward leaking.

SELF-CONTAINED ATMOSPHERE-SUPPLYING RESPIRATORS

Self-contained atmosphere-supplying respirators provide clean air from an uncontaminated source rather than filtering contaminants from the environment. There are three types: self-contained breathing apparatus (SCBA) devices (air is provided by a tank worn on the user's back), supplied air respirators (SARs) (air is provided by a tether), and closed-circuit SCBA systems known as "rebreathers" (the user's own air is scrubbed, oxygen enriched, and recycled). Firefighters often use SCBAs in oxygen-deficient conditions.

A comparison of many APRs can be found in the Department of Homeland Security's *Guide for the Selection of Personal Protective Equipment for Emergency Responders.*[191] If you decide to purchase a respirator, it is highly recommended that you first consult this document.

Avon Protection SCBA

ESCAPE RESPIRATORS

Avon Protection escape respirator

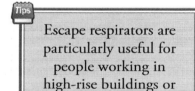

Escape respirators are particularly useful for people working in high-rise buildings or traveling on subways.

Escape respirators are relatively low-cost emergency devices used for rapid evacuation of respiratory hazard areas, such as burning buildings or contaminated subway tunnels. These devices are grouped into two categories: air-purifying escape respirators (APERs) and self-contained escape respirators. APERs are negative pressure systems that use filter canisters to remove contaminants, whereas self-contained escape respirators have an attached air tank. Escape respirators are designed to be used for a short duration, perhaps 5 to 60 minutes. Both types seal around the neck and are capable of providing the wearer with respiratory protection from CBRN threats.

For most people, the purchase of an air-purifying respirator cannot be easily justified. However, escape respirators may make sense if you work in a high-rise building or travel on the subway each day. Also, many people decide that a simple particulate filtering respirator (e.g., N95) is sufficient because it provides a modest level of protection from airborne contaminants. As with every aspect of disaster preparation, the decision is ultimately yours to make. Assess your risk and take the appropriate actions.

If you decide that a air-purifying respirator is needed, consider the following list of questions when making your selection:[192]

1. What contaminants does the escape hood protect against?
2. Is there more than one size available? How do you ensure that the gas mask or escape hood will fit properly?
3. What type of training do you need to use the respirator?
4. Is data available to show that the escape hood has been tested against various contaminants, such as biological agents, chemical warfare agents, toxic industrial chemicals, and radioactive dust particles?
5. Who performed the testing? What were the tested levels and test durations? Are they representative of your concerns?
6. Is the escape hood certified by an independent laboratory or government agency?
7. Are there any special maintenance or storage requirements? What is the shelf life?
8. Will you be able to talk or drink while wearing the respirator?
9. Does the hood restrict vision or head movement in any way?
10. How would you carry the device to where it might be needed? In the trunk of your car? In your purse or brief case? Or would it be left in a drawer at work?
11. Is a training respirator available from the vendor to practice with?
12. Can you use the respirator more than once? Does it have replaceable filter cartridges? If so, where are they purchased?

Respirators are just one part of a complete ensemble of personal protective equipment needed to adequately protect you from biological or chemical threats. To be fully protected, clothing, gloves, and boots should also be considered. For a detailed discussion of CBRN protective clothing and respirators, refer to the *Guide for the Selection of Personal Protective Equipment for Emergency Responders*.[191]

DP PLAN EXAMPLE

Table 9-5 Sample DP Plan Entry

Need: Air			
Danger	**Goals**	**Needs**	**Implementation**
Air pollution	Reduce the air pollution levels in our community	Steps to conserve resources and reduce emissions	Follow Individual Action pollution checklist.
		A way to motivate public officials	Write local and national representatives to express support for Clean Air Act legislation.
Contamination	Protect family from airborne hazards	An area that can be sealed	Use the large bedroom closet for sheltering in place. Stock it with emergency supplies.
			Pre-cut and label plastic to cover windows and doors.
		Air purification devices	Use a 250 CADR HEPA air filter in the safe room.
			Equip everyone with appropriate fitting N95 respirators.
	Detect and respond to high levels of radiation	A portable device that warns of radiation threats	Keep a NukAlert on keychain.
		Medicine that mitigates radiation poisoning	Stock iOSAT KI tablets and administer if radiation threat is detected.
		A dosimeter that monitors total exposure	Place a RADSticker on the back of driver's license.

Quick Summary - Air

➢ Air pollution is a serious problem facing our world. Over 60% of the U.S. population lives in areas where air pollution reaches unhealthy levels.

➢ In the long term, a potentially irreversible disaster is looming if pollution is not brought under control. Everyone needs to take steps to improve the situation.

➢ Numerous disasters can affect air quality, including a volcanic eruption, chemical release, widespread fire, biological attack, nuclear event, or airborne pandemic.

➢ Air hazards can be mitigated by sheltering in place and through the use of air purification devices, such as HEPA filters and respirators.

➢ Sheltering in place involves sealing an internal room of your house or workplace using heavy plastic sheeting and duct tape.

➢ HEPA air purifiers can remove very small contaminants from the air, including radiological and biological.

➢ Radioactive contamination occurs when you come into physical contact with radioactive materials. This can be external, such as on the skin, or internal (inhaled or ingested). Potassium iodide and potassium iodate are both effective preventive measures against thyroid damage caused by radioactive iodine.

➢ Particulate filtering facepiece respirators (e.g., N95, N99) are low-cost protective devices that offer limited protection from airborne contaminants.

➢ Advanced respiratory equipment includes air-purifying respirators (APRs), powered air-purifying respirators (PAPRs), and self-contained atmosphere-supplying respirators. Due to cost, these devices may not be practical for many families.

➢ Escape respirators are particularly useful when trying to escape a burning building or CBRN-contaminated area.

Recommended Items - Air

❑ Shelter in place
    a. Roll of 4 mil thick plastic sheeting
    b. Roll of 10 mil thick duct tape
    c. Portable HEPA air filter

❑ Protection from radioactive contamination
    a. iOSAT potassium iodide pills or ThyroidShield liquid
    b. Optional: NukAlert™ radiation detector
    c. Optional: Dosimeter (e.g., RADSticker™)

❑ Individual air filtration
    a. Particulate filtering facepiece respirators (e.g., N95) for each family member, *or*
    b. Air-purifying respirators
    c. Optional: Escape respirators

CHAPTER 10

SLEEP

Challenge

The stock market is crashing. Worrying over what might happen to your family's nest egg is keeping you awake at night. The lack of sleep is affecting you both physically and emotionally. What can you do to get more sleep?

Sleep is one of life's simple pleasures that isn't fully appreciated until you don't get enough of it. Up to 70 million Americans suffer from some form of insomnia.[193] It affects people regardless of status or profession, from entertainer (e.g., Michael Jackson) to political leader (e.g., Benjamin Franklin) to scientist (e.g., Thomas Edison).

During times of crisis, even otherwise sound sleepers can experience difficulties. People react to stress in varying ways. For many, stress can keep them awake at night worrying. For others, sleep comes quickly as a means of escaping from the pressures and decision making. It is no wonder that when your roof has a large tree poking through it, your car is floating in rising waters, or your children are coughing from the latest flu outbreak, sleeping may be unusually difficult. However, it is during these times of great stress that your body and mind need sleep the most.

There is strong evidence to support that getting enough sleep is important not only to your performance of routine tasks, such as driving or working, but also to your health. Heart disease, diabetes, and obesity have all been linked to lack of sleep.[194] Getting an adequate amount of quality sleep is akin to exercising and eating a balanced diet—they all lead to a longer, healthier life.

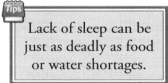
Lack of sleep can be just as deadly as food or water shortages.

This chapter is unique in that there's nothing to buy, no skill to learn, no community or government resources to draw upon. Sleep is a natural part of life, one of the few things not explicitly taught to humans. That is not to suggest that sleep shouldn't be given consideration in your preparations. Lack of sleep will affect you emotionally, physically, and mentally, and those effects can lead to lapses in judgement that could prove just as deadly as shortages of food or water.

RECOMMENDATIONS

There are countless reasons for having difficulty sleeping, from anxiety, to bodily pain, to uncomfortable sleeping conditions. Unless you are a master yogi, trying to force yourself into sleep usually only exacerbates the problem. The truth is that sleep is not something you can control; the sooner you accept that fact, the better. The most you can do is work to establish a comfortable environment that helps you to relax and take your mind off your troubles.

Below is list of recommendations shown to help people get a good night's sleep. Some of these suggestions are from reputable insomnia experts, while others are mere personal observations. Certainly not all of them apply to every situation, but together they present an overview of important considerations to overcoming sleeping difficulties.

- Establish a sleep schedule. Go to bed and wake up at the same time each day—even on the weekends.
- Don't drink alcohol or caffeine within three hours of bedtime.
- Invest in a comfortable bed and pillow.
- Keep the temperature in your bedroom comfortable; experts recommend 65° to 72°F.
- Put socks on your feet to keep them warm (if necessary).
- Turn off the TV or computer at least one hour before bedtime.
- Read a good book to get your mind on something else.
- Limit the size of any late night snack. Warm milk or caffeine-free tea may be better choices.
- Take a warm bath to help you relax.
- Exercise routinely—but not within two hours of bedtime.
- Enjoy sex with your partner.
- Budget more than enough time to sleep, relieving the feeling that you must fall asleep quickly.
- Manage stress by making a checklist of worries and the actions you will take to resolve them.
- Experiment with a sound machine. Most devices generate white noise, but some will also emit natural sounds, such as a trickling stream, rainfall, or chirping birds.
- If you are experiencing pain or sore muscles, consider taking an over-the-counter pain reliever. As a longer term solution, try stretching or yoga.
- If necessary, consult your physician for sleep aids, such as zolpidem (Ambien) and eszopiclone (Lunesta). Many people feel that drug-assisted sleep is not as restful as natural sleep, but medicines may be helpful when nothing else seems to work. Be aware that sleep aids may have side effects, including headaches, diarrhea, and dizziness, and can limit your sleep to perhaps only 4 to 6 hours.

For more information on insomnia and its treatments, review the many sleep-related articles and resources at *www.webmd.com* and *www.health.com*.

The good news is that if you get tired enough, you *will* sleep. There are no exceptions to this rule—save for the unfortunate people who suffer from a deadly hereditary disease, familial insomnia. The best thing you can do to facilitate healthy sleep is to give yourself a comfortable environment and adequate time to rest.

CHILDREN

Children are particularly susceptible to the effects of insufficient sleep. They need sleep not only to function, but also to grow and develop. It affects their mood, view of the world around them, alertness, social behavior, and learning ability. If you have children, you can surely testify to this if you consider how difficult your children become when they haven't had adequate rest.

> Ensuring that your children get adequate quality sleep will help your family to run smoother in times of crisis.

Perhaps this topic relates only peripherally to disaster preparedness, but consider how much more difficult a challenging situation can become if your children are bickering, misbehaving, and pushing all your buttons. Keeping your children well rested helps your entire family to function smoother and thus be better prepared for whatever challenges any given day might bring.

Children need significantly more sleep than adults. Table 10-1 lists the recommended number of hours of sleep required for children of different ages.[195,196] Every child is different, some requiring less than the recommended amount, some more. One of the most important things you can do for your children is to provide them with adequate sleep.

Children need adequate, restful sleep

Table 10-1 Hours of Sleep Needed by Age

Age	Sleep / Night (hours)
1-4 weeks	15½ to 16½
1-4 months	14½ to 15½
4-12 months	14 to 15
1-3 years	12 to 14
3-6 years	10¾ to 12
7-12 years	10 to 11
12-18 years	8¼ to 9½
19+	7-8

Not only is it important to ensure that children sleep sufficient hours, but also that they have quality, uninterrupted sleep. This permits their developing brains to go through the various sleep cycles. Children should not be expected to outgrow sleeping problems. Sleeping issues must be resolved before they become a part of a child's routine. Consider a few observations regarding the correlation of sleep quality and child behavior:[197]

- Children who get adequate sleep have longer attention spans.
- Babies and toddlers who sleep more hours are less fitful and socially demanding.
- Even small sleep deficits can accrue and lead to long-term effects on the brain. Children with higher IQs have been shown to sleep longer.
- Children suffering from attention deficit problems, such as ADHD, can benefit dramatically from improved sleep. In fact, many learning and behavioral problems are improved with healthy sleeping.

TOGETHERNESS

During a crisis, you may find that having the entire family sleep together in the same room helps everyone to sleep better. Children in particular often benefit because they feel a heightened sense of security when close to their parents and siblings. With parents, this tactic tends to be more hit and miss. Some parents feel relief when close to their children; others may feel anxiety over the lack of privacy. Sleeping in close proximity also helps to keep your family members warm, which can be important in many situations.

Personal aside: I sleep better when I know that my children are nearby and safe. On the other hand, if I am under great stress, I sometimes have trouble sleeping in the same bed as my wife. I have no logical explanation for this, only the observation.

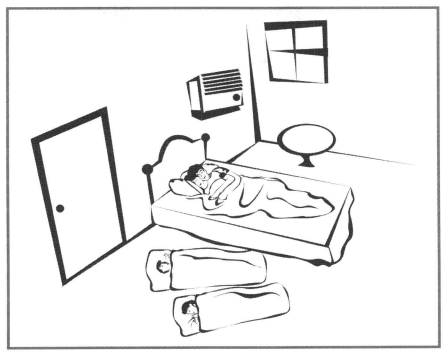

Reducing anxiety by sleeping in the same room

HOW LONG CAN YOU GO?

When everything is falling apart, it is easy to push too hard, telling yourself that you will sleep later. However, not getting the necessary rest can lead to irrational decision-making, perhaps with deadly consequences. A good night's sleep not only maintains your mental acuity, but also refreshes your body, preparing you for the physical challenges ahead. With that said, there are times when you may be forced to interrupt your regular sleep schedule, such as during an emergency evacuation or a particularly dangerous weather event that requires vigilant monitoring.

You may wonder how long a person can go without sleep. In December of 1963, Randy Gardner, a 17-year-old Californian boy, went without sleep for a record 11 days. His 264 hours of being awake is the longest verified period anyone has ever voluntarily gone without sleep.[198] Gardner suffered from mood swings, memory lapses, coordination difficulties, slurred speech, and hallucinations. He did not take stimulants, but people were constantly interacting with him.

Without constant interaction, staying awake for more than 36 hours is difficult, and more than 48 hours is nearly impossible. When you become very tired, your mind will force sleep on you in tiny bursts, some of which you may not even be aware. These sleep bursts can occur with your eyes open or closed. Testing on rats has proven that if you were forced to remain awake for a long period of time, it would kill you in less time than it would take for you to starve to death.

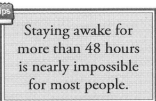

Staying awake for more than 48 hours is nearly impossible for most people.

There are medications, such as modafinil, to help combat excessive daytime sleepiness often associated with obstructive sleep apnea or other sleep disorders. These drugs have also been used by militaries around the world to help soldiers remain alert for long periods of time. It is not difficult to imagine situations where you might need to stay alert for a long time, perhaps during an evacuation or a period of intense danger. However, stimulants of any kind should always be considered a last resort because they have side effects and may become physically addictive. Usually working with family members to sleep in shifts is a much more judicious choice.

DP PLAN EXAMPLE

Table 10-2 Sample DP Plan Entry

Need: Sleep			
Danger	**Goals**	**Needs**	**Implementation**
Insomnia	Get adequate sleep, especially during stressful times	7-8 hours sleep per night for adults; 10-11 hours for three children	Budget one hour more than needed for sleep. Monitor the children for any signs of inadequate sleep. Don't exercise within three hours of bedtime. Don't drink any coffee or alcohol after supper. Make a to do list of any pressing items, with plan of action noted. Take ibuprofen at bedtime if suffering from aches or pain.
Sleepiness	Stay alert in the presence of any threat	Have someone remain alert during a lengthy evacuation or extended threat	When dangers are threatening, establish a sleeping schedule with wife or other DP network members.

Quick Summary - Sleep

➤ People of every class and profession suffer from insomnia.

➤ Sleep is important not only to remain mentally alert, but also to maintain good health. Many diseases have been connected to inadequate sleep, including diabetes, heart disease, and obesity.

➤ Sleep is not something you can control. Rather, you should establish a comfortable environment and consistent bedtime routine.

➤ Consider taking over-the-counter pain relievers to help with aches and pains, or seeing your doctor about sleep aids if absolutely necessary.

➤ Children need much more sleep than adults. Inadequate or poor quality sleep can cause children to develop slowly, have shorter attention spans, feel irritated, and experience behavioral problems. Their misbehavior can make a bad situation much more stressful.

➤ When a situation is particularly threatening, consider sleeping with the entire family in one room. Consolidating helps to keep everyone warm and can relieve worry and stress—especially in young children.

➤ Going without sleep for more than 36 hours is very difficult and may eventually lead to mood swings, poor decision making, loss of coordination, and even hallucinations. Resist the temptation to push yourself too hard.

➤ Even when staying alert for a long period of time is required, sleeping in shifts is preferred to using stimulants.

Recommended Items - Sleep

❏ An environment that facilitates sleep
  a. A comfortable bed, pillow, and sheets
  b. Adequate temperature control of the room

❏ Medicine
  a. Optional: Pain killers (e.g., ibuprofen, acetaminophen)
  b. Optional: Sleep aids (e.g., Ambien, Lunesta)

CHAPTER 11

MEDICAL / FIRST AID

Challenge

Your spouse awakens you during the night complaining of chest pain and difficulty breathing. Should you administer aspirin, and if so, how much? Should you have him/her sit down, lie down, or stand up? Should you call emergency services or rush your spouse to the hospital? If he/she suddenly collapses and stops breathing, do you know how to correctly perform CPR?

Medical concerns are important every day, but even more so during times of disaster because services and supplies may be limited. Hospitals can quickly become overloaded, or worse yet, inaccessible. Doctors and pharmacists may be compelled to close their practices in order to handle their own family emergencies. This shortage of staff and supplies might force your family to rely on its own first aid abilities and existing stockpile of medicines. With that said, you should avoid the temptation to feel that you must become an emergency room doctor when a disaster strikes. There are a few books, such as David Werner's *Where there is No Doctor*, that offer valuable advice for treating medical conditions when doctors are unavailable. However, for the vast majority of disaster situations likely to occur in the U.S., some level of professional medical care will remain available—albeit, perhaps without the same expedience that we expect under normal conditions. Therefore, when a serious medical emergency occurs, you should call upon expert medical services as expeditiously as possible. Your goals should be to learn how to recognize a medical emergency, correctly stabilize the victim, and seek the appropriate care. Additionally, you should develop basic first aid skills that enable you to treat a wide variety of minor medical situations.

MEDICINE

If anyone in your family has a serious existing medical condition, maintaining a stockpile of medicine may be critical to their survival. When a disaster occurs, you may be unable to gain timely access to a doctor or pharmacist, meaning that whatever supplies you have on hand must last until the situation improves. A few of the many possible concerns include: insulin for diabetics, respiratory inhalers for asthmatics, opioids for those suffering chronic pain, and nitroglycerin for people with heart conditions.

> Treat medicines similar to food; maintain a minimum of a 30-day supply.

Recent regulations permit prescriptions up to a 90-day supply for nearly all medications. A reasonable approach to preparing is to handle medications in a manner similar to food—stockpiling a minimum of a 30-day supply. If your customary prescriptions don't support this, explain your preparedness rationale to your doctor to receive the necessary supplies. Just as with food, rotate the newest medicine to the back and use that which is oldest. If a crisis causes you to experience a shortage of medicine, make every effort to inform your doctor and family of your predicament *before* the issue becomes serious.

MEDICAL EMERGENCIES

Medical emergencies are to be expected when times are especially challenging. This is in part due to the immediate dangers posed by the threats but also from having to adapt to difficult living conditions. Connecting a generator, foraging for water, cooking with portable stoves, and shoring up damaged shelters are all examples of activities that introduce additional risk of injury.

To better handle medical emergencies, make the following preparations:

- Teach everyone in the family how to call for emergency medical assistance (see *How to Call for Help*). This includes all children old enough to hold and dial a phone.
- Ensure that all adults in the household learn and practice basic first aid.
- Compile well-stocked first aid kits for your home and automobiles.
- Share information about any serious, existing medical conditions with family and friends.
- Memorize driving directions to at least two emergency medical facilities.
- Investigate which hospitals offer the best services and equipment. For example, some hospitals have specialized cardiac care centers, improving a heart patient's chances for survival.

HOW TO CALL FOR HELP

The United States uses 911 as its three-digit emergency number. This number can be called without charge from any landline, payphone, or cell phone (active or not). Once received, an operator or dispatcher requests that the caller specify the nature of the call: police, fire, or medical services. In the case of an operator, the call is then forwarded to the correct emergency service center. A dispatcher is specially trained to dispatch emergency services as well as provide urgent lifesaving advice.

> **Tips**
> Calling 911 from a landline helps the dispatcher to more quickly locate you.

Note that 911 is not a universal emergency number. Every country has one or more of its own emergency phone numbers. For example, much of Europe uses 112—see *www.911dispatch.com* for a listing of emergency numbers.

When placing an emergency call, it is better to use a conventional landline rather than a cell phone. A landline enables the emergency operator to immediately identify your location. Location determination is slower and less precise when you call from a mobile phone.

The most important thing to remember when calling for emergency service is to remain calm. It may feel like the world is tumbling down around you, but staying calm improves your chances of a successful outcome, whether that is saving someone's life or getting the police to your home as quickly as possible. Explain your situation clearly, follow the dispatcher's directions, and do not hang up until directed to do so.

WHEN TO CALL FOR HELP

There are many health conditions that require emergency medical services. Recognizing those conditions is critical to increasing the victim's chances of survival. Most life-threatening conditions can be recognized by one or more basic warning signs. Consider the following list of symptoms that often indicate a medical emergency.[199] If you witness or experience any of these symptoms, immediately call for emergency medical assistance.

> ### Symptoms of a Medical Emergency
> ➢ Loss of consciousness
> ➢ Chest or severe abdominal pain
> ➢ Sudden weakness or numbness in face, arm, or leg
> ➢ Sudden changes in vision
> ➢ Difficulty speaking
> ➢ Severe shortness of breath
> ➢ Bleeding that does not stop after ten minutes of direct pressure
> ➢ Any sudden, severe pain without an obvious cause
> ➢ A major injury such as a head trauma or broken limb
> ➢ Unexplained confusion or disorientation
> ➢ Bloody diarrhea with weakness
> ➢ Coughing or vomiting blood
> ➢ A severe or worsening reaction to an insect bite, food, or medication
> ➢ Suicidal feelings

APPLYING FIRST AID

First aid symbol

First aid is defined as the initial medical assistance given to someone who is ill or injured. It may be required at home, in the workplace, or while traveling the roadways; medical emergencies can happen anywhere. In the case of minor illnesses or accidents, treatment can usually be rendered without any outside medical assistance. More serious conditions require that you stabilize the patient and call for emergency medical assistance.

As a caregiver, your role begins with an understanding of the six first aid priorities:[200]

1. Assess the situation quickly but calmly.
2. Don't put yourself or the patient in additional danger.
3. Prevent cross contamination by cleaning your hands, using sterile supplies, and equipping yourself with protective clothing.
4. Provide comfort and reassurance to the patient.
5. Administer life-saving treatment, such as stopping the bleeding, clearing the airway, and administering chest compressions, before taking any other actions.
6. Never hesitate to call for emergency medical assistance even if the victim is reluctant to have you do so.

First aid training can be acquired from your local Red Cross or medical "how to" manuals. With that said, there is no substitute for hands-on experience. Reading about injuries is one thing; feeling the flow of warm blood over your fingers as you try to stop the bleeding from an arterial wound is something altogether different. Beyond professional training, the single best way to become proficient at first aid is to practice. For example, to become proficient at applying bandages, take every opportunity to apply bandages over a variety of wound types. Learn by doing.

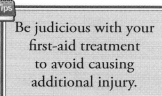

Be judicious with your first-aid treatment to avoid causing additional injury.

Above all, follow the doctor's dictum, *primum non nocere* . . . first do no harm. The human body does a remarkable job of healing itself, so be judicious when deciding your course of action. Perhaps the most important rule to remember is that if you are unsure about what to do, stabilize the person and seek professional medical assistance.

Step-by-step treatments for a variety of first aid situations are outlined in this chapter. Recognize that gaining proficiency in administering first aid is very different from becoming a self-prescribed emergency room doctor. For serious medical situations, you should always seek a trained physician in a fully-equipped facility. With that said, there are times when you may have to handle medical emergencies on your own—even if it is only to stabilize the patient until help arrives.

This chapter should serve as a good start to your first aid preparations. The bulk of this material describes symptoms and treatments for a host of common first aid situations. For each situation, recommended actions are provided based on a number of reputable medical sources.[200,201,202,203] A list of supplies necessary to treat each condition is also given. The lists of supplies have been combined into a comprehensive first aid kit at the end of the chapter. The goal of organizing it this way is to ensure that you know how to use everything in your first aid kit.

Note: For ease of writing, the terms "him" and "he" are used throughout the descriptions of medical conditions, but they are meant to refer to both male and female patients.

FIRST AID CONDITIONS AND TREATMENT

ANAPHYLAXIS

Anaphylaxis is a severe allergic reaction that can result in respiratory distress and circulatory shut-down. Anaphylactic shock can occur immediately or take several hours from the time of exposure. In an anaphylactic reaction, blood vessels dilate, causing the victim's blood pressure to fall and air passages to constrict. Additional symptoms may include hives; swelling of the lips, eyes, tongue, and throat; wheezing and gasping; dizziness; abdominal cramping; nausea; vomiting; and diarrhea. There exists a wide variety of allergens, but some of the most common are nuts, insect stings, shellfish, latex, and eggs.

Recommended Actions:

1. Call 911 for emergency medical assistance.
2. Check to see if the patient is carrying an epinephrine auto-injector.
3. If epinephrine is available and the patient is conscious, help him to use the auto injector.
4. If the patient is unconscious or nonresponsive, administer epinephrine as directed:
   * Remove the safety cap.
   * Hold the injector in your fist and place the tip firmly against the patient's thigh—the shot can be given through most clothing.
   * Release the medicine and hold the injector in place for at least 10 seconds.
   * Massage the injection site to increase absorption.
5. Have the patient sit in whatever position feels most comfortable for his breathing. If he begins to appear pale or weak, have him lie on his back with his feet elevated about 12 inches.
6. Administer an over-the-counter antihistamine pill with a little water if the patient is able to swallow it without choking.
7. Loosen any tight clothing and cover the patient with a blanket. Do not give him anything else to drink.
8. If the patient begins vomiting or bleeding from the mouth, turn him on his side to prevent choking.
9. If the patient stops breathing, perform CPR.

Supplies needed:

* Epinephrine auto-injector—requires a doctor's prescription
* Antihistamine pills (e.g., Benadryl)

ANIMAL/HUMAN BITES

Animal bites

Domestic pets are the source of most animal bites. Dog bites occur more often, but cat bites are more likely to cause infection. Any bite from a wild or non-immunized animal should be immediately treated by a medical doctor due to the increased risks of infection, tetanus, and rabies. Bites from immunized animals that break the skin should also be treated. Rabies is most common in raccoons, skunks, bats, and foxes, and less likely in rodents, rabbits, and squirrels. Reptiles, fish, and birds do not carry rabies. Human bites can also cause infection or transfer potentially deadly viruses, such as HIV/AIDS, and should therefore be treated immediately by medical personnel.

Recommended Actions:

1. If the bite is from a wild animal, or if the immunization of the animal is unknown, seek medical care immediately. Rabies is a fast-acting, deadly disease that requires immediate diagnosis for the greatest chance of survival.
2. If the bite doesn't break or barely breaks the skin *and* is from an immunized, domesticated animal, wash the wound with soap and water (or Betadine antiseptic if water is unavailable), apply an antibiotic cream, and cover with a clean bandage.
3. If the bite is a deeper wound (i.e., punctured or torn), apply pressure with a clean cloth or bandage to stop the bleeding, and then seek emergency medical care.
4. As the wound heals, if you notice signs of infection, such as swelling, redness, or oozing, seek medical attention.

Supplies needed:

- Soap and water, or Betadine
- Antibiotic cream
- Adhesive bandages

BLACK EYE

A black eye is caused by bleeding under the skin around the eye. Most black eye injuries are not serious and can be handled without assistance.

Recommended Actions:

1. To reduce swelling, gently apply a cold pack or ice-filled cloth to the area. Apply it as soon as possible, and continue using the cold pack a few times a day for 24 to 48 hours.
2. If the patient experiences vision problems (e.g., blurring, double vision), severe pain, bleeding within the eye, or bleeding from the nose, seek emergency medical care.
3. If both eyes become black, seek emergency medical care since it may indicate a skull fracture.

Supplies needed:

- Cold pack

BLISTERS

Blisters are typically caused by friction (e.g., your foot rubbing against your shoe) or heat (e.g., grease splashing onto your hand from a hot pan). Most blisters are not serious and can be handled without medical assistance.

Recommended Actions:

1. Wash the blister and surrounding area with clean water. Gently pat dry with a soft, clean cloth.
2. If the blister isn't too painful, try to avoid popping it. Keeping the skin intact will help to prevent infection.
3. If the blister is small, cover it with an adhesive bandage. Larger blisters can be covered with porous, non-stick gauze pads.
4. If the blister must be drained due to its size or location, follow these steps:
   a) Wash your hands with soap and water.
   b) Apply iodine or rubbing alcohol to the blister.
   c) Sterilize a clean needle with alcohol.
   d) Use the needle to puncture the blister at several places near the blister's edge. Let it drain, but try to keep the overlying skin intact.
   e) Apply an antibiotic cream, and cover with an adhesive bandage or porous, non-stick gauze pad.
   f) After several days, it may be possible to use sterile scissors and tweezers to gently cut away the top skin. After doing so, apply more antibiotic ointment and a fresh bandage.
   g) Seek medical attention if signs of infection develop, such as pus, redness, increased pain, or warm skin.

Supplies needed:

- Antibiotic cream or ointment
- Adhesive bandages
- Porous, non-stick gauze pads
- Rubbing alcohol
- Needle

BRUISE

Bruises are pools of blood just beneath the skin, usually caused by a physical blow of some sort. They fade within a few days and are not a serious medical concern. Generally, the older a person is, the more easily bruising occurs.

Recommended Actions:

1. If the skin isn't broken, there is no need to cover a bruise with a bandage. However, some patients may prefer to cover particularly unsightly bruises.
2. To help a bruise heal faster:
   a) Elevate and rest the injured area if possible.
   b) Apply a cold pack to the bruise several times a day for 24 to 48 hours. Do not apply ice directly to the skin.
3. After about 48 hours, apply a warm washcloth for 10 minutes, twice a day, to help increase blood flow. This allows the skin to reabsorb the blood more quickly.
4. Seek medical attention if the bruise is unusually large or painful—particularly if the patient can't remember suffering any trauma to the region.
5. Seek emergency medical attention if the patient is experiencing abnormal bleeding elsewhere, such as from the nose, gums, or eyes, or if blood is present in the stool or urine.

Supplies needed:

- Cold pack
- Warm washcloth

BURNS

Burns are described by three classifications: first degree, second degree, and third degree. Each increase in degree indicates an increase in severity (i.e., third degree is more serious than second degree). The classification is based on the depth of the burn. It is important to recognize the symptoms of each type of burn because their respective treatments are different.

First-degree burns (a.k.a. superficial burns) damage only the outermost layer of skin, the epidermis. The affected area is usually red, swollen, and painful. Mild sunburn is the most common type of superficial burn. First-degree burns are generally treated as minor burns as outlined below. However, if the burn covers a large portion of the hands, feet, face, groin, buttocks, or a major joint, seek medical attention.

Second-degree burns (a.k.a. partial-thickness burns) damage down to the second layer of skin, the dermis. Symptoms include blisters, intensely red and splotchy skin, pain, and swelling. If the burn is smaller than three inches across, treat it as a minor burn as outlined below. If the burn is larger than three inches across, or is located on the hands, feet, face, groin, buttocks, or over a major joint, treat as a major burn.

Third-degree burns (a.k.a. full-thickness burns) are serious burns that involve all layers of the skin and sometimes even muscle and bone. The area may appear waxy, pale, or charred. Full-thickness burns require immediate emergency medical care and should be treated as major burns.

Recommended Actions (Minor Burns):

1. Immediately cool the burn by holding the area under cool running water for 10 to 15 minutes. If this is difficult to do because of the burn's location, soak the affected area in cool water or apply cold compresses. Do not apply ice directly to the burn.

2. Cover or wrap the burn loosely with a non-stick gauze bandage. Try to avoid breaking any blisters.
3. Administer an over-the-counter pain reliever as needed.
4. Watch for signs of infection such as increased pain, redness, fever, swelling, or oozing. If infection occurs, seek medical attention.
5. To minimize pigmentation color changes, apply sunscreen on the burned area for at least one year.

Recommended Actions (Major Burns):

1. Call for, or seek, emergency medical assistance.
2. Cool the injury by flooding the area or entire body with cold water. Be careful about overcooling the patient, especially when dealing with babies and the elderly.
3. Do not touch the injured area. If possible, remove any burned clothing as well as any watches or jewelry that might become constricting if the limb should swell. If there is a risk of injury, leave the clothing in place, but be sure that it is fully extinguished.
4. Cover the burned area with a non-stick gauze or gel-soaked bandage, such as those from Water-Jel. If bandages are not available, cover the wound with plastic wrap to keep out infection.
5. Reassure the patient and treat for shock if necessary.
6. If the patient stops breathing, administer CPR.

Three degrees of burns *(Wikimedia Commons/K. Aainsqatsi)*

Supplies needed:

- Non-stick gauze or gel-soaked burn bandage
- Cold compress
- Pain reliever

CARDIOPULMONARY RESUSCITATION (CPR)

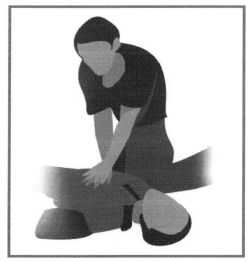

Administering chest compressions

Cardiopulmonary resuscitation (CPR) is a lifesaving treatment used on a person whose breathing or heartbeat has stopped. If you are interested in receiving formal CPR training, contact the American Heart Association at 1-800-AHA-USA1 or your local American Red Cross.

CPR consists of chest compressions and mouth-to-mouth rescue breathing. For adult patients, experts now recommend that untrained people only administer chest compressions, rather than the combination of compressions and rescue breathing. The reason for this change is that for the first five minutes after an adult's heart stops, the blood remains oxygen rich, and chest compressions are the highest priority for survival. The chest compressions help keep oxygenated blood flowing to the brain and organs until emergency personnel arrive.

For infants and children, however, both chest compressions and rescue breathing are necessary. The differences in treatment are outlined in the steps below.

Recommended Actions:

1. Check to see if the person is unconscious by touching him and asking, "Are you okay?" Be sure he isn't just sleeping!
2. If he doesn't respond, have someone call for emergency medical assistance while you start CPR. If you are the only person on the scene and have immediate access to a phone, make the call before beginning CPR. However, if the patient is unconscious due to suffocation or drowning, or if the patient is a child, administer CPR for one full minute before calling for emergency assistance.
3. If you suspect the patient has suffered a heart attack, and an Automatic External Defibrillator (AED) is available, retrieve the unit, push the ON button, and follow the voice prompts. Otherwise, administer manual CPR as outlined below.
4. Before administering CPR, check to be sure that the patient's airway is clear and that he is not currently breathing.
   a) Lay him on his back.
   b) Kneel next to his shoulders.
   c) Tilt back his head using his chin and forehead.
   d) Check for breathing using the *look, listen, feel* technique. Look for chest motion; listen for breathing; feel for breath on your cheek. If he is breathing, wait for emergency personnel to arrive while monitoring his condition.
   e) If the patient is not breathing, look to see if his airway is obstructed. If it is visibly obstructed, or you know the patient choked, follow the steps outlined in *Choking*. Otherwise, begin administering CPR.

5. ***Adolescent/Adult Patient:*** Administer chest compressions.
   a) Place the heel of one hand over the center of the patient's chest directly between his nipples. Place your other hand on top of the first.
   b) Keeping your elbows straight, use your upper body weight to quickly compress the chest about two inches. Repeat at a rate of 100 compressions per minute.
   c) Continue chest compressions until the patient revives or medical personnel arrive.
6. ***Child Patient (age 1-8):*** Administer chest compressions and rescue breathing.
   a) Administer chest compressions using the heel of one hand placed directly between the child's nipples. The compression should compress the child's chest by about one third. Administer 30 compressions at a rate of 100 compressions per minute.
   b) Administer rescue breathing by tilting the child's head back, pinching his nose closed, and placing your lips around his mouth. Blow steadily for one second; the chest should rise. Remove your mouth, take a breath, and administer a second rescue breath.
   c) Continue repeating 30 chest compressions followed by two rescue breaths until the child revives or rescue personnel arrive.
   d) If alone, repeat the cycle five times before calling for emergency medical assistance.
7. ***Infant Patient (<1 year old):*** Administer chest compressions and rescue breathing.
   a) Administer chest compressions using the two fingers of one hand placed directly between the infant's nipples. The compression should compress the child's chest by about one third. Administer 30 compressions at a rate of 100 compressions per minute.
   b) Administer rescue breathing by tilting the infant's head back, and placing your lips around the infant's mouth *and* nose. Blow a gentle breath for one second; the chest should rise. Remove your mouth, take a breath, and administer a second rescue breath.
   c) Continue repeating 30 compressions followed by two rescue breaths until the infant revives or rescue personnel arrive.
   d) If alone, repeat the cycle five times before calling for emergency medical assistance.

Supplies needed:

- Optional: rescue breathing masks or face shields

CHEMICAL SPLASH IN EYE

Chemical splashes in the eye can cause serious injury if not treated quickly. It is important to thoroughly flush the eye with water or saline solution as rapidly as possible.

Recommended Actions:

1. Turn the patient's head sideways under a gentle running faucet (or shower), flushing the eye thoroughly with water for up to ten minutes. Orient the patient so that the affected eye is down toward the sink, preventing the contaminants from washing over into the uninjured eye. With a young child, it might be easier to have him lay back while you gently pour water onto his forehead or the bridge of his nose.

2. After cleaning his eye, wash your hands and have the patient wash his hands to remove any possible chemical contamination.
3. Have him remove any contact lenses if they haven't been removed already.
4. Try to prevent any rubbing of the eye, as this may cause damage.
5. If pain or irritation continues, seek emergency medical care. Take the bottle of chemical contaminant with you.

Supplies needed:

- Water or saline solution

CHEST PAIN

Chest pain can be an indication of many different medical conditions, ranging from minor to deadly. An accurate diagnosis is often very difficult for a layperson to make. A few of the many possible causes for chest pain are given below:

- Heartburn—stomach acid rises into the esophagus
- Sore muscles, injured ribs, or pinched nerves
- Heart attack—an artery providing oxygen to the heart becomes blocked
- Angina—plaque buildup causes reduced blood flow in the heart
- Pulmonary embolism—a blood clot lodges in an artery of the lungs
- Aortic dissection—a tear develops in the inner layer of the aorta (the heart's large blood vessel)
- Costochondritis—pain in the chest wall causes tenderness around the cartilage that connects the ribs to the sternum
- Pleurisy—membrane lining the chest wall becomes inflamed, causing sharp, localized chest pain, especially when inhaling

Recommended Actions:

1. If someone experiences unexplained chest pain that lasts for more than a few minutes, seek emergency medical care. Either call for emergency medical assistance and wait for an ambulance, or immediately drive the patient to the nearest emergency room. A patient should not be allowed to drive himself in case his condition worsens.
2. If you suspect the chest pain is due to a heart attack, follow the steps given under *Heart Attack*.

Supplies needed:

- None

CHOKING

Choking is usually the result of something lodged in the throat, partially or fully blocking the airflow. With adults, the cause is often a piece of food, but with children it can also be a toy, coin, or other small object. Teach everyone in your family the universally accepted sign for choking—clutching their throat with both hands. Symptoms of choking can include inability to talk or cough, difficulty or noisy breathing, skin and lips turning blue or dusty-looking, and loss of consciousness. Choking is a potentially deadly condition that requires immediate attention.

Recommended Actions (for Adult Patient):

1. Ask the patient if he is choking. If he is able to talk or cough forcefully, allow him to dislodge the food himself. Do not administer abdominal thrusts if the patient is coughing.
2. If you determine that his airflow is blocked, administer standing abdominal thrusts—see below.
3. If the patient collapses unconscious, examine his airway to see if the obstruction can be removed. If so, attempt to remove it, but be careful not to push it further into the airway.
4. If you cannot remove the object, administer prone abdominal thrusts—see below.
5. If you fail to clear the obstruction, call for emergency medical assistance, and administer CPR until emergency personnel arrive.

Administering abdominal thrusts (a.k.a. Heimlich maneuver):

Stand behind the patient and bend him slightly forward at the waist. Wrap your arms around his waist, making a fist with one hand and grabbing it with the other. Place your fist slightly above the patient's navel. Press hard up into the abdomen with a quick thrust—as if you are trying to lift him off his feet. In the case of an obese or pregnant patient, position your hands slightly higher.

If the patient is unconscious, deliver abdominal thrusts by straddling the patient and placing the heel of your hand just above his navel. Place your other hand on top of the first. Keeping your elbows straight, give several quick upward thrusts until the object is expelled.

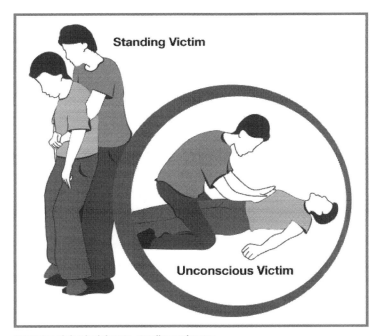

Delivering abdominal thrusts, standing and prone

Recommended Actions (for Infant Patient):

1. Sit down, and place the infant face down on your forearm, resting the supporting arm on your thigh. The infant's head should be lower than his trunk.
2. Using your heel of your palm, thump the infant five times in the middle of his back.
3. If back blows fail to relieve the choking, roll the infant over and check his mouth and throat. Remove any obvious obstructions with your fingertips—don't sweep the object, since that may push the object further down the throat.
4. If the obstruction cannot be removed with your fingers, keep the infant face-up on your forearm with his head lower than his trunk. Using two fingers, give up to five quick chest compressions at the center of the infant's breastbone. The purpose of the compressions is to dislodge the blockage.
5. If chest compressions do not free the obstruction, call for emergency medical assistance, and repeat this routine of five back blows followed by five chest thrusts until the object clears or emergency personnel arrive.
6. If the infant loses consciousness, administer CPR.

Supplies needed:

• None

CUTS AND SCRAPES

Minor cuts and scrapes occur frequently and can usually be treated without medical assistance. However, if you can't stop the flow of blood, or if it seems to spurt out, seek emergency medical care. Likewise, if the wound is deeper than about ¼ inch or is gaping open, seek emergency medical care to receive stitches or other treatment.

Recommended Actions:

1. Most small cuts and scrapes will stop bleeding on their own, but if they don't, apply gentle pressure with a clean bandage or cloth for 20 to 30 minutes. If possible, elevate the wound above the patient's heart.
2. Clean the wound by rinsing it with clean water. You can also use soap (or alcohol wipes) to clean the area, but try to keep the soap or alcohol out of the wound. If dirt or debris remains in the wound after rinsing, use tweezers dipped in alcohol to remove the debris.
3. Apply a thin layer of antibiotic cream or ointment to the cut or scrape.
4. Cover the wound with a clean bandage.
5. Change the bandage at least once daily or when it becomes wet or dirty.
6. Once the wound has healed sufficiently to make infection less likely, remove the bandage and expose it to the air to speed healing.
7. If you see signs of infection, such as redness, increasing pain, drainage, or swelling, seek medical attention.

Supplies needed:

- Bandages
- Antibiotic ointment
- Soap and water
- Rubbing alcohol
- Tweezers

DISLOCATION

Bones can dislocate as a result of a blow or fall. This type of injury occurs frequently in contact sports, such as football or hockey. Dislocation can occur in major joints, such as the shoulder or knee, or in smaller joints, such as the fingers and toes. Torn ligaments, damage to the synovial membrane, and fractured bones can also accompany dislocated joints.

Recommended Actions:

1. Do not move the bone or try to reposition it back in the joint. Keep the limb stationary in the position that causes the patient the least amount of pain. If possible, use a splint or sling to prevent movement.
2. Apply a cold pack to the injured joint to reduce swelling.
3. Seek emergency medical care immediately.

Supplies needed:

- Splint, sling
- Medical tape
- Cold pack

ELECTRICAL SHOCK

Minor electrical shock (i.e., one that causes no noticeable symptoms) doesn't usually require medical treatment. However, a major electrical shock can cause serious complications, including cardiac arrest, respiratory failure, burns, seizures, and loss of consciousness. A patient who receives a shock by alternating current (AC) may be unable to let go of the electrified object. Direct current (DC) tends to produce a single violent muscular contraction that throws the patient away from the electrified object.

If you witness someone being shocked or encounter someone who you suspect has been shocked, use extreme caution to prevent injury to yourself. Electricity can arc through the air from person to person or from object to person if you get too close. High-voltage power lines, for example, can arc through the air up to 60 feet.

Recommended Actions:

1. Carefully inspect the situation. Do not touch the person until you are certain that he is no longer being shocked.

2. If the patient is being shocked by a high-voltage power line, do not approach him. Call for emergency assistance immediately.

3. If the patient is being shocked by a house appliance or outlet, unplug the source or throw the circuit breaker. If that is not possible, use a non-conducting object, such as a wood broomstick, to push the source away from the person.

4. Check to see if the patient is breathing. If he isn't breathing, begin administering CPR and call for emergency medical assistance.

5. If the patient is injured but still breathing, treat for shock by laying him down and elevating his feet about 12 inches. Then call for emergency medical assistance.

Supplies needed:

- None

FAINTING

Fainting is a brief loss of consciousness caused by a temporary reduction in blood flow to the brain. It can be brought on by dehydration, lack of food, exhaustion, reaction to pain, or emotional stress. It may also occur when standing up quickly after sitting for a long period of time because blood may pool in the legs and be unable to reach the brain fast enough. A person who faints is typically unconscious for only a brief time (less than one minute) until blood flow to the brain improves.

Recommended Actions:

1. If someone says he feels faint, have him sit or lie down. Do not have him put his head between his legs while sitting because he may fall and become injured. If he chooses to lie down, elevate his feet slightly.

2. If someone faints, first check to see if he is breathing. If he is not breathing, immediately begin administering CPR and call for emergency medical assistance.

3. If he is unconscious but breathing, lay him on his back and elevate his feet above his heart level by about 12 inches. Loosen any constrictive clothing, and monitor his breathing.

4. Once he regains consciousness, have him rest a moment before slowly standing up.

5. If he does not regain consciousness within one minute, call for emergency medical assistance.

Supplies needed:

- None

FEVER

Fever is a symptom of many different medical conditions but is most often caused by a bacterial or viral infection. Every person's normal body temperature is slightly different, with the average being 98.6°F. A moderate fever is not usually harmful to adults, but may introduce seizures in young children.

Many doctors recommend not treating fevers below 102°F with medications. They argue that the body's elevated temperature will help fight off infection, but only at the expense of added discomfort for the patient.

Recommended Actions:

1. Keep the patient cool and comfortable, preferably remaining in bed.
2. Provide cool beverages to drink (non-alcoholic, caffeine-free).
3. Treat with fever reducers or pain medicine as needed.
   - Adult fevers can be treated with acetaminophen, ibuprofen, or aspirin (assuming there are no allergies).
   - Children over 6 months of age can be treated with ibuprofen or acetaminophen. Do not administer aspirin because it might trigger Reye's syndrome—a rare but deadly disorder.
   - Children under 6 months of age should be treated only with acetaminophen.
4. Seek medical care if any of the following conditions apply:
   - A baby younger than 3 months has a rectal temperature of 100.4°F or higher. A rectal reading is generally about 1 degree higher than oral readings
   - A baby older than 3 months has a temperature of 102°F or higher
   - A newborn has a rectal temperature below 97°F
   - A child younger than age 2 has a fever longer than one full day, or a child older than age 2 has a fever longer than three days
   - An adult has a temperature of more than 103°F or has a fever for more than three days

Caution: Be very careful to give the correct dose of acetaminophen. Overdosing of acetaminophen can lead to liver failure and death. Tens of thousands of Americans are rushed to emergency rooms every year from acetaminophen overdose.[204] Be aware that acetaminophen may also be present in other medicines, such as cough suppressants, that when combined with fever-reducing medicine, can lead to a dangerous overdose.

Supplies needed:

- Fever-reducing medicine
- Thermometer

FOREIGN OBJECT IN EAR

A foreign object can become stuck in the ear and cause pain or hearing loss. Common objects include cotton swabs and insects. In most cases, foreign objects can be safely removed without medical assistance, but care must be taken not to injure the patient's ear or push the object deeper into the ear canal.

Recommended Actions:

1. If the object is clearly visible, attempt to gently remove it using tweezers. Do not try to remove the object using a cotton swab, matchstick, or other ill-suited tool; you risk pushing it further into the ear and causing damage.
2. If you can't easily pull it out, try tilting the patient's head such that his ear faces the ground, and then gently try to dislodge it.

3. If the object is an insect (alive or dead), fill the ear with warm water, mineral oil, baby oil, or olive oil to float the insect out. Do not use this method if there is any pain, bleeding, or discharge from the ear.
4. If these simple methods fail to remove the object, seek medical attention.

Supplies needed:

- Tweezers
- Mineral, baby, or olive oil

FOREIGN OBJECT IN EYE

Most eye injuries involve the cornea—the clear protective coating at the front of the eye. Dust, dirt, sand, wood shavings, or other particles can scratch, cut, or become lodged in the cornea. Corneal abrasions are painful, but most are not serious unless they become infected or inflamed. Symptoms may include tearing, blurred vision, headache, and eye redness.

Recommended Actions:

1. Have the patient sit facing a bright light so that you can more easily inspect his eye.
2. Wash your hands with soap and water.
3. Using your thumb and forefinger, gently separate his eyelids (pulling the top one up and the bottom one down). Have the patient look left, right, up, and down, while you are examining the eye for foreign objects.
4. If you see a foreign object on the surface of the eye, rinse the eye with saline solution (or clean water if saline isn't available).
5. If rinsing is not successful, try to lift the object out by gently dabbing with a clean moist cotton swab. Try not to rub the eye, which can cause additional scratching.
6. You can also have him pull the upper eyelid over the lower one to help dislodge anything stuck on the undersurface of the upper eyelid.
7. Don't try to remove anything that is embedded in the eyeball.
8. If you are unable to remove the foreign object, or if it is embedded in the eyeball, cover the eye with sterile eye pad or gauze, and seek emergency medical care.

Supplies needed:

- Saline solution or eye wash
- Eye pad
- Bright light
- Cotton swab

FOREIGN OBJECT IN NOSE

Children are especially likely to lodge a small object in their noses. Common obstructions include marbles, candy, toys, or batteries. If the object is sharp, it can cause tissue damage, which can lead to infection. Likewise, batteries can cause tissue burns. Small objects lodged in the nose can usually be removed without medical assistance; however, care must be taken to ensure that the object isn't pushed further into the airway.

Recommended Actions:

1. Have the patient breathe through his mouth, preventing him from inhaling the object.
2. Have him blow his nose gently. If the obstruction is affecting only one nostril, plug the other nostril when blowing.
3. If the object can be easily grabbed, carefully use tweezers to remove it. Do not use a cotton swab or other ill-suited tool to try to pry the object out.
4. If the object is a marble and it will not blow out, try gently pinching the nose from the top down, pressing the marble down and out of the nose.
5. If you are unable to remove the object, seek emergency medical care.

Supplies needed:

- Tweezers

FOREIGN OBJECT IN SKIN

Splinters of wood, metal, glass, or fiberglass can become lodged in the skin. It is important to remove the splinter to prevent infection. Getting a splinter out can usually be done without medical assistance.

Recommended Actions:

1. Wash your hands and the affected area with soap and water. Sterilize a needle and tweezers using isopropyl alcohol.
2. If the splinter juts out above the skin, use tweezers to draw the object out in a straight line opposite the direction that it went in.
3. If the splinter is completely embedded in the skin, use a small sterilized needle to break the skin above it. Then gently lift the tip of the splinter out, and use tweezers to remove the object.
4. Gently squeeze the area to stimulate a little bleeding, which helps to flush out the wound.
5. Wash and dry the area again.
6. Apply a thin layer of antibiotic ointment and a small adhesive bandage as needed.

Supplies needed:

- Tweezers
- Needle
- Isopropyl alcohol
- Antibiotic ointment
- Adhesive bandage

FOREIGN OBJECT SWALLOWED

Most small foreign objects that are swallowed will usually pass through the digestive system without complications. However, sharp objects may become lodged in the esophagus (the tube connecting the throat to the stomach). Also, button batteries used in watches, calculators, and toys, can cause tissue damage and should be removed by a medical professional immediately. Finally, swallowing multiple magnets may cause internal complications and requires immediate medical consultation.

Recommended Actions:

1. If the object is small and doesn't cause choking, contact your family doctor. He will likely recommend that you allow the item to pass through the body.
2. If the object is sharp, lodged in the esophagus, or a battery, seek emergency medical care.
3. If the object causes choking, treat the patient as outlined under *Choking*.

Supplies needed:

- None

FRACTURES (BROKEN BONES)

A fracture is a break or crack in the bone. Open (compound) fractures are those where the broken bone penetrates the skin, whereas closed (simple) fractures leave the skin intact. Open fractures are particularly susceptible to infection. Fractures may be characterized by bleeding, swelling, deformity, bruising, pain, and difficulty in moving the limb. Fractures require professional medical attention.

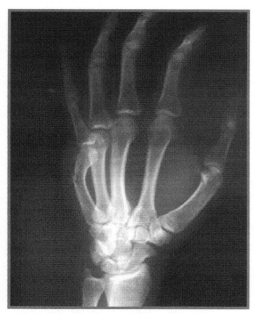

Fractured hand *(Wikimedia Commons/Roberto J. Galindo)*

Recommended Actions:

1. If the fracture is on the arm, hand, or foot, and there is no bleeding, immobilize the limb using a sling or structural aluminum malleable (SAM) splint. Apply a cold pack to the area, and seek emergency medical care.
2. If the fracture is elsewhere on the body, such as the neck, back, leg, or hip, call for emergency medical assistance and have the patient remain still until emergency personnel arrive.
3. If the fracture is bleeding, carefully apply pressure to the wound with a clean cloth or bandage, and call for emergency medical assistance. Keep the area immobilized, and do not try to realign or push a protruding bone back in.
4. If the patient feels faint, treat him for shock by laying him on his back and elevating his feet.

Supplies needed:

- SAM splint
- Sling
- Tape
- Cold pack
- Clean cloth or bandage

FROSTBITE

Frostbite is the result of the skin being exposed to very cold temperatures. Extremities such as hands, feet, nose, and ears are most likely to be affected. If left unchecked, frostbite can cause loss of sensation and tissue death. Symptoms include the skin appearing white or grayish-yellow, being very cold to the touch, and having a hard or waxy feel. Itching, burning, or numbness may also be present. As the skin thaws, it becomes red and painful. Frostbite is often accompanied by hypothermia, which must also be treated (see *Hypothermia*).

Recommended Actions:

1. Immediately get the patient out of the cold. Try to keep him from walking on frostbitten feet or toes.
2. Warm the affected area. You can warm hands or feet by placing them under your armpits. Hands and feet can also be warmed by placing them in lukewarm (not hot) water. Cover ears, nose, and face with warm hands.
3. Raise the affected limb to reduce any swelling.
4. Do not rub the affected area as this can cause skin and tissue damage.
5. Administer over-the-counter pain medicine as needed.
6. Seek emergency medical assistance if numbness remains after warming or if skin damage is present.

Supplies needed:

- Over-the-counter pain medicine

GASTROENTERITIS

Gastroenteritis is a general term used to describe the inflammation of the stomach and intestines. There are many possible causes, including viruses, foodborne or waterborne bacteria, and stress. Symptoms may include nausea, vomiting, diarrhea, abdominal cramps, and occasionally a low-grade fever. The symptoms may last from 24 hours to more than a week. See *Chapter 3: Food* and *Chapter 4: Water* for additional information on food poisoning and waterborne contamination.

Recommended Actions:

1. Have the patient rest and drink plenty of liquids to prevent dehydration.
2. Have him gradually ease back into eating, starting with bland foods, such as crackers, toast, gelatin, bananas, and chicken soup. Have him avoid dairy products, caffeine, alcohol, nicotine, and fatty foods for a few days.

3. Generally, do not administer anti-diarrheal medications early in the sickness because they may slow the elimination of contaminants from the digestive system.
4. Seek medical care for any of the following conditions:
   - Vomiting persists for more than two days or turns bloody
   - Diarrhea persists for more than a week or turns bloody
   - High fever develops
   - Confusion, lightheadedness, or fainting occurs
5. Additionally, seek emergency medical care for a child for any of the following conditions:
   - Child becomes unusually drowsy
   - Child shows signs of dehydration such as marked thirst, sunken eyes, crying without tears, if the soft spot on an infant's head starts to become sunken, or if diapers remain dry for 8 or more hours
   - Fever remains longer than one day for child under age 2, or fever remains longer than 3 days for child over age 2

Supplies needed:

- Thermometer

HEADACHE

A headache may accompany many illnesses, particularly those with fevers. Headaches may also be the result of stress, fatigue, or excessive alcohol consumption. Most headaches are minor and can be treated with over-the-counter pain medications. However, headaches can also signal dangerous medical conditions, such as meningitis or a stroke.

Recommended Actions:

1. Have the patient sit or lie down in a quiet room. Provide a cold compress for his head.
2. If the headache is minor, consider treating with over-the-counter pain medication:
   - Treat adult headaches with acetaminophen, ibuprofen, or aspirin (assuming there are no allergies).
   - Treat a child's headache only with acetaminophen or ibuprofen. Do not give aspirin to children because it might trigger Reye's syndrome—a rare but deadly disorder.
   - Treat infants younger than 6 months of age with acetaminophen. Be very careful to administer the correct dose.
3. Seek emergency medical care for any of the following:
   - A severe headache develops suddenly
   - The headache is accompanied with a fever, stiff neck, rash, confusion, seizures, vision changes, dizziness, weakness, paralysis, or difficulty speaking
   - The headache is the result of a head injury or fall
   - The headache progressively worsens throughout the day or persists for several days

Supplies needed:

- Over-the-counter pain reliever
- Cold compress

HEAD TRAUMA

Minor abrasions and cuts to the head are treated as they would be to any other part of the body (see *Cuts and Scrapes*). However, in the case of a more severe head injury, the patient requires careful monitoring because head trauma may cause more serious medical conditions, including a skull fracture, concussion, and bleeding in the brain.

Recommended Actions:

1. If the head trauma is minor, treat the abrasions as detailed in *Cuts and Scrapes*. An ice pack can also be applied to reduce localized swelling.
2. If any of the following symptoms occur at the time of the injury, or become evident within 24 hours, call for emergency medical assistance immediately:
   - Severe head or facial bleeding
   - Bleeding from the nose or ears (other than from a minor injury to the nose or exterior of the ears)
   - Severe headache
   - Loss of consciousness for more than a few seconds
   - Black and blue discoloration below the eyes or behind the ears
   - Cessation of breathing
   - Confusion
   - Loss of balance
   - Weakness or inability to use an arm or leg
   - Unequal pupil size
   - Vomiting
   - Slurred speech
   - Seizures

3. While awaiting emergency personnel, keep the patient still, stop any bleeding by applying pressure using a clean cloth or bandage, and stay alert for any changes in breathing.

4. If the patient stops breathing, administer CPR.

Supplies needed:

- Bandages
- Antibiotic ointment
- Cold pack

HEART ATTACK

A heart attack occurs when the artery supplying blood to the heart becomes partially or completely blocked. The loss of blood flow causes damage to the heart muscle. Symptoms can include any of the following: chest pressure; pain in the upper abdomen or back; pain spreading from the chest to shoulders, neck, jaw, and one or both arms; shortness of breath; dizziness; sweating; and nausea. The earliest warning sign may be episodes of chest pain that occur during physical exertion but are relieved by rest. However, a heart attack can also occur without any warning signs.

Recommended Actions:

1. If you suspect that someone is having a heart attack, call for emergency medical assistance.
2. Have the patient sit down, rest, and try to remain calm.
3. Have him chew one regular strength aspirin tablet (assuming no allergy), or if he has been prescribed nitroglycerin or other heart medication, have him take it as directed. Do not administer nitroglycerin to someone who has not had it prescribed.
4. If the patient stops breathing, administer CPR while waiting for emergency personnel.
5. If you are in a public building that has an automatic external defibrillator (AED), have someone else retrieve it while you administer CPR. Training is certainly desired when using an AED, but even if you are not trained, don't be afraid to use it on a patient who has stopped breathing. It is their best chance for survival.

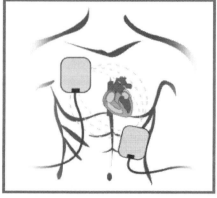

AED pad placement

Using an AED: An AED analyzes a patient's heart rhythm and determines if an electrical shock is required. After switching the unit on, attach the two pads to the correct location: one just under the patient's right collarbone, and the other along his left ribcage. Get everyone to stand clear while the AED analyzes his heart rhythm. The AED will then advise whether a shock is needed. If advised to do so, press the *Shock* button on the AED control box to administer the shock. The AED will then direct you to continue administering CPR for two minutes before it reanalyzes the patient's heart rhythm again. If at any time the patient starts to breathe, stop using the AED (but leave it attached), and help him to be comfortable while awaiting emergency personnel.

Supplies needed:

- Aspirin
- AED (if available)

HEAT CRAMPS

Heat cramps are the first stage of hyperthermia (see *Chapter 8: Heating/Cooling*) and consist of painful involuntary muscle spasms. They are most often the result of overexertion in hot environments. Inadequate fluid intake also contributes to heat cramps.

Recommended Actions:

1. Move the patient to a cool environment to rest.
2. Have him drink cool liquids that do not contain caffeine or alcohol.
3. Gently stretch and massage the affected muscle group.

4. Advise the patient to wait a few hours after the heat cramps subside before resuming any strenuous activity.
5. Seek medical care if the cramps don't go away within an hour.

Supplies needed:

- Cool liquids

HEAT EXHAUSTION

Heat exhaustion is the second stage of hyperthermia (see *Chapter 8: Heating/Cooling*). It is typically a result of overexertion in hot conditions. Symptoms often begin suddenly and may include heavy perspiration, dizziness, nausea, a rapid but weak heartbeat, low blood pressure, cool skin that may look pale, low-grade fever, heat cramps, headache, dark urine, and fatigue.

Recommended Actions:

1. Get the patient out of the sun and into a cooler location.
2. Lay the patient down and elevate his feet about 12 inches. Loosen any tight clothing.
3. Have him drink cool liquids that do not contain caffeine or alcohol.
4. Cool the patient by gently spraying or sponging him with water.
5. Monitor his temperature and condition for signs of heatstroke.
6. If any of the following symptoms occur, call for emergency medical assistance:
   - Fever greater than 102°F
   - Fainting
   - Confusion
   - Seizures

Supplies needed:

- Cool liquids
- Thermometer

HEATSTROKE

Heatstroke is the third and most severe stage of hyperthermia (see *Chapter 8: Heating/Cooling*). It is often a result of continued overexertion in hot conditions combined with inadequate fluid intake. Young children, older adults, and obese people are at higher risk of suffering from heatstroke. Heatstroke can also result from the use of some drugs, such as Ecstasy. The main symptom of heatstroke is an elevated body temperature, generally higher than 104°F. This condition may result in personality changes, confusion, coma, and even death. Other symptoms can include rapid heartbeat, shallow breathing, cessation of sweating, dizziness, headache, nausea, and fainting.

Recommended Actions:

1. Move the patient out of the sun to a cooler location, and remove as much of his outer clothing as possible.
2. Call for emergency medical assistance.
3. Have him drink cool liquids that do not contain caffeine or alcohol.
4. Cool the patient by covering him with damp sheets, or gently spraying with cool water and then fanning him.
5. Once the patient's temperature returns to normal, replace the wet sheets with a dry one. If his temperature rises again, repeat the cooling process.
6. If the patient loses consciousness, continue cooling while monitoring his pulse and breathing.
7. If the patient stops breathing, administer CPR.

Supplies needed:

- Cool liquids
- Sheets

HYPOTHERMIA

Hypothermia is a condition that occurs when the body temperature falls below 95°F, caused by prolonged exposure to the cold (see *Chapter 8: Heating/Cooling*). Severe hypothermia, in which the core body temperature falls below 89°F, is usually fatal. Becoming wet or not covering your head in cold weather increases your chances of suffering hypothermia. Symptoms include shivering, slurred speech, slowed breathing, cold pale skin, loss of coordination, fatigue, and confusion. Symptoms usually develop slowly, with a gradual loss of mental acuity and physical ability. Older adults, infants, young children, and people who are very lean are at higher risk of hypothermia. Alcohol and drugs can exacerbate the condition. If a person is submerged in cold water, hypothermia can develop very quickly.

Recommended Actions:

1. Move the patient out of the cold, preferably into a warm, sheltered location. If you must lay him on the ground, place a pad or other insulating material beneath him.
2. Call for emergency medical assistance.
3. Remove his wet clothing and replace it with something warm and dry. Handle the patient very gently, since rubbing frostbitten skin can cause severe damage.
4. Warm the patient gradually. Do not apply direct heat, such as from hot water, a heating pad, or heat lamp. Instead, warm slowly with compresses to the trunk of the body. Do not attempt to directly warm the arms and legs because that can cause a fatal condition in which cold blood flows back to the heart, lungs, and brain.
5. Give the patient something warm to drink as well as a small quantity of high-energy food, such as chocolate, to eat.
6. While awaiting medical help, monitor the patient's breathing. If he stops breathing, administer CPR.

Supplies needed:

- Warm liquids
- High-energy food, such as chocolate
- Dry clothing

INSECT BITES AND STINGS

Most insect bites and stings cause only mild symptoms, such as itching, stinging, or swelling, and can be safely treated without medical assistance. Only a small percentage of people develop severe reactions to insect venom (see *Anaphylaxis*).

Recommended Actions:

1. Move the patient away from the area to avoid additional stings.
2. If visible, remove any stinger with tweezers, or scrape it out using a playing card or other flat object. Be careful not to squeeze the stinger, which might inject additional venom.
3. Wash the area with soap and water.
4. Apply a cold pack or cloth filled with ice to reduce swelling and pain. Raise the infected limb above heart level if possible.
5. Have the patient take an antihistamine containing diphenhydramine (e.g., Benadryl).
6. Apply hydrocortisone cream, calamine lotion, or baking soda paste (3 teaspoons baking soda to 1 teaspoon water) several times a day until the symptoms subside.
7. For severe reactions, call for emergency medical assistance and follow the steps under *Anaphylaxis*.
8. If the sting occurs on the throat, give the patient ice water to sip. If swelling starts to occur, seek emergency medical assistance.

Supplies needed:

- Tweezers or flat card
- Cold pack
- Hydrocortisone cream, calamine lotion, or baking soda
- Antihistamine with diphenhydramine

NOSEBLEEDS

Most nosebleeds are not serious and are easily treated without medical assistance. In older adults, however, some nosebleeds originate from deeper within the nose and may be a result of hardened arteries or high blood pressure. These deeper nosebleeds can be difficult to stop and may require medical assistance.

Recommended Actions:

1. Have the patient sit upright and lean forward to avoid swallowing blood.
2. If bleeding is the result of an injury that may have broken the nose, seek emergency medical assistance.

3. Have him pinch his nose for 5 to 10 minutes to stop the bleeding.

4. If re-bleeding occurs, have him blow out forcefully to clear his nose of blood clots. Then spray nose with a decongestant nasal spray containing oxymetazoline (e.g., Afrin). Then have him pinch his nose for another 5 to 10 minutes.

5. If bleeding persists for more than 20 minutes, seek emergency medical assistance.

Supplies needed:

- Decongestant nasal spray containing oxymetazoline

POISONING

Poison symbol

Poisons are toxic substances that once introduced into the body, can cause sickness and death. They can be ingested, inhaled, splashed into the eyes, or absorbed through the skin. The specific effects of each poison are unique, but common reactions are: vomiting and nausea (ingested poisons), breathing difficulties (inhaled poisons), blurred vision and watering eyes (splashed poisons), and skin irritation (absorbed poisons). Other effects, including seizures, dizziness, and pain are also possible.

Recommended Actions:

1. If you suspect poisoning of any kind, immediately phone the **Poison Control Center at 1-800-222-1222.** If the patient is already unconscious, call 911 for emergency medical assistance.

2. If a poison was ingested, remove anything still remaining in the patient's mouth. If possible, ask him what was swallowed. If he is unable to answer, look for clues, such as leaves, berries, pill bottles, or containers. Do not give the patient anything to induce vomiting unless directed to do so by the Poison Control Center.

3. If the patient was exposed to poisonous fumes, such as carbon monoxide, immediately move him to an area with fresh air.

4. If the poison spilled onto the patient's clothing or skin, or splashed into his eyes, remove the clothing and flush the skin and eyes with water.

5. If the patient stops breathing, administer CPR.

Supplies needed:

- None

PUNCTURE WOUNDS

A puncture wound, such as a nail through the foot, usually does not cause excessive bleeding but can be serious due to the depth of the wound and risk of infection. The puncturing object may have spores of tetanus or other bacteria, especially if exposed to the soil. Puncture wounds from mammals carry the additional risk of rabies (see *Animal Bites*).

Recommended Actions:

1. Apply gentle pressure to help stop the bleeding as needed. If bleeding persists, seek emergency medical assistance.
2. If the object remains embedded, seek emergency medical attention.
3. If the object is not embedded, flush the wound with clean water. Use tweezers disinfected in alcohol to remove any superficial particles. Clean around the wound with soap and water. If debris remains in the wound, seek emergency medical assistance.
4. Apply a thin layer of antibiotic ointment.
5. Cover the wound with a clean bandage.
6. Change the bandage at least daily or when it gets wet or dirty.
7. Watch for signs of infection including redness, drainage, or swelling. If infection occurs, seek medical care.
8. If the puncture is deep, or from an animal or human bite, or in the foot, seek emergency medical assistance. The doctor may recommend a tetanus booster or a series of rabies vaccinations.

Supplies needed:

- Tweezers
- Antibiotic ointment
- Isopropyl alcohol
- Bandages
- Soap and water

SEVERE EXTERNAL BLEEDING

Trauma can result in severe external bleeding, which can be distressing for both you and the patient. Keeping calm is critical to administer lifesaving first aid successfully. If the bleeding is not stopped quickly, the patient will lose consciousness and die.

Recommended Actions:

1. Call for emergency medical assistance.
2. If possible, wash your hands and put on rubber or latex gloves.
3. Have the injured person lie down, and cover him to prevent loss of body heat.
4. Treat for shock by elevating his legs about 12 inches.
5. Remove or cut away clothing to expose the wound.
6. Remove any obvious dirt or debris from the wound. Do not try to remove any deeply embedded objects.
7. Using a clean cloth or thick sterile bandage, such as a trauma pad or bloodstopper compress, apply direct pressure to the wound until bleeding stops (at least 20 minutes). If a bandage or cloth is not available, use your hands directly (preferably with gloves on). If there is an object embedded in the wound, press to either side of it.
8. If possible, elevate the wound above the patient's heart while applying pressure.

9. If the bleeding continues and seeps through the bandage, do not remove it. Instead, add absorbent material over it.

10. If the bleeding doesn't stop, use one hand to compress a main artery that feeds the limb by pressing it against the bone. Use the other hand to maintain pressure on the wound:
    - Artery pressure points on the arm are on the inside of the arm, one above the elbow and one below the armpit.
    - Artery pressure points on the leg are behind the knee and in the groin.

11. Once the bleeding has stopped, secure the bandage with tape or gauze, and immobilize the injured part until emergency personnel arrive.

Supplies needed:

- Thick, sterile bandages such as trauma pads, bloodstopper compress bandages, or multi-trauma dressings
- Conforming gauze rolls
- Rubber or latex gloves

SHOCK

Shock is a life-threatening condition most often caused by a significant loss of blood (i.e., more than two pints). It may also be a result of trauma, heatstroke, allergic reaction, infection, poisoning, electric shock, heart failure, hypothermia, hypoglycemia, or severe burns. When in shock, a person's vital organs, such as his brain and heart, aren't getting enough oxygen. This deprivation can lead to organ damage and death. Initial symptoms of shock may include cool, clammy skin; weak, rapid pulse; and sweating. As shock develops, additional symptoms may include nausea, thirst, confusion, weakness, gasping for air, and loss of consciousness. If not treated, the patient's heart will eventually stop.

Recommended Actions:

1. If you suspect that a person is going into shock, call for emergency medical assistance.
2. Have the patient lie on his back and remain still. Elevate his feet about 12 inches.
3. Reassure the patient. Keep the patient warm and comfortable by loosening clothing and covering with a blanket.
4. If the patient begins to vomit or bleed from the mouth, turn him on his side to prevent choking.
5. Monitor the patient's breathing and heart rate. If he stops breathing, administer CPR.

Supplies needed:

- Blanket

SNAKEBITES

Fortunately, most North American snakes are not poisonous. Obvious exceptions are the rattlesnake, coral snake, water moccasin, and copperhead, all of which can deliver fatal bites. Snakes do not carry rabies.

Recommended Actions:

1. Call for emergency medical assistance immediately.
2. Do not try to capture the snake, but do try to remember its color and shape so you can describe it to emergency personnel. Also, note the exact time of the bite.
3. Have the patient sit upright and remain calm. Reassure him as needed.
4. Keep the bite below the level of the patient's heart if possible.
5. Remove any jewelry on the limb, and cover the wound with a clean, dry bandage.
6. Immobilize the bitten limb using a splint if possible. If a splint is not available, wrap the limb with heavy gauze up past the joint to prevent bending.
7. Do not use a tourniquet, apply ice, cut the wound, or try to suck out the venom.

Supplies needed:

- Splint
- Bandage
- Gauze roll

SPINAL INJURY

Injuries to the spine may involve the vertebrae, the disks of tissue separating them, the surrounding muscles and ligaments, and the spinal cord and associated nerves branching off from it. The most serious risk exists when the spinal cord is damaged, which can lead to temporary or permanent paralysis. Spinal injuries are often the result of an impact, such as falling off a ladder or horse, landing awkwardly on a trampoline, diving into a shallow pool, or suffering a misplaced football tackle. Symptoms often include pain in the neck or back, irregularity in the normal curve of the spine, paralysis, abnormal sensations, and loss of bladder or bowel control.

Recommended Actions:

1. If spinal injury is suspected, call for emergency medical assistance.
2. Keep the patient still. Brace his head on both sides using rolled up clothes or towels. Do not roll him over or straighten him out; instead, leave him in the position that you found him.
3. If the patient is wearing a helmet, do not remove it.
4. Provide emergency first aid if needed, such as stopping any bleeding. But try to do so without moving the patient's head or neck.
5. Monitor the patient's vital signs. If he stops breathing, administer CPR—chest compressions only.

Supplies needed:

- Towels

STRAIN OR SPRAIN

A strain is an injury to the muscles or tendons, and is the result of the muscle being overstretched. The muscle can be partially torn or fully ruptured. The injury often occurs at the junction between the muscle and the tendon that joins it to the bone.

A sprain is an injury to the ligaments connecting bones together or holding joints in place. The ligament can be stretched, partially torn, or completely separated. Ankle and knee sprains are the two most common types. Sprains cause rapid swelling and can be very painful. Generally, the greater the pain, the more severe the sprain. Most minor sprains can be treated without medical assistance by keeping the swelling down and treating the pain.

Recommended Actions:

1. Help the patient sit or lie down. Support the injured limb in a comfortable raised position using pillows or other soft materials.
2. Apply a cold pack to the area immediately after injury.
3. Compress the area with an ACE bandage or neoprene wrap.
4. Elevate the injured limb above the patient's heart whenever possible to prevent swelling.
5. Use over-the-counter anti-inflammatories/pain relievers to help with swelling and pain.
6. The injured person may require the use of crutches or splints for two to three days. During this time, apply a cold pack for 10 to 15 minutes at a time, four times a day.
7. If the sprain isn't improving after three days, consult a doctor.
8. If a fever develops, or if the joint feels unstable, seek emergency medical assistance.

Supplies needed:

- Cold pack
- ACE bandage or neoprene wrap
- Over-the-counter anti-inflammatories, pain relievers
- Crutches

STROKE

Strokes are the third most common cause of death in the United States. The condition is more common in older people and is often associated with circulatory system disorders such as high blood pressure.

A stroke is caused from bleeding in the brain or when the blood flow into the brain is interrupted (usually by a clot in a blood vessel). Within minutes, brain cells start dying—a process that may continue for several hours. The best ways to prevent brain damage are to recognize the symptoms and quickly seek emergency medical treatment.

Symptoms of a stroke may include:

- Sudden weakness or numbness in the face, arm, or leg on one side of the body
- Sudden dimness, blurring, or loss of vision, particularly in one eye
- Trouble talking or understanding speech
- Sudden and severe headache
- Dizziness

Recommended Actions:

1. Ask the patient to smile while you are looking at his face. If he is having a stroke, one side of his mouth may droop.
2. Ask the patient to raise both his arms out to his sides. If he is having a stroke, he may be able to lift only one arm out straight.
3. Ask the person several questions and see if he responds intelligibly. If he is unable to understand you or reply coherently, he may be suffering from a stroke.
4. Based on your assessment, if you suspect a person is having a stroke, call for emergency medical assistance.

Supplies needed:

- None

SUNBURN

Overexposure to the sun can cause sunburn. Numerous medicines, including some antibiotics and chemotherapy drugs, can trigger hypersensitivity to the sun, making the skin burn very easily. Sunburn is prevented by staying in the shade or applying protective sunscreen. Symptoms of sunburn usually appear within a few hours and may include pain, redness, swelling, and blistering. If sunburn covers a large portion of the body, it can also cause headache, fever, and fatigue.

Recommended Actions:

1. Cover the burn with a cool wet cloth, or have the patient take a cool bath.
2. Apply aloe vera, calamine, or moisturizing lotion several times a day. Do not use petroleum jelly, butter, or other home remedies.
3. Leave any blisters intact to speed healing and avoid infection. Apply antibiotic ointment to any that burst.
4. Administer over-the-counter pain reliever for pain as needed.
5. If complications develop, such as rash, fever, or blistering, seek medical care.

Supplies needed:

- Aloe vera, calamine, or moisturizing lotion
- Over-the-counter pain reliever

TOOTH LOSS

If an adult tooth is knocked out, it may be possible to replant it in the socket, either at home or at the dentist's office. Proper handling of the tooth is critical to improving the chances of a successful replanting.

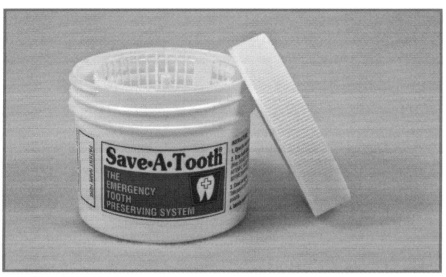

Save-A-Tooth preserving system

Recommended Actions:

1. Retrieve the tooth, handling it only by the top, not the roots.
2. Do not rub or scrape the tooth to remove debris since this might damage the roots.
3. Rinse away any dirt with milk (or lukewarm water if milk is unavailable).
4. Try to replant the tooth back into the socket. If it doesn't fully reseat, have the patient bite down slowly and gently on a gauze or moistened tea bag. Have him hold the tooth in place while traveling to the dentist or hospital.
5. If you can't replace the tooth in the socket, put it in a plastic bag or container along with some milk or the patient's saliva. If you have a Save-A-Tooth preserving kit, use it.
6. Regardless of whether replanting is successful or not, seek immediate care from a dentist or hospital.

Supplies needed:

- Save-A-Tooth preserving system
- Gauze
- Milk

TOOTHACHE

Toothaches are often the result of tooth decay. The first signs might be a painful sensation when eating something sweet, hot, or cold. When these early symptoms become evident, the patient should seek out dental care to address the underlying cause. If an injury to the mouth causes tooth pain, the patient should see his dentist immediately.

Recommended Actions:

1. Have the patient rinse his mouth with warm water, and then use dental floss to remove any food particles that might be wedged between teeth.
2. Administer over-the-counter pain reliever as needed. You can also apply an over-the-counter antiseptic containing benzocaine directly to the irritated tooth or nearby gum.
3. Call the dentist or seek emergency care if:
   * There are signs of infection such as swelling, pain, red gums, or foul discharge
   * The pain persists for more than a day
   * A fever is present
   * The patient has trouble breathing or swallowing
   * The toothache is a result of injury to the mouth

Supplies needed:

* Over-the-counter pain reliever
* Antiseptic containing benzocaine

FIRST AID KIT

A good first aid kit is one stocked with items that you know how to use. There is no point in having a huge stockpile of medical supplies if you can't put them to use safely and effectively.

The following first aid kit is compiled to treat the first aid conditions described in this chapter (as well as many others). For the most part, the supplies are readily available from your local drug store or through online medical supply stores. The quantities listed are completely subjective, based on family size, likelihood of injury, and types of expected injuries. You should stock your kit with enough supplies to meet your family's needs. It is also recommended that you build up a smaller first aid kit, primarily focused on trauma, for inclusion in your roadside emergency kit (see *Chapter 14: Transportation*).

To keep your medical supplies fresh, as well as maintain familiarity with your kit's contents, use your first aid kit for your family's daily medical needs. As long as you are vigilant at replacing supplies as they are consumed, daily use in no way compromises your preparation for a more significant first aid emergency.

First aid Kit		
Qty	Item	Use
1	Large first aid bag with individual compartments	Contain your first aid supplies
1	Bottle of alcohol or alcohol wipes	Disinfect tweezers, needles, or around wounds
1	Bottle of Betadine or hydrogen peroxide	Clean wounds when soap and water is unavailable
1	Bottle of hand sanitizer or sanitizer wipes	Sanitize hands when water is not available

1	Bottle of mineral or baby oil	Float insects out of ear
1	Bottle of saline solution or eye wash	Flush contaminant from eye
1	Bottle of decongestant spray	Clean blood clots from nose
1	Tube of antiseptic containing benzocaine	Apply for mouth pain
10	Individual doses of burn gel (e.g., Water Jel)	Treat burns, sunburn
1	Bottle of aloe vera lotion or gel	Treat sunburn
1	Bottle of calamine lotion	Treat poison ivy, sunburn
1	Tube of hydrocortisone cream	Treat insect bites or itchy rashes
1	Tube of antibiotic cream or ointment	Apply to wounds or broken blisters to prevent infection
2	Pairs of rubber or latex gloves	Protect against infection
1	Tweezers	Remove foreign objects
1	Needle in protective case	Remove splinters
1	Penlight	Examine eyes, ears, throat
1	Bandage scissors	Cut gauze, tape
1	Rescue shears	Cut away clothing
1	Magnifying glass	Examine wounds, foreign objects in eye and skin
6	Safety pins or bandage clips	Secure bandages
1	Digital thermometer	Measure temperature
1	Small plastic bag	Dispose of trash, bloody bandages
1	Plastic measuring spoon	Administer correct dosages of liquid medicines
1	Roll of medical tape, 1 in. × 10 yds.	Secure bandages and splints
1	Bulb syringe, 3 oz.	Remove congestion from nose; irrigate wounds
1	Small package of cotton swabs (Q-tips)	Clean around wounds; remove foreign object from eye
3	Instant, disposable cold packs	Reduce swelling; relieve pain
1	SAM splints, 1 finger, 1 large (36 in.)	Immobilize limb

1	Roll of duct tape	Immobilize limb
1	Rescue blanket	Treat for shock
1	Epinephrine auto-injector	Administer for anaphylactic shock
1	Save-A-Tooth storage system	Transport tooth to dentist or hospital
1	Pocket mask	Protect against infection when administering rescue breathing
1	Bottle of acetaminophen or ibuprofen tablets	Relieve pain in adults
1	Bottle of acetaminophen or ibuprofen liquid	Relive pain in children
1	Bottle of aspirin	Treat heart attack
1	Bottle of diphenhydramine antihistamine pills	Treat allergic reaction
1	Package of pink bismuth tablets (or bottle of liquid)	Treat upset stomach, diarrhea, and indigestion
50	Adhesive bandages, assorted sizes	Cover minor scrapes, cuts, and punctures
20	Gauze pads, assorted sizes	Cover wounds; clean around wounds; insert lost tooth
20	Non-stick gauze pads, assorted sizes	Cover burns, blisters, wounds
2	Conforming gauze rolls, 4 in. wide	Secure bandages; compress joints
2	Eye pads	Protect injured eye
10	Trauma pads, 5 in. × 9 in and 8 in. × 10 in.	Stop bleeding of deep wounds
1	Multi-trauma dressing, 10 in. × 30 in.	Protect and pad major wounds
2	Bloodstopper compress dressings	Stop bleeding of deep wounds
2	Water-Jel burn dressings, 4 in. × 4 in., 4 in. × 16 in.	Treat burns
20	Fingertip and knuckle bandages	Protect wounds on fingers and toes
1	Triangle bandage, 40 in.	Cover large wounds; secure limbs
25	Butterfly wound closure strips, assorted sizes	Hold wound edges together
1	Notepad and pen	Write down patient information, vital signs
1	First aid manual	Guide your actions

DP PLAN EXAMPLE

Table 11-1 Sample DP Plan Entry

Need: Medical / First Aid			
Danger	Goals	Needs	Implementation
Loss of access to medical care and medications	Establish an emergency supply of medicines and first aid supplies	30-day supply of medications	Explain preparedness rationale to family doctor and request the necessary prescriptions.
			Stock additional insulin and glucose for diabetic daughter.
		First aid kit capable of meeting emergency medical needs	Assemble a first aid kit capable of meeting the needs of a family of five.
			Assemble a smaller kit with trauma supplies; keep it in the car.

Quick Summary - Medical / First Aid

➢ Keep a minimum of a 30-day supply of medicine for everyone in your family. Give special consideration to any critical medicinal needs, such as insulin, nitroglycerin, or pain medicine.

➢ Everyone in your family should learn first aid. Critical life-saving first aid includes: administering CPR, stopping heavy bleeding, recognizing the signs of a stroke, treating for shock, responding correctly to poisoning, and delivering abdomen thrusts to a choking victim.

➢ When placing an emergency call, try to use a landline so the dispatcher can immediately determine your location.

➢ First aid care is not meant to replace professional emergency medical treatment.

➢ An effective first aid kit is one that contains medical supplies that you know how to use.

Recommended Items - Medical / First Aid

❑ Minimum of a 30-day supply of medicines

❑ A well-stocked first aid kit for your home

❑ A smaller first aid kit for your car, primarily focused on trauma injuries

❑ A comprehensive first aid manual

COMMUNICATION

Challenge

Your city is attacked by terrorists, causing widespread panic and mayhem. Local television stations are off the air. Where else can you obtain information? Do you have a backup method of requesting help if the telephone lines become overloaded?

When a disaster threatens, your ability to communicate can mean the difference between life and death. Without communication, you are cut off from information as well as unable to request emergency assistance—both extremely important to survival during a widespread crisis.

Communication can be divided into two broad functions: *incoming* and *outgoing*. Incoming communications consist of radio, television, or internet broadcasts for the purpose of providing warnings or situational updates. Critical information, such as evacuation orders, food and water restrictions, and early warnings of dangerous weather events, can be quickly relayed to large populations. Information may be your single greatest need during a disaster. Numerous questions require answers: How bad is the situation? What are the specific dangers? What should your family do to remain safe? Is evacuation necessary? If so, which are the best routes? How long will the threat last? What areas are affected? Are more threats expected?

Outgoing communications consist of radio transmissions, telephone, written, or even informal person-to-person messaging. Obvious uses are to request assistance from emergency services, provide updates to friends and family, and relay situational updates from the affected area.

Communication Needs

Incoming:
> ➤ Warn of approaching dangers
> ➤ Receive evacuation instructions and routes
> ➤ Receive timely situational updates

Outgoing:
> ➤ Request help from emergency services
> ➤ Contact family, friends, and DP network members
> ➤ Relay first-person updates out of the affected area

The benefits of a communication system are not only practical but psychological. Being cut off from the rest of the world can quickly cause people to grow fearful and perhaps make irrational conclusions about the state of things. These conclusions can lead to decisions that might ultimately prove dangerous or costly.

For many situations, a TV set, radio, computer, or a couple of well-placed telephone calls can answer your questions and help you to make informed decisions. During serious disasters, however, a more robust communication system may be required. In this chapter, different types of communication systems are discussed. Some allow you to receive information, while others enable you to reach out for help.

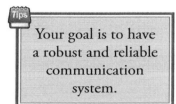

Your goal is to have a robust and reliable communication system.

Your goal should be to have a communication system that is robust and reliable—capable of working under nearly any set of circumstances. Unfortunately, there is no way to know which communication methods will be dependable because reliability tends to be a strong function of the surrounding conditions. For this reason, redundancy is especially important. Design your communication system with a fallback plan in mind, and perhaps another fallback to that fallback. If you prepare in this way, it is unlikely that *all* of your methods will become inoperable.

COMMERCIAL BROADCASTS

In this modern age of technology, information comes from a variety of broadcasters, including those who operate using radio, television, newspapers, cable, satellite, and the internet. The internet in particular serves as an immense and diverse information system capable of providing news in many forms, including text, audio, photographs, and video.

Together, these commercial services represent the primary methods of keeping citizens informed in times of emergency. However, depending on the type and severity of the disaster, some or all of your commercial services may become inoperable. This leaves you to gather your own information through less conventional services, such as shortwave radio or emergency service broadcasts.

EMERGENCY SERVICES

The United States government provides three important emergency notification services: the National Weather Service, the Emergency Alert System, and the National Terrorism Advisory System (replacing the Homeland Security Advisory System). The government also maintains a host of internet websites with valuable disaster preparedness information (see *Internet Resources*).

NATIONAL WEATHER SERVICE

Americans live in the most severe-weather prone country on earth.

Americans live in the most severe-weather prone country on earth with an average of 10,000 thunderstorms, 5,000 floods, 1,000 tornadoes, and 2 landfalling hurricanes each year.[205] To help combat these numerous threats, the National Oceanic Atmospheric Administration's National Weather Service (NWS) provides weather, flood, and climate forecasts and warnings.

The NWS is organized as a collection of national and regional weather centers as well as 122 Weather Forecast Offices (WFOs). The WFOs report local weather forecasts and conditions, such as temperature, humidity, and chance of precipitation, which are then combined in the National Digital Forecast Database. The NWS website, *www.weather.gov,* contains a great deal of useful information, including weather forecasts and warnings, weather radar, satellite imagery, water levels, and air quality maps. The NWS also provides information on sky conditions (for aviators), water warnings and advisories (for boaters), and fire conditions (for those concerned about wildfires).[206]

The NWS solicits help from over 290,000 SKYWARN® severe weather spotters, whose primary responsibility is to report local storm conditions. These volunteer weather reporters include police and fire personnel, EMS workers, public utility workers, and private HAM radio operators. If you are interested in becoming a SKYWARN® weather spotter, contact your local Warning Coordination Meteorologist.[207]

Wildfire *(FEMA photo/Liz Roll)*

EMERGENCY ALERT SYSTEM

The Emergency Alert System (EAS) is a nationwide public warning system (audio only) that replaced the Emergency Broadcast System (EBS) in 1994. Radio and TV broadcasters, cable TV systems, and satellite providers are required to make their communications systems available so that the president can address the American public during times of a national emergency. The system is also used by local and state authorities to deliver emergency information, such as weather warnings and missing children (AMBER) alerts.[208]

EAS messages are composed of four parts: a digitally-encoded SAME (Specific Area Message Encoding) header, an attention signal, an audio announcement, and a digitally-encoded end-of-message marker. The SAME header indicates who originated the message (e.g., the president, state or local authorities, NWS, or the broadcaster), a description of the event, the counties or states affected, the expected duration of the event, and the date and time it was issued. This information is translated by special receivers that decode the information (see *Weather Radios*).

Every broadcaster is required to test their EAS equipment on a weekly basis. As a courtesy to their listeners and viewers, broadcasters will typically announce that a test of the EAS is underway.

It is worth noting that the federal EAS (or its predecessor, the EBS) has never been officially used. This is in part due to the nearly universal access to news media throughout the country, making the EAS of questionable value. There are plans to eventually replace the EAS system with the Integrated Public

Alert and Warning System (IPAWS), which will provide warnings through personal communication devices such as cell phones and computers.

HOMELAND SECURITY ADVISORY SYSTEM

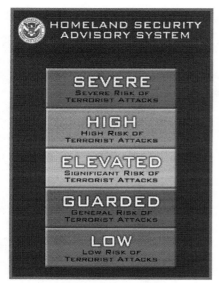

DHS's Color-coded Threat Level System
(no longer in use)

The Department of Homeland Security (DHS) was formed following the terrorist attacks of 2001 and consists of twenty-two agencies, including the U.S. Customs Service, Coast Guard, Secret Service, and Immigration and Naturalization Service.[209] DHS's purpose is to protect the nation and its critical infrastructure from dangerous people and goods as well as to strengthen the country's preparedness and emergency response capabilities.[210] In 2002, DHS established the Homeland Security Advisory System (HSAS) that consisted of Homeland Security Threat Advisories, Homeland Security Information Bulletins, and the Color-coded Threat Level System.[211] As you may recall, the color-coded system indicated graduated threat levels by progressive colors (e.g., red indicating imminent danger, green indicating very low risk). However, the lack of specificity and general uncertainty of how to respond to the threats led experts and citizens alike to question the system's effectiveness. For these reasons, the HSAS was discontinued in 2011.

NATIONAL TERRORISM ADVISORY SYSTEM

In April of 2011, the Homeland Security Advisory System was replaced with the National Terrorism Advisory System (NTAS). The goal of NTAS is to more effectively communicate information about terrorist threats by providing timely, detailed information to the public, government agencies, first responders, airports and transportation hubs, and the private sector.[253]

The Secretary of Homeland Security can use NTAS to issue two types of alerts:

> **NTAS Alerts**
> ➤ Imminent: Warns of a credible, specific, and impending terrorist threat
> ➤ Elevated: Warns of a credible terrorist threat

The alerts must include a concise summary of the potential threat, information about actions being taken to ensure public safety, and recommended steps that citizens, communities, businesses, and governments can take to help prevent, mitigate, or respond to the threat. Additionally, some alerts will be sent directly to law enforcement or affected areas of the private sector, while others may be more broadly announced to the American people through official and media channels as well as social media networks, such as Twitter (*www.twitter.com/NTASAlerts*) and Facebook (*http://facebook.com/NTASAlerts*). Alerts will also be posted at the official DHS NTAS website at *www.dhs.gov/alerts*.[254]

> **Tips**
> "If You See Something, Say Something™" was adopted to encourage the reporting of suspicious behavior.

Alerts will contain a "sunset provision" that indicates a specific date when the alert expires. There will no longer be open-ended alerts that indicate an ever present general threat since that type of alert has been routinely criticized as being ineffective and leading to confusion and complacency.

As part of a public awareness campaign, DHS adopted the slogan "If You See Something, Say Something™." The slogan was originally used by New York's Metropolitan Transit Authority and is designed to encourage citizens to be vigilant in watching for, and reporting, potential terrorist activity. Beyond encouraging vigilance, the slogan is worded to emphasize behavior, rather than appearance, in an effort to respect the civil liberties of all individuals.

The "If You See Something, Say Something™" campaign was launched in conjunction with the Nationwide Suspicious Activity Reporting Initiative (NSI). The NSI was designed to help America's law enforcement "connect the dots" of suspicious activity, thereby improving their effectiveness in combating crime and terrorism. The NSI essentially establishes a process for gathering, processing, analyzing, and sharing suspicious activity reports across all levels of law enforcement.

Both the "If You See Something, Say Something™" and the Nationwide Suspicious Activity Reporting Initiative were designed to underscore the importance of hometown security, where an alert public is a key element in keeping our nation safe. If you see something that you believe to be potential terrorist activity, you should immediately notify your local law enforcement officials.

INTERNET RESOURCES

The internet is the worldwide, publicly accessible network of interconnected computers that transmit data by packet switching using the standard Internet Protocol (IP). It consists of more than a billion computers on domestic, academic, business, and government networks, which together carry various information and services, such as electronic mail, online chat, file transfer, and the interlinked webpages and other documents of the World Wide Web.[214]

Individual connection to the internet is accomplished by connecting to dedicated servers using phone lines (dial-up or DSL), cable, fiber-optic, wireless, or satellite links. Vast amounts of information, only some of which is accurate, can be found on the internet on every conceivable topic. Keep in mind that anyone can post anything, so information from personal webpages or organizations with particular agendas should be viewed with skepticism.

Nationwide news outlets, such as Reuters (*www.reuters.com*) and the Associated Press (*www.ap.org*), provide real-time updates of world events and are likely to be valuable resources of information in times of crisis. Public bulletins, disaster preparedness information, and contact information can also be found on the internet, both on official government and non-government sites.

Below is an alphabetical listing of useful disaster-related public education websites.[212] Each has its own unique focus, such as infectious diseases (CDC), community readiness (Citizen Corps), and emergency preparedness (Ready.gov). The specific content of the websites changes frequently, and there is a great deal of overlap. If you want to change your perspective not only about disaster preparedness, but also about important issues facing the country, spend a few hours visiting these websites. Links to the websites are also available at *http://disasterpreparer.com*.

Table 12-1 Useful Disaster Preparedness Websites

Government Websites	
Agency for Toxic Substances and Disease Registry Provides information regarding toxic substances, including exposure registries, medical education, emergency response, risk assessments for contaminated sites, and comprehensive information about toxic substances of all types.	www.atsdr.cdc.gov
Air Now Displays air quality index ratings, ozone, and particulate levels for every state in the United States.	**www.airnow.gov**
Be Ready Campaign Provides information to help people prepare for and respond to emergencies. Offers advice on assembling a simple emergency supply kit and creating a family emergency plan. Furnishes links to state and local emergency management services.	www.ready.gov
Centers for Disease Control and Prevention Offers information on many health safety topics, including diseases, healthy living, injuries, travelers' health, and environmental health. Provides weekly updates of influenza outbreaks and publications that discuss infectious diseases.	**www.cdc.gov**
Citizen Corps A grassroots organization of over 2,000 local councils focused on citizen awareness and preparation. Offers online training and discusses Community Emergency Response Team (CERT) training. (The listing of Citizen's Corps Councils is a good way to find other *preppers* in your area.)	**www.citizencorps.gov**
Department of Health and Human Services Contains extensive information on preparing for natural and man-made disasters. Provides links to FEMA's daily National Situation Update, the National Weather Service, and NOAA weather radio broadcasts. Lists state emergency management and health agencies and Red Cross offices.	**www.hhs.gov/disasters**

Department of Homeland Security Provides a broad range of information relating to homeland security, including the national terror alerts, travel alerts and procedures, emergency preparation guidelines, immigration policies and border initiatives, school safety, and other security-related topics.	**www.dhs.gov**
Department of Interior Responsible for the conservation and management of federal land and administration programs relating to native Americans, Alaskans, and Hawaiians. Provides many useful links, including USGS, National Park Service, and U.S. Fish and Wildlife Service.	**www.doi.gov**
Department of Justice Provides information on federal law enforcement, including the country's Most Wanted and Missing Persons lists. Furnishes links to report a crime, report or identify a missing person, locate an inmate or sex offender, or provide a tip to the FBI.	**www.justice.gov**
Disability Preparedness Provides information targeted to the physically disabled. Has discussions regarding evacuation, preparedness, and emergency planning.	**www.disabilitypreparedness.gov**
Environmental Protection Agency Discusses environmental issues, including water, air quality, climate, waste/pollution, and ecosystems. Provides regional and national environmental news, and a listing of environmental laws and regulations. Contains links to report environmental violations or spills.	**www.epa.gov**
Federal Emergency Management Agency Provides extensive information about preparing for, and recovering from, numerous types of emergencies and disasters. Gives instructions on applying for disaster assistance; displays flood maps and declarations of disaster areas.	**www.fema.gov**
Food and Drug Administration Provides information concerning food safety, nutrition, bioterrorism, food bacteria outbreaks, animal-related illnesses, drug approvals and alerts, radiation-emitting products, and vaccines.	**www.fda.gov**
Food Safety Lists information pertaining to food handling, cooking temperatures, food illnesses, inspections, and product recalls.	**www.foodsafety.gov**
National Oceanic and Atmospheric Admin. Focuses on ocean and atmospheric conditions. Provides links to forecasts and weather alerts, atmospheric research, conservation activities, and environmental concerns.	**www.noaa.gov**

National Weather Service Gives up-to-date weather forecasts, warnings, radar, air quality, flooding, and weather safety information, including hazard assessments and weather radios.	**www.weather.gov**
Nuclear Regulatory Commission Oversees issues relating to nuclear energy production and safety, including reactors, materials, and waste. Provides recommendations for radiological emergency preparations and maps of active nuclear reactors and waste disposal sites.	**www.nrc.gov**
National Traffic and Road Closure Information Displays maps and associated links to traffic conditions and road closures across the United States.	**www.fhwa.dot.gov/trafficinfo**
U.S. Department of Agriculture Develops and executes policies on agriculture, farming, and food. Topics include pest control, weather, animal health, food aid, a plant database, food and nutrition, farm bills, and agricultural news.	**www.usda.gov**
U.S. Fire Administration Manages programs relating to fire fighting and fire safety; collects statistics on incidents and deaths. Also provides fire safety and prevention information to the public.	**www.usfa.dhs.gov**
U.S. Geological Survey Provides scientific information about the country's landscape, resources, and natural hazards, including earthquakes, floods, droughts, wildfires, climate change, volcanoes, and invasive species.	**www.usgs.gov**
U.S. Office of Personnel Management Maintains disaster preparedness information specifically targeted to federal employees.	**www.opm.gov/emergency**
The White House The official government website details policy, achievements, and challenges facing the United States. Also includes press briefings, presidential addresses, proclamations, and cabinet appointments.	**www.whitehouse.gov**
National Wildfire Programs Database A national database of state and local wildfire mitigation programs.	**www.wildfireprograms.com**
Pandemic Flu Furnishes extensive information about pandemic and flu outbreaks, vaccinations, and preparations around the world.	**www.pandemicflu.gov**

Non-government Websites	
American Red Cross Lists links to disaster preparation, including for those with special needs (children, elderly, and disabled). Also provides information on how to receive assistance after a disaster and take first aid or other training classes.	**www.redcross.org**
American Veterinary Medical Association Offers emergency preparedness information relating to pets and livestock.	**www.avma.org/disaster/saving_family.asp**
Equipped to Survive Furnishes information on first aid kits, position locator beacons, survival equipment, and other emergency gear.	**www.equipped.org**
Google Flu Trends Tracks influenza outbreaks around much of the world. Compares current year's flu activity to previous years'.	**www.google.org/flutrends/**
Institute for Business and Home Safety Offers practical information on preparing for earthquakes, floods, freezing weather, hail, high winds, hurricanes, tornados, and wildfires.	**www.disastersafety.org**
Kids Health Provides medical and first aid information relating specifically to children's needs. Contains useful instruction sheets for dealing with a host of medical and first aid issues.	**www.kidshealth.org/parent**
National Fire Protection Association Provides fact sheets and safety tips regarding prevention, fire proofing your home, and evacuation.	**www.nfpa.org**
Web MD An online resource that offers medical condition information, symptom-based diagnoses, and health-related news.	**www.webmd.com**

COMMUNICATION EQUIPMENT

The effectiveness of different communication methods will largely depend upon the type, duration, proximity, and extent of a particular disaster. Each method utilizes specific communications hardware, from commonplace TVs and AM/FM radios, to more specialized shortwave radios and scanners. Below are brief discussions of many communication devices designed to receive and/or transmit vital information during an emergency.

AM/FM RADIO AND TELEVISION

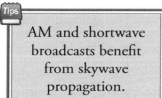

AM and shortwave broadcasts benefit from skywave propagation.

Tried and true, the AM/FM radio and television are very valuable tools to receive updates during a crisis. Not only will local stations broadcast the most relevant information, but the nation's Emergency Alert System also uses radio and TV broadcasts to inform the public of dangerous events.

One distinct advantage of AM radio over other broadcasts is that, in some weather conditions, it can transfer very long distances. This is known as skywave propagation, or more simply as "skip," and refers to the amplitude-modulated wave refracting back from the ionosphere and returning to the surface at a faraway place. The skipping phenomenon is much less likely to occur with TV or FM signals. AM skip can be especially beneficial during large scale disasters. One disadvantage of AM, however, is that its signals don't penetrate well. For this reason, reception in an underground shelter is very limited without external antennas.

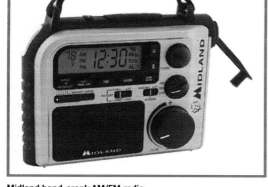

Midland hand-crank AM/FM radio

At a minimum, you should own at least one high-quality, portable, battery-powered AM/FM radio (e.g., Grundig, Sangean, Sony). Determine your radio's battery use, and stock enough spare batteries to run the radio for a minimum of 84 hours. If you listen to the radio for six hours per day, this represents two full weeks of listening. If you opt for rechargeable batteries, be sure you have the appropriate equipment to recharge them from your backup electrical source (see *Chapter 7: Electrical Power*).

Battery-powered radios with hand cranks or solar panels are another viable option. However, treat the hand crank or solar recharging as a backup capability only. Cranking the radio every few minutes can grow tedious, and the durability of the cranking system is often the radio's weakest link. Likewise, recharging by solar means can be slow and unpredictable (see *Chapter 7: Electrical Power*). Several companies, including Sangean, Freeplay, and Eton, offer quality hand-crank and solar units. Read numerous consumer reviews before purchasing a solar or hand-crank radio. Many units are not well built and won't stand up to continued use.

Portable, battery-operated televisions are also becoming fairly commonplace. If you plan to use a TV to pick up local stations without the use of a cable or satellite box, select a unit with a built-in-digital tuner that supports digital TV (DTV). Analog broadcasts are no longer available.

WEATHER RADIO

Weather radios are special-purpose radios designed to receive weather-related announcements, forecasts, and emergency broadcasts. For many years, NOAA's National Weather Service has broadcasted continuous weather information, including forecasts, watches, and warnings on seven frequencies from 162.4 MHz to

162.55 MHz. Recently however, they have also begun working in conjunction with the Federal Communications Commission's Emergency Alert System to broadcast warning and post-event information for natural disasters, such as earthquakes, volcanoes, and tornadoes, and technological emergencies (e.g., chemical releases, oils spills).[215] This "all hazards" capability makes weather radios critically important to disaster preparedness.

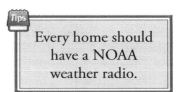

Every home should have a NOAA weather radio.

It is strongly recommended that you have a weather radio at home and another at work. These radios sound a loud tone as well as provide critical information when dangerous weather or other events threaten your area. In case the threat occurs overnight, place the radio somewhere in your home where it will awaken you. Early warnings of this sort can mean the difference between life and death. Modern radios are also equipped with the Specific Alert Message Encoding (SAME) feature. SAME allows you to designate the particular area for which you want to receive alerts, specifying your local county, for example.[216] This reduces the number of alarms you receive to only those of greatest interest.

Weather radio *(courtesy of Midland)*

Additional information regarding weather radios and the SAME messaging feature can be found at NOAA's NWS website: *www.weather.gov/nwr.*[217]

SCANNERS

Scanners are radio receivers that allow you to listen to emergency service broadcasts, including police, fire, and rescue, as well as broadcasts by airports, trains, utility companies, the National Weather Service, and amateur radio operators. A scanner sweeps through the frequency ranges that you specify until it finds a broadcasted signal. Once that signal subsides, the scanner continues the cyclic scan, repeating continuously.

Previous generation scanners were bulky, expensive, and had very limited channel capacity. Modern units, however, can be handheld, mobile, or stationary, and are available in nearly every price range. Much like computers, scanners are now equipped with microprocessors and solid-state memory that enables them to store thousands of channels, monitoring hundreds of them in a single second.

Before purchasing a scanner, it is vital that you match your requirements to the receiver's capabilities. Otherwise, you may be sorely disappointed when you are unable to receive the signals in your area. All scanners enable you to easily dial in and listen to single discrete frequencies—like a conventional radio station. However, many agencies have replaced analog transmissions with APCO-25 digital transmissions. This means that broadcasts are transmitted using binary modulation (1's and 0's), which cannot be decoded by basic analog receivers. Note that digital receivers can receive both analog and digital transmissions.

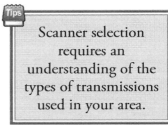

Scanner selection requires an understanding of the types of transmissions used in your area.

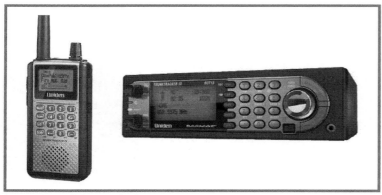

Handheld and automotive scanners *(courtesy of Uniden)*

Trunked systems, such as EDACS and LTR, are also becoming very common. Trunking refers to the sharing of a small set of frequencies by a number of broadcasters. Each time a user attempts to broadcast, he is assigned one of the frequencies from the set, and the receivers on his user group are automatically notified to tune into that new frequency. This frequency hopping is very efficient in using a small number of frequencies, but it also makes it impossible for someone to listen to the broadcasts without a receiver that also automatically hops between frequencies.

It is important to understand whether your local area emergency services use basic analog, trunked analog, digital, or trunked digital communications, so you can select a scanner that supports those transmissions. Most cities appear to be migrating to trunked digital technology, so it is a safer bet, albeit more expensive, to purchase a unit capable of this type of reception.

If you are interested in learning more about scanners, start by visiting *www.scannermaster.com* and *www.radioreference.com*. Between the two websites is a wealth of information on scanners and the transmissions in your local area. You may also post questions on their forums to solicit advice from scanner experts. If you enjoy listening to emergency broadcasts and have the time to become familiar with this hobby, scanners can be a very valuable emergency preparedness tool.

TELEPHONES

Telephones are broadly classified as conventional, internet, cell, or satellite. Conventional phones rely on hardwired telephone cables and remain the primary means of telephone communication due to their many benefits, including moderate cost, high reliability, and good voice quality. Internet calling uses conventional phones or computers to transmit voice data through internet connections. There are many voice-over-internet products, including magicJack, Skype, and Vonage. The specific features and equipment for each differ, but the products all offer low-cost, internet-enabled, point-to-point communications.

Tips

When voice services fail, it may be possible to send and receive information using text messages.

For many households, cellular phones are replacing conventional phones. This migration is in part due to the "always connected" society in which people now live. It is estimated that over 80% of the U.S. population has a cell phone.[218] The distinct advantage of cell phones is their portability. Disadvantages include cost and poor voice quality. Cell phones can also be used to text messages, which can be especially important because text messages can often get through when cell service is congested and otherwise unavailable.

Satellite phones use orbiting satellites to relay information, and thus offer communication to and from remote areas where other services are unavailable. Satellite phones are not part of the mainstream commercial market due to prohibitively high costs.

Depending on the type and severity of a disaster, some telephone services may no longer operate or may be overwhelmed by users. Hardwired landline phones are preferable to cordless home phones because they will operate even when your power is out—assuming electricity is available to the phone company. For this reason, it is recommended that you have at least one hardwired landline phone in your home.

Compile a list of emergency phone numbers, and keep a copy near the phone as well as in your safe room. An emergency contact worksheet is included in the *Appendix* for this purpose. Below are some phone numbers that might be useful:

Service or Agency	Phone Number
Emergency Services	911
Nationwide Traveler Information (available in some states)	511
Poison Control Center	1-800-222-1222
Federal Emergency Management Administration (FEMA)	1-800-621-FEMA
Centers for Disease Control and Prevention (CDC)	1-800-311-3435
Federal Bureau of Investigation	1-800-CALL-FBI

EMAIL

Due to its speed and convenience, email has largely replaced written correspondence, both for work and personal communications. Email can be a valuable method of communicating in times of crisis because it is less likely to be overwhelmed than telephone services and can be used to send out messages to multiple recipients simultaneously. However, email depends on the availability of internet or cellular phone service as well as electrical power, making the loss of either a single point of failure.

If you are part of a disaster preparedness network, it is a good idea to create an email group to which you can quickly broadcast information in times of crisis. This allows your group to keep everyone up to date without the worry of accidentally forgetting someone in the network.

SHORTWAVE RADIOS

Shortwave radios operate in the frequency range from 3 to 30 MHz. There are several different types of transmissions, including amplitude modulation (for general broadcasting), continuous wave (for Morse code transmission), single sideband (for long distance communication), and narrow band FM (for military communications).

In the United States, you do not need a license to own or operate a shortwave receiver. However, the privilege of operating a shortwave radio transmitter for non-commercial two-way communications (known as amateur radio) is granted through a licensing process by the Federal Communications Commission (FCC). Amateur radio operators often make themselves available to transmit emergency communications when normal communications channels have failed. In the case of a disaster, it is a good bet that shortwave operators will find a way to relay important information across the country. As with AM, shortwave radio signals also benefit from skywave propagation, meaning they can often transmit vast distances—even around the globe under certain favorable weather conditions.

There are many types of shortwave broadcasts, but most fall into the following categories:

- Amateur radio ("HAM") operators, many of who help local communities during times of crisis
- Domestic broadcasting from countries with widely dispersed populations
- World band radio broadcasting internationally to foreign audiences
- Political, religious, and conspiracy theory radio networks
- Numbers stations, believed to be operated by government agencies to communicate with clandestine operatives working in foreign countries
- Rebel and insurgent force communications

Of all the shortwave broadcasters, amateur radio operators should have a special place in the hearts of disaster preparers. HAM radio operators often serve as an information lifeline to affected communities when conventional broadcasts have gone offline. If you have an interest in becoming a HAM radio operator, consult the internet and you will find many user groups that can help you get started. Even if

Kaito shortwave radio receiver

you are not interested in becoming an operator, you should consider having a radio or scanner capable of receiving shortwave transmissions. This only allows you to function in a "listen only" mode, but at least you will become familiar with local area operators.

CB RADIOS, WALKIE-TALKIES

Walkie-talkies and CB radios are inexpensive, short-range communication devices that can be useful when cell phone service is unavailable. Don't rely on the mileage claims put forth by retailers. Actual effective operating distances are a strong function of the environment, but in general, walkie-talkies and CB radios should always be considered line of sight devices.

Cobra CB radio

CB Radios—The Citizen Band is a 40-channel radio system operating in the 27 MHz band. CB radio technology is a bit dated, going back to its inception in 1945. Since the transmission does not propagate well through objects, CB radios are used predominantly outdoors. In the United States, these radios are used extensively by truckers communicating traffic information along the major roadways.

Walkie-Talkies—Modern walkie-talkies operate using the Family Radio Service (FRS) and the General Mobile Radio Service (GMRS):

- **FRS**—The Family Radio Service is a frequency-modulated (FM), 14-channel walkie-talkie system authorized for use in the United States since 1996. It operates in the UHF band around 462 and 467 MHz and thus does not suffer from the interference effects found with CB radios. FRS radios are limited by FCC regulations to a maximum of 0.5 watts. Channels 1 to 7 of FRS radios are shared with the GMRS. A license is required for those channels only if the power output exceeds the FRS 0.5 watt limit.
- **GMRS**—The General Mobile Radio Service is a 15-channel, FM, UHF radio service available for short-range two-way communications. GMRS operates in the same frequency band as FRS, but with a higher maximum allowed transmit power. By law, GMRS use requires an FCC license (costing about $85), although the vast majority of users are unlicensed. GMRS permits transmission on eight dedicated channels at up to 50 watts, but typical GMRS walkie-talkies transmit at only a few watts. GMRS licensees are also permitted to use the first seven FRS channels (the "interstitial" frequencies) but only at a 0.5-watt maximum output power level.

Uniden walkie-talkies

Recently, hybrid FRS/GMRS consumer radios were introduced that include all 22 channels (the 14 FRS and 8 dedicated GMRS). On this type of radio, only channels 8 through 14 are strictly license-free FRS channels. Transmitting on channels above 14 requires a license, although many operators are either unaware of the requirement or simply unwilling to register. Illegal GMRS radio operators are referred to as "bubble-pack pirates," referring to the fact that these radios are often sold in bubble packs. GMRS radios can be very useful short-range communication devices because they transmit at a higher output power than FRS systems.

ONSTAR

OnStar Corporation, a subsidiary of General Motors, offers a subscription-based communications service for use in your vehicle. The benefits of OnStar are quite significant. Below is a brief list of some of the important features:[219]

- Automatic Crash Response—Sensors monitor your car to determine if you are involved in an accident and then automatically dispatches emergency services to your location.
- Remote Door Unlock—Operators can remotely unlock your car in case you have lost your keys or locked them in the vehicle.
- One-button Call for Help—Press a single button to call for emergency or roadside services.
- Vehicle Diagnostics—Sensors monitor the health of your vehicle, reporting any anomalies or maintenance needs.
- Optional add-ons include voice-activated calling, turn-by-turn navigation, and stolen vehicle assistance.

One significant advantage over a cell phone is that OnStar transmits at three watts of power, about five times more than a typical cell phone. This extra transmit power enables you to call for help in areas

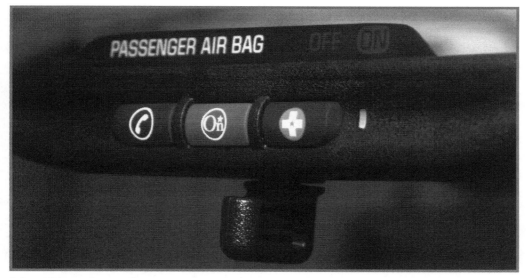

OnStar *(Wikimedia Commons/BlackHawkTraffic)*

where cell service is weak. OnStar also allows an operator to locate your car using embedded GPS. If you should become injured or ill, or break down in an unfamiliar area, the operator can guide emergency or repair services directly to you. Likewise, if you are reported missing, OnStar representatives can help rescue services to quickly locate you.

It should be obvious from the benefits discussed that having a vehicle equipped with OnStar is an excellent way to be prepared on the road. The most significant disadvantage of OnStar is that it is only equipped in newer vehicles from General Motors. You cannot currently add OnStar to a vehicle that didn't have it factory installed as original equipment. Some people are also concerned with privacy issues relating to the tracking ability of the service. Finally, there is a cost associated with the service (currently $199-$299 per year). However, all things considered, if you have a car currently equipped with OnStar, it is highly recommended that you keep the service active.

LAST DITCH CALL FOR HELP: THE WHISTLE

One final form of personal communication device that shouldn't be overlooked is the whistle. A loud whistle can be useful in calling for help, such as when lost outdoors or trapped in a collapsed building. Blowing a whistle takes much less energy than screaming, and the sound travels farther. Keeping a whistle in your safe room is a simple, perhaps life-saving precaution. Should your home ever collapse on top of you, a whistle can help you signal for help. Finally, everyone should always carry a whistle when hiking outdoors. Lost children and adults alike stand a much greater chance of surviving if they follow one simple rule: hug a tree and blow your whistle.

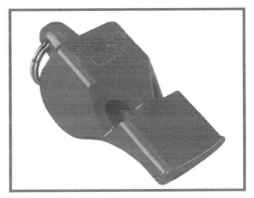

Fox 40 Classic Pealess Whistle—loud and reliable

CRISIS JOURNAL

It is often helpful to create a crisis journal during an extended disaster. The journal can have many uses, including recording events, tracking the passing of time, listing worries and their potential solutions, documenting injuries or illnesses, and keeping up with supply lists. Creating a crisis journal can help you gain a sense of organization and control that is often critical not only to making sound decisions, but also to keeping a positive attitude.

First person accounts also serve as historical records for generations to come. A journal can be kept with pen and paper (both very portable), or with a word processor if electricity is available.

DP PLAN EXAMPLE

Table 12-2 Sample DP Plan Entry

Need: Communication			
Danger	**Goals**	**Needs**	**Implementation**
Loss of telephone	Request emergency assistance, and inform family of situation	Backup telephone systems	Use cell phone service as primary backup.
			Use magicJack plugged into computer to serve as a backup VOIP phone system.
		Electronic mail and texting	Set up DP network email group to quickly broadcast updates and warnings.
			Use cell phone text message capability to relay information when phones are overwhelmed.
	Communicate across short ranges with DP members	A two-way radio system	Use a GMRS walkie-talkie system for neighborhood and car-to-car communication.
Loss of radio and TV broadcasts	Receive news and emergency broadcasts	Online news	Use existing internet connection to monitor online news channels.
		Emergency weather and other hazard news	Put NOAA weather radios in bedroom and on desk at work.
		Terrorist alerts	Monitor www.dhs.gov/alerts from computer and cell phone.

Quick Summary - Communication

➢ Timely information is critical to making the correct decisions during a disaster. There are many possible sources of information, including radio and TV broadcasts, internet websites, government emergency service announcements, and first person accounts.

➢ Commercial services, namely AM/FM radio and TV, will continue to be the primary methods used to distribute information from government agencies and media. More modern systems are underway that include using cell phones, text messages, and the internet.

➢ Governmental emergency services include the National Weather Service, Emergency Alert System, and the new National Terrorism Advisory System.

➢ The internet offers access to numerous resources, including real-time news, evacuation routes, traffic conditions, medical advice, weather forecasts, terror alerts, disaster preparedness advice, and crisis assistance.

➢ Every home should have a weather radio. Modern weather radio broadcasts provide early warnings not only for dangerous weather events but also for a variety of natural and man-made disasters.

➢ Specialty radios, including shortwave and scanners, allow you to listen in on amateur broadcasts and emergency services. Before purchasing a scanner, be sure to understand the radio systems that your local emergency services use.

➢ Every home should be equipped with at least one hardwired landline phone that can operate without electrical service.

➢ Walkie-talkie and CB radios should be considered short-range, line of sight devices. GMRS radios have more transmit power than FRS radios but require a license to operate.

➢ A crisis journal can help you make better decisions and track supplies. It can also serve as a historical record for generations to come.

Recommended Items - Communication

❑ Receive information and situational updates
   a. AM/FM or shortwave radio (hand-cranked or battery operated)
   b. Television capable of receiving local and national news
   c. Computer with internet access
   d. NOAA weather radio
   e. Optional: Scanner

❑ Request assistance from home or while on the road
   a. Hardwired landline home phone (not cordless)
   b. List of emergency phone numbers
   c. Cell phone, preferably with text messaging
   d. Whistle
   e. Optional: OnStar
   f. Optional: GMRS walkie-talkies

❑ Keep a record of events
   a. Notebooks and pens

CHAPTER 13
FINANCIAL PREPAREDNESS

Challenge

The nation's financial situation suddenly worsens. A run on the banks forces them to close their doors, limiting access to money. Businesses begin laying off workers in huge numbers. The stock market takes a nose dive, devastating retirement savings, pension plans, and personal nest eggs. How will your family survive this serious national crisis?

You may question the relevance of money management in disaster preparedness. Certainly it is not a topic covered in most handbooks. Its importance, however, is easy enough to see. Consider that your finances play three distinct roles in disaster preparedness:

1. Extra money enables you to purchase emergency supplies.
2. An emergency fund helps you to handle unexpected financial pressures.
3. A financial safety net protects your assets during times of loss.

Each of these three roles is vitally important and will be discussed in detail. Realize, however, that money is not the most important element of preparation. What you know is always more valuable than what you own. You should never feel that you are not properly preparing because of budget constraints. Do what you can when you can. Being prepared is as much a frame of mind as it is a closet of supplies. There are many steps to becoming better prepared that cost nothing more than your time, including learning first aid, forming a community preparedness group, and ensuring that your home is protected from hazards.

For now though, let's turn our attention to money.

PRIORITIZING

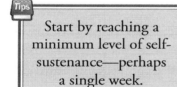
When it comes to preparing, patience and persistence win the day.

By this point in the book, you have probably come to realize that preparing can be rather expensive. Throughout the chapters, every effort has been made to limit the recommended supplies to include only the most important items. Even with this limitation, costs can be significant. With all the various recommendations, you may wonder where to begin. Everyone has a limited budget, so becoming fully prepared will take time—perhaps months or even years. Patience and persistence win the day.

It would be satisfying to possess a simple linear checklist to run down to get prepared—buy item A, perform action B, learn skill C, etc. Unfortunately, such a general approach is impractical. Every family's priorities will differ because each is exposed to different types of threats. If you live on the East Coast and experience powerful nor'easters, you will want to prioritize shoring up your home to better weather those events. Likewise, if you live in the Northwest where heavy snowfall is common, a fully stocked pantry and secondary heat source are likely high priorities. The point is that every family's situation is unique.

The best way to start is to identify and prioritize *your* family's needs. This process begins with completing a DP Plan Worksheet (see *Appendix*) for each of the fourteen needs (e.g., food, water, shelter, etc.). To make it easier, sample DP plan entries have been furnished at the end of every chapter. The goal of using the worksheets is to map out the specific actions you will take as well as the supplies you will purchase to prepare for various types of threats.

Start by reaching a minimum level of self-sustenance—perhaps a single week.

Preparation is easier when accomplished in stages. For example, you might wish to first identify everything you need to reach an arbitrary minimum level of self-sustenance, that is, able to provide all of your family's needs for a brief time (perhaps for a single week). This might require stocking the cupboards with extra food, purchasing a water filter, shoring up your home, getting a few lanterns, and other basic preparations.

After reaching this minimum level, you can then expand your capabilities based on the likelihood of the need and the cost of the preparation. Increasing your food stockpile to a 30-day supply and your water to a 14-day supply are often reasonable starting points for the second phase of preparation. The order of your preparations is dependent on the particular threats your family faces, their likely impact, and the extent of your ability to prepare for them. For example, a generator is very expensive, but if you frequently find yourself without electrical power, it might be a high priority purchase for your family.

Once again, the idea is to methodically work through your DP Plan worksheets while considering which threats are most likely to affect your family.

ARE YOU FINANCIALLY PREPARED?

People come from all walks of life, each with his or her unique financial situation. Some people are unquestionably wealthy while others struggle to pay the rent. One commonality spanning all classes, however, is that we spend too much and save too little. Consider that Americans carry a revolving debt in excess of $900 billion (as of June of 2007). Add in loans, but still exclude mortgages, and the personal debt rises to a whopping $2.46 trillion.[220] These figures don't consider the trillions of dollars that the country's wasteful politicians have overspent.

One of the goals of this chapter is to provide you with common sense approaches to becoming better financially prepared. But the darker truth is that every citizen should be deeply concerned with the financial condition of our country. Both as individuals and as a country, we are overspending. The solution to the nation's coming insolvency problems will surely involve government and citizens alike, but you should prepare for this coming debacle by getting your own financial house in order now.

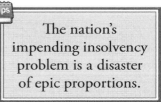

The nation's impending insolvency problem is a disaster of epic proportions.

Complete immunity to economic downturns would require that you essentially live outside the economy, providing your own food, water, shelter, electricity, medicine, etc. While some hardcore survivalists would argue that this is the only way to be truly prepared, such extremism is not practical for the vast majority of people. Not only is it impractical, but for many of us, it would prove to be extremely disruptive to enjoying a healthy, happy life in our modern society. With complete economic independence ruled out as impractical, the next logical step is one of self-reliance. This refers to establishing the necessary safeguards so that even under the worst economic conditions, you do not have to rely on the generosity of others to survive. For most people, this is the optimal level of financial preparedness.

Perhaps you haven't given much thought to your financial health. If that's true, use this foray into disaster preparedness to get financially fit. Some may shy away from the uncomfortable topic of money because it forces them to acknowledge some poor decisions they may have made in the past. Rest assured, from billionaires to paupers, *everyone* has made poor financial decisions at one time or another. Learn from your mistakes and move on.

A good place to start is to informally assess your own personal level of financial preparedness. Take a moment to answer the following questions as honestly as possible.

1. Are you living paycheck to paycheck?
2. Do you have an emergency fund set aside to handle a significant, unexpected expense?
3. Can you cope with the financial difficulties caused by an extended illness in your family?
4. If you or your spouse dies suddenly, will your family be able to maintain the same standard of living as it does today?
5. If your home or car is damaged or destroyed, can you afford to repair or replace it without excessive financial hardship?
6. Do you have a reasonable plan for retirement? (Hoping to win the lottery is *not* a plan!)

Your answers should give you insight into your current level of financial preparedness. Maybe your financial preparations are rock solid, in which case, you should stay the course. However, if your financial readiness is less than ideal, now is the time to turn things around. Sound money management will not only help you

Unexpected financial hardship

become better prepared, but also grant you a sense of control in your life. You are encouraged to read this chapter carefully, follow it up with your own research, and then put the principles into practice.

THE MONEY PLAN

To accomplish anything difficult generally requires a plan, and becoming financially prepared is no exception. Readying your finances for the challenges that hard times are almost certain to bring begins by adopting a sound money management plan (a.k.a. the *Money Plan*).

Designing an effective money management plan is painfully simple. Putting it into practice is the hard part! Getting financially healthy is analogous to getting physically fit. On paper, it all sounds great . . . eat more vegetables, exercise 30 minutes a day, get plenty of sleep, etc. But once you see that cheesecake in the fridge, the best laid plans go out the window! This chapter will do its part to lay out a reasonable plan. You must do your part to resist the cheesecake.

> ### The *Money Plan*
> 1. Stay employed.
> 2. Spend less than you make.
> 3. Save more.
> 4. Limit your debt.
> 5. Get the most bang for your buck.
> 6. Be adequately insured.
> 7. Don't get scammed.

The *Money Plan* is broken down into seven steps (see tip box). Each and every step is crucial to your success. Together, they act as the foundation to your financial house. If any part is missing, the house quickly collapses.

As you will see, achieving financial preparedness will likely require you to adopt new spending and saving habits. This may lead to some dramatic changes in your way of living. For that reason, it must be a family commitment. Involving your children is especially important because they are the next generation either doomed to repeat our mistakes or to learn from them.

STAY EMPLOYED

If previous recessions and depressions have taught us anything, it is that those who stay employed will weather the storm fairly well, while those who don't will suffer terribly. Maintaining a job during tough economic times is arguably *the* most important financial preparedness goal. Unemployment may skyrocket, and what were previously lucrative and stable jobs, might quickly disappear due to cost cutting. There is no surefire way to guarantee that you won't lose your job during a financial downturn, but a few actions can help:

1. Assess your job stability and change jobs if necessary. Try to position yourself in a career that is relatively insensitive to economic conditions. Some service sectors are highly dependent on the amount of disposable income people have. For example, salons, luxury spas, and construction can all suffer when money gets tight. Manufacturing and retail sales can also suffer, especially if the goods are not considered necessities. Jobs likely to be relatively unaffected might include health care, government civil service, teachers, and critical services (such as heating/air repair, plumbers, automotive repair, etc.).

2. Make yourself invaluable to your employer. Look for opportunities to learn unique skills or accept important responsibilities. Not only might this help your salary, but it also makes it less likely that you will be let go when times get tough.

3. Become a dream employee. Do good work, show up on time, and burn the midnight oil when necessary. Let your employer know that you appreciate your job and will do everything possible to help the company succeed.

4. If it looks like your company may have to conduct layoffs, consider offering to take a

Unemployment lines *(FEMA photo/Michael Raphael)*

pay cut until conditions improve. Let your employer know that you believe in the company and are willing to tighten your belt to stay employed.

5. Establish some backup sources of income. Most people are good at something. Consider finding your niche and establishing a small business. Business ideas might include becoming a yard care provider, scrap booking supplier, daycare provider, author, tutor, artist, party planner, freelance beautician, or photographer. An extensive list of home-based business ideas is given at *www.ahbbo. com/ideas*. If you're not the small business kind of person, look for a second job that could serve as a safety net.

SPEND LESS THAN YOU MAKE

The second step of the *Money Plan* is certain to be the most difficult. But without taking it, you simply cannot succeed. Simply put, you must learn to live below your means. Not *above* your means. Not even *at* your means. You must spend less than you make.

If we lived in a cash-driven society, it would be much easier to keep from overspending. With cash, you can only spend what you have in hand. With the advent of debt, however, many people have adopted lifestyles that are beyond their means. People observe friends and neighbors living to a certain standard and assume that they should live to that level too (a.k.a. *Keeping up with the Joneses'* syndrome).

One simple way to determine if you are living beyond your means is to answer this question:

Do you have short-term debt, such as credit cards or loans, that rolls over from month to month?

If the answer is yes, then you have borrowed money from your future to pay for your past. Said another way, you bought things you couldn't afford at the time. If it makes you feel any better, you are certainly not alone in this predicament. Debt is as seductive as the song of the Sirens.

Living below your means implies that you have extra money each month, thereby eliminating the need for short-term debt. By doing so, you will achieve a level of financial freedom. The word *freedom*

---

may seem a bit extreme, but taking on debt is akin to placing yourself in the servitude of a company, organization, or individual. Is that really where you want to be? Being in debt is neither a natural nor desirable state for anyone.

HOW TO LIVE BELOW YOUR MEANS

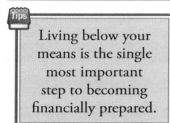

Living below your means is the single most important step to becoming financially prepared.

Learning to live below your means sounds like such a simple thing. All you have to do is spend less than you make. What could be easier? In reality, living below your means will likely be a tremendous challenge, one that will require significant changes to your current lifestyle.

The process is straightforward. Begin by determining your monthly gross and net incomes. Gross income is the amount you earn before taxes and withdrawals. Net income is your take-home pay—the number of dollars you actually put into your wallet after the government takes its share. If your pay is not the same each month, then use the last 12 months to establish averages.

Next, determine the amount that you would like to save. Most financial experts recommend that you save 10% of your gross income. As you become an experienced saver, you may wish to increase this amount. For now though, 10% is doing fantastic. Subtract that amount from your net income to get your available income—this is the money you have available to pay your bills, buy groceries, and keep gas in the car.

Once you determine your available income, calculate your monthly expenses. Determining your expenses is a bit trickier because some bills vary from month to month, and others may come due only once or twice a year. For those, determine the yearly totals and divide by twelve to determine average monthly expenses. With a bit of effort, you can get estimates that are close enough for budgeting purposes.

Finally, see how it all adds up. Any extra cash remaining each month is your available income minus your expenses. If this number is positive, congratulations are in order. You can live below your means without any appreciable changes to your income or spending. Unfortunately, for most people, the difference is more likely a negative number. This means that your monthly expenses are in excess of your available income. Your challenge is to increase your available income, decrease your monthly expenditures, or both.

Recognize that doing either one is difficult. Raising your income may require working extra hours, taking on a second job, or perhaps even making a job change. Lowering your expenses may require sacrifices that you simply don't want to make. Remind yourself that your financial readiness depends on working this out.

Numerous suggestions for increasing your income and reducing your expenditures are given in this chapter. They represent a smorgasbord of ideas, not all of which will apply to you. But don't dismiss them out of hand without consideration. Remember, the first and most important step to becoming financially prepared is to spend less than you make. Consider all the options, decide how you will make this happen, and then do it!

Increasing Money Flowing In

Below are suggestions to increase or supplement your income. As you take steps to raise your income, never lose sight of what you are trading for the additional money, whether it is simple "down time" or something more important like quality time with those you love. Make the appropriate choices.

- **Upgrade your skills.** Get the training or education you need to advance to a more highly paid position. Remember, if you have skill, energy, and a decent attitude, you are an employer's dream.
- **Become more valuable to your employer.** Put in the extra hours; volunteer for additional responsibilities; do exceptional work. Your employer *will* notice.
- **Take on a second job.** Burn the midnight oil and work a second job—at least until you can get some of your debt knocked out.
- **Maximize your investments for the best returns.** Put your money in the highest yielding accounts. Money under your mattress is losing its value in an inflationary economy.
- **Start up a small business.** Do something that you enjoy and are good at, such as a lawn cutting service, math tutoring, scrapbooking, etc.
- **Sell some stuff.** Look through your attic and closets and sell some of the things collecting dust. You can do this through consignment shops, online auctions, or garage sales.

When looking to increase your income, don't underestimate your abilities. The difference between a fast food restaurant manager and the CEO of a Fortune 500 company is not so great. Both handle daily business pressures, deal with difficult people, and are always on the lookout to raise revenue and cut costs. If you see jobs that are more rewarding than the one you currently have, take a practical approach to making the transition. Believe in yourself.

Decreasing Money Flowing Out

Below are suggestions to reduce your monthly expenses. Everyone has different priorities. You may feel that some of the suggestions are too great a sacrifice, but once again, give each item some honest consideration. The goal is to assess where your money is going each month, and then figure out ways to reduce or eliminate those expenses.[221]

- **Make a budget.** Manage your money using a formal budget. Start by planning how you will spend every dollar of your paycheck. Bills, savings, grocery money, and other necessities are budgeted first. Any additional money is then mapped out for leisure activities—movies, dinner out, amusement parks, etc. Numerous budget worksheets are available for free online.
- **Downsize.** It may be particularly painful to think about moving to a smaller house or replacing that fancy new car with a used one, but these steps can free up a significant amount of money.
- **Get rid of high-interest debt.** Shop around for favorable rates on credit cards, car loans, home mortgages, and home equity lines of credit.
- **Stop smoking.** Quitting will not only reduce your monthly expenses but also help you to obtain more affordable life insurance—not to mention live longer.
- **Go out for bids on insurance.** Shop around for the best rates on your home, car, and life insurance.
- **Cut back on services.** Try living without cable, internet, cell phone, caller ID, and other techno-luxuries.

Look for this book and others at the library

- **Brew your own coffee.** Make coffee at home instead of hitting the coffee shop.
- **Learn to love water.** Drink tap water when eating out, rather than sodas, wine, or beer. Also, use a home water purifier instead of bottled water—this saves money and the environment.
- **Avoid impulse buying.** Limit your purchases to real needs, rather than impulse buys. Always ask "Do I really need this?"
- **Carpool.** Establish or join a carpool to work.
- **Sell that second car.** Get by with only one car. It will reduce your debt as well as cut insurance and tax expenses.
- **Save energy.** Run the heater/AC less; weatherproof your house; turn down the water heater; cutoff unused lights; run appliances less frequently.
- **Drop memberships.** Discontinue memberships if you are not getting your money's worth (e.g., gym, golf club, hunting club).
- **Avoid using ATMs.** Eliminate ATM fees by keeping a little money in your sock drawer.
- **Limit trips to town.** Plan your trips to town to accomplish as much as possible.
- **Stay healthy.** Getting sick can cost you bundles. Take care of yourself, and it will fatten your wallet.
- **Buy store brands.** Try out those generic store brands. For many products, you will never notice the difference.
- **Thrift stores.** Check out the local thrift or consignment stores for great deals on secondhand products, particularly in well-to-do areas.
- **Generic or online drugs.** Generics can save 30% or more over brand-name drugs. Likewise, drugs can sometimes be ordered online for additional savings (e.g., *www.drugstore.com*).
- **Online shopping.** Compare local prices to those of online stores. Many online stores offer lower prices and have the advantage of not charging tax.
- **Avoid eating out.** Cook at home. Not only is it cheaper, but it is often healthier.
- **Buy clothes off season.** Never pay the full retail price for clothing. Even designer clothes can be purchased at huge savings when out of season.
- **Use coupons.** Use coupons for groceries, restaurants, services, etc. Check out newspaper inserts as well as online coupon websites, such as *www.couponchief.com, www.fatwallet.com, www.retailmenot.com,* and *www.ultimatecoupons.com*.
- **Loyalty cards.** Use grocery store loyalty cards to receive special savings on promotional items.
- **Watch the register.** Monitor the checkout register for pricing errors.
- **Join a warehouse club.** Join Costco, Sam's Club, or other large warehouse stores to receive significant savings on bulk purchases.
- **Keep your car as long as possible.** Resist the urge to replace your car when the new wears off. Try to get at least 10 years (or 150,000 miles) out of a car.
- **Buy used cars.** When you must replace your vehicle, buy one that is a couple of years old to save on depreciation.
- **Learn to love the library.** Quit buying what you can get for free. Discover the treasure trove of books, CD's, and movies that libraries offer.

- **Rent instead of going to the movies.** Stay at home to watch movies using low-cost services, such as Netflix, Blockbuster, or Redbox.

The single most important thing you can do to cut your expenses is to learn to curb your consumerism. Quit accumulating stuff! When is enough really enough? The answer is different for each person, but even recognizing that there is a logical limit to just how much you want to accumulate in life is eye opening.

SAVE MORE

The third step to becoming financially healthy, and thus better prepared for unexpected financial burdens, is to save more. If you are like most Americans, you don't save enough. How much is enough? A respectable goal is to save 10% of your gross income. That means if your gross family income is $80,000, you should be saving $8,000 a year.

There are several important benefits to saving:

- Savings allow you to deal with unexpected hardships using an emergency fund.
- As your money grows, it serves as a new income stream.
- Savings help you to have a more comfortable life after you retire.
- With savings can come generosity, whether it is in the form of donations to your local animal shelter, paying for your grandchildren's college costs, or helping your church to build a new gymnasium. Prosperity can also be passed down to your children and other loved ones.

There are many people who don't save a dime. Either they argue that they are living life to the fullest—mistakenly correlating spending to happiness, or more likely, they convince themselves that they simply can't afford to save.

> Your goal is to save 10% of your family's gross income.

Those in the first category will almost certainly come to regret their choices as they grow older and are forced to work to their dying day—assuming they are healthy enough to do so. As for the people who feel that they don't have enough income to save, it is generally more a matter of willingness to sacrifice. If your income was suddenly cut by 10%, would you still survive? Almost certainly you would. You might have to make some adjustments to how you live, but you would adapt and survive.

Also, recognize that savings is a sliding scale. If you make a lot, then you need to save a lot—enabling you to handle larger financial burdens and ultimately retire at the same standard that you have today. Likewise, if your income is small, then it is possible to save fewer dollars and still achieve the same results. Remember, 10% of your gross income is your savings goal.

A STRATEGY FOR SAVINGS

An important objective of saving is to become financially prepared—out of debt, socking away money each month, and prepared to deal with unexpected challenges. The obvious first step to achieving this goal is to establish an emergency fund.

However, there is another equally important goal of saving. Getting ready for the huge financial challenge nearly everyone faces. It's called retirement! There will very likely be a point in your life when you will be unable (or perhaps just unwilling) to work. If you have properly prepared, then retirement can be a wonderful and comfortable phase of life. However, if retirement is forced upon you or arrives before you have fully prepared, it can become a disaster like any other. A thorough savings plan should therefore meet both emergency *and* retirement needs.

Before anything else, you need to prepare for unexpected financial challenges. Establishing a strong emergency fund comes first. If you don't have a suitable emergency fund, you will be forced to resort to using credit cards with the first unexpected financial setback—potentially undoing all your hard work.

After establishing your emergency fund, take advantage of any employer matching of retirement funds. The return on investment is simply too good to pass up. Apply what's left of your monthly savings to paying off short-term debt, starting with the accounts that have the highest rates and fees.

Finally, once all your short-term debt is knocked out, put the full 10% savings into your retirement accounts. If you have more than 10% available, apply the additional money toward other savings goals. Don't make the common mistake of saving for your kid's college needs at the expense of your retirement. There are many types of assistance available to pay for college (e.g., grants, loans, scholarships), but there is no one to help you with retirement. This four-step savings strategy is summarized below:

1. First, establish your emergency fund.
2. Once the emergency fund is fully funded, begin contributing to retirement accounts that offer employer matching. Invest enough to take full advantage of any employer matching. With any remaining money, pay off short term debt, such as credit cards.
3. When short-term debts are paid off, put the full 10% into retirement accounts, even if it is more than what your employer matches.
4. If you have more than 10% available to save, either supplement your retirement accounts, or put the extra aside for other savings goals, such as the kids' college accounts, travel, a new car, or leisure items.

Let's apply the plan to an example situation.

Example: Assume that you have an income of $60,000 gross per year, $9,000 of credit card debt, and work for a company that offers 5% retirement matching. Through sacrifice and hard work, you have found that you can save 10% of your gross income, or $6,000 annually ($500 per month). You determine that your emergency fund should be $7,000. The savings plan would be applied as follows.

1. *Save the full $500/month for 14 months to fully fund your emergency fund.*
2. *After your emergency fund is established, begin contributing enough into your retirement account to take advantage of your employer's matching. In this case, your employer will match up to 5% of $60,000, which is $3,000 annually ($250/month). Therefore, invest $250 per month in your retirement account to receive the full matching. The other $250/month of the 10% savings goes toward paying down short-term debt.*
3. *Once the $9,000 in short-term debt is eliminated (taking about 36 months at $250/month), put the entire $500 each month into your retirement accounts.*
4. *If you find that you have additional money beyond the 10% to save, put it toward specific goals. These could include college savings accounts, annual family travel, new car planning, or simply things that you would really enjoy.*

This example shows that in a little over 4 years, you will have established a $7,000 emergency fund, paid off $9,000 of short-term debt, and begun establishing a retirement nest egg. Not a bad start to improving your financial preparedness!

EMERGENCY FUND

Establishing an emergency fund should be your first priority in savings. This fund will serve many important purposes, including handling the costs of unexpected car/house repairs, health care expenses, and weathering a job loss. It should be drawn upon only for emergencies, and only when no other options exist.

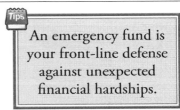

An emergency fund is your front-line defense against unexpected financial hardships.

Financial experts have differing recommendations about how big an emergency fund should be. A reasonable goal is to establish an emergency fund large enough to pay all your bills for at least three months. If your bills come to $3,000/month, then you should have a minimum of $9,000 in an emergency fund. Keep the bulk of your emergency fund liquid, not locked up in a CD or other account that isn't easily accessible. It is usually best to put the money in an FDIC-insured savings account or money market fund. The yields will be low from these accounts, but safety comes first when dealing with your emergency fund.

Check into using an online bank, since they frequently offer higher interest savings rates than your local bank or credit union. When you establish an online savings account, it links directly to your local bank account, making it very easy to transfer money between the accounts. Saving in an account like this also helps to keep the money out of sight, making it less likely to be used for things other than true emergencies.

The importance of an emergency fund cannot be overemphasized. It is the safety net that catches you when times get tough. If you draw money out of your emergency fund, immediately begin replenishing it with the money you set aside each month for savings. Maintaining the emergency fund is always your first priority, even if it means temporarily not putting savings into other accounts.

RETIREMENT

Planning for retirement is an exercise in wisdom. Unfortunately, people don't tend to gain this wisdom until later in life. Many people don't get serious about building their retirement savings until they see retirement on the horizon, perhaps when they are in their 40s or 50s. This is unfortunate because the most benefit can be had if you begin saving for retirement when in your 20s (or even earlier).

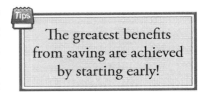

The greatest benefits from saving are achieved by starting early!

Compound Interest

Albert Einstein is attributed with saying that the most powerful force in the universe is compound interest. Regardless of whether that is true or simply urban legend, it does highlight a very important savings phenomenon. In layman's terms, compound interest simply means that interest can earn more interest.

The equation for calculating compounded interest is given below:

$$FV = PV\left(1 + \frac{i}{k}\right)^{kt}$$

Where FV = future value, PV = present value, i = interest rate (e.g., 10% = 0.1), k = number of times the interest is compounded per year, and t = number of years. If you are not particularly math savvy, the equation may appear confusing. Don't worry over it. The coming conclusion is what's important.

Let's look at two examples. First, assume that you have $1,000 in an investment account that earns 10% annually and is compounded monthly. Even if you never added another penny to the account, how much money would you have at the end of 10 years?

$$FV = \$1000\left(1 + \frac{0.10}{12}\right)^{12(10)} = \$2,707$$

At the end of 10 years, your $1,000 has grown to over $2,700. Almost a tripling of your money—not bad!

Now, what if you put the same $1,000 into an account when you were age 20 and let it sit until you were ready to retire 45 years later?

$$FV = \$1000\left(1 + \frac{0.10}{12}\right)^{12(45)} = \$88,354$$

When you were ready to retire, that single $1,000 investment would have grown to over $88,000!

These examples demonstrate that there are two significant impacts you can have on how much your money will grow: invest in an account with the largest rate of return, and invest as early as possible.

There is a very useful rule that can be used to estimate how long it takes money to double through interest accumulation. It is called the "Rule of 72." The rule states that if you divide the number 72 by the interest rate, it will indicate the number of years required to double an investment. For example, if you invest in an account with an interest rate of 6%, it takes about $72/6 = 12$ years for that money to double. Likewise, if the interest rate is 10%, the money will double about every $72/10 \approx 7$ years.

GOLD

One issue frequently discussed when considering financial preparedness is the buying of gold or other precious metals. The concern is that the value of assets tied to paper currency will drop if inflation occurs. Gold is seen as a "hard asset" offering protection from inflation and wild stock market swings. If you listen to the radio and television commercials, you might think that gold is a prepper's financial salvation.

> Tips
>
> Investing in gold has many drawbacks and is not the only way to hedge against inflation.

In reality, there are many drawbacks to buying physical gold. The worst of which is that you have to pay a premium both when you buy and sell the gold—typically losing 5 percent of the market price each way.[222] This means that gold has to rise 10% in value just for you to break even. To make matters worse, the IRS considers gold a collectible and taxes gains at much higher tax rates than stock gains. Finally, you have to store gold in special segregated vaults to guarantee its purity. If you keep it at home in a safe or in a bank's safety deposit box, you must have the gold re-assayed before selling it—yet another expense.

If you want to invest in gold while avoiding the many drawbacks, buy it through exchange-traded funds like SPDR Gold Trust (GLD) or iShares COMEX Gold Trust (IAU). The shares are backed by gold bars stored in certified vaults in New York and London. Admittedly, it is not

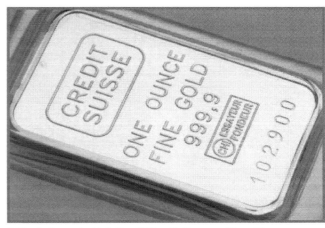

Gold, a hedge against inflation *(Wikimedia Commons/Olegvolk)*

quite the same as holding a gold Krugerrand in your hands, but it is a better investment. Either way, if you decide to own gold (or any other precious metal), most experts advise that you limit the investment to no more than 10% of your portfolio.

A better option all around might simply be to put your money in inflation-protected index funds (available through all major investment brokers). Gold is far less predictable. It has been a good investment over the past few years, but if you look at it over the past few decades, the yield has not been very impressive. Even with the recent spike in prices, gold has risen only about 400% over the last 50 years. That corresponds to an average compounded yearly rate of less than 3%.

LIMIT YOUR DEBT

The fourth step to becoming financially prepared is to limit your debt. Consider the definition for debt as provided by Merriam-Webster Online:[223]

> ***Debt*** *- sin, trespass; something owed, an obligation; a state of owing*

Wow. Does that cause you to pause for a second? Are these the words you want used to describe your financial condition?

Debt is so prevalent that it is taken for granted as a necessity in modern society. It enables you to purchase expensive items and amortize that cost over time. The problem arises when people use debt to fund everyday expenses. Worse yet is when debt is used to fund living expenses, essentially borrowing from the future to pay for the present. As the world's governments are now learning, this method of borrowed living is simply unsustainable.

Credit cards—friend or foe?

CREDIT CARDS

The personal debt problem is largely the result of the ease of credit. Each year, the credit card industry mails out approximately six billion credit card offers across the United States.[224] Why do these companies go to the expense of such widespread solicitation? The answer is obvious—there is a lot of money to be made!

There are countless complaints about the unethical practices of credit card companies, including tacking on exorbitant penalty fees, delaying payments to incur late fees, charging rates higher than originally promised, targeting college kids with the expectation that their parents will pick up the expenses, and raising interest rates to outrageous levels when a borrower falls behind. Credit card companies are the modern day equivalent of the loan shark business but on an unprecedented scale. Fortunately, the recently adopted Credit CARD Act legislation is taking steps toward protecting the consumer, but many questionable practices still remain.

From the tone of the condemnation, you may think that credit cards should always be avoided. This is far from the truth. When used correctly, credit cards provide very real benefits:

- They provide buyer protection by limiting your liability to fraud.
- Many offer extended warranties on card-purchased items.
- Many offer rewards, such as cash back or frequent flyer miles.
- They are easily replaced if lost or stolen.
- They are necessary for certain purchases and rentals.
- They help you build a higher credit score, which is needed for mortgages or other large purchases.
- They can help you categorize your purchases for purposes of budgeting.

Therefore, despite questionable company practices, you may still wish to enjoy the many benefits of credit cards. The key to credit of any type is using it correctly. How do you do that with credit cards? Follow three simple steps and you can't go wrong.

1. Find a credit card with the best benefits. Cards can be compared online at *www.creditcards.com* and *www.cardweb.com*.
2. Use the credit card to make your purchases.
3. Pay off the credit card balance each month.

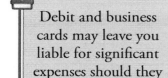

Debit and business cards may leave you liable for significant expenses should they ever be lost or stolen.

If you are unable to balance your checkbook well enough to guarantee your ability to pay off your credit card balance each month, then don't use credit cards. The bottom line is that credit cards can be a convenient tool or an irresistible temptation to overindulge.

Many people use debit cards—cards that charge directly against funds in your bank account. The advantage of a debit card is that you can't use it to put yourself into debt. The drawback is that debit cards don't provide the user the same protection that credit cards do. For example, when using VISA or MasterCard, you are limited to a $50 liability should the card be stolen and used fraudulently. The same is not generally true for debit cards. Likewise, business credit cards are often pushed on people (even those without a business). Why do credit card companies offer better deals on business credit cards? The answer once again comes back to consumer liability. If a business card is lost or stolen, the credit card company can claim that you were negligent with the handling of the card and hold you fully responsible for all fraudulent charges.

OTHER DEBT

There are of course many other types of debt besides credit cards. Car loans are one good example of this. The ideal way to purchase anything that depreciates in value is using cash. To borrow money to invest in something that depreciates is financially unwise. With that said, there is an argument for purchasing automobiles on credit. Cars are used to gain and keep employment, which in turn brings in income. Therefore, the case can be made for borrowing money to purchase a car *if* it is used to support employment. However, this does not suggest that you borrow money to purchase a new Lexus to get back and forth to your job. If the cost of the debt prevents you from meeting your savings goals, then you can't afford it. Also, you will save significantly on depreciation by buying a car that is at least two years old.

Home mortgages are a unique class of debt. Most people are unable to purchase a home outright, so either they must rent for the majority of their lives, or they must surrender to this form of debt and take out a mortgage. Of all the types of debt, home mortgages are the arguably the best. There are several reasons for this. First, homes have historically appreciated in value, making them a reasonable investment. Second, they replace the expense of renting. The interest on home mortgages is also tax deductible, effectively lowering the borrowing costs. Finally, a mortgage gives an immediate return on the investment since you are able to enjoy living in the house. However, the same rules of affordability still apply. You can only afford a house if the mortgage, utilities, taxes, maintenance, association fees, and insurance costs don't prevent you from meeting your savings goals.

GET THE MOST BANG FOR YOUR BUCK

If you give two people one hundred dollars and ask them to go out and purchase a set of clothes, they will come back with very different results. A business man might come home with a pair of trousers and a tie. A frugal grandmother on the other hand might come home with two dresses, a pairs of shoes, some pantyhose, a new purse, and a pocket full of change. The difference being that one knows how to bargain shop. This is a classic example of getting the most bang for your buck. Learn to buy things off season or when they are falling out of fashion. Even high-end clothing can be found at huge discounts if you are willing to do end-of-season shopping. Online stores are also an excellent place to do comparison shopping and find bargains (especially on electronics).

One word of caution . . . if something sounds too good to be true, then *it is* too good to be true. There are rarely any exceptions to this rule. Be very careful when parting with your money! Remember what Thomas Tusser said, "A fool and his money are soon parted." Scams target everyone, rich and poor alike. The most common elements of scams are discussed a little later in this chapter.

BE ADEQUATELY INSURED

A very important step to financial preparedness is being adequately insured. There are countless disasters that can cause property damage, injury, and loss of life. Having appropriate safety nets in place is vital to your family's financial security.

HOME INSURANCE

Your home is likely to be your largest financial investment, both the structure and the contents. If you are a renter, then the property contained in your apartment, townhouse, or rental home is probably a large part of your assets. Having adequate homeowner's or renter's insurance is therefore imperative.

There are many companies that offer homeowner's or renter's insurance (e.g., State Farm, Allstate, Prudential). All of these companies have good and bad reputations depending on who you ask. It probably doesn't really make much difference which of the major insurers you choose. What does matter is that you select adequate coverage and sufficiently document your belongings.

The definition of "adequate coverage" is enough insurance to fully rebuild your home and replace all of your belongings with new items should they be lost to a major disaster, such as a fire, flood, tornado, or earthquake. Be sure to clarify that you want replacement cost insurance, not coverage that is pro-rated based on the age of the item. You may also need special endorsements known as riders to cover specialty items such as jewelry, guns, electronics, collectibles, or antiques.

> **Tips**
> Document your belongings using a video camera.

Homeowner's insurance generally comprises two parts: property protection and liability protection. Property protection covers the dwelling, detached structures, personal contents, and costs associated with loss of use. Liability protection covers personal liability, such as damage to other people's property and medical expenses associated with an accident on your property. Do you know the details of your current policy? If not, take a few minutes to read through it. If your policy doesn't cover the worst case scenarios, make the appropriate changes.

Importance of home insurance *(FEMA photo/Dave Gatley)*

One major problem that often arises when a catastrophic event completely destroys a home is convincing the insurer of what was actually lost. Insurers are all too familiar with customers claiming that every room was filled floor to ceiling with Gucci handbags and Rolex watches. It is your responsibility to document what you actually possess and want insured. For this reason, you should create a home inventory video. Simply walk through your house with a camcorder (or camera), documenting every room and closet, as well as noting any high-value or irreplaceable items. Give a copy of the recording (or photos) to your insurer, and keep a copy for yourself somewhere other than inside your house. Repeat the process annually or when you have significant changes to your contents. Video documentation will help to protect you as well as make claims much easier to process.

AUTO INSURANCE

Auto insurance coverage is another financial necessity. Whether you drive a new BMW or a true junkyard classic, you need car insurance. At a minimum you will need liability insurance to pay for damages or injuries to others, as well as comply with the law. If your vehicle is of any significant value, you will also want collision insurance, which pays for your car to be repaired in case of an accident. The only recommendation here is to find the best price on the best policy. It may make sense to use the same provider as do for your home since companies typically offer multi-policy discounts. Also, find an insurer with a local office and get to know the agent and staff.

DISABILITY INSURANCE

Disability insurance is likely to be the most difficult insurance decision you will have to make. This type of insurance provides a monthly stipend in case of disability. Depending on the type and coverage, it has a specific payout (paying a fixed amount each month) and duration (lasting a given number of years). If you opt for this coverage, be sure to ask numerous

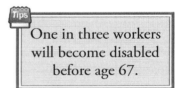

One in three workers will become disabled before age 67.

questions, such as how disability is determined, how insurance payments are affected by government disability payments, how long the payments will keep coming, and how long must you have to be out of work before the first payment is made.

It is estimated that one in three workers will become disabled for an extended period before age 67.[225] In a perfect world, every worker should therefore have disability insurance. In reality, few do. Disability insurance is relatively expensive and often does not provide a comparable replacement income. It can, however, soften the financial impact of a disability.

MEDICAL INSURANCE

Medical insurance is another of life's necessities. Whether you are age 6 or 60, health problems can arise forcing you to seek medical services. It is truly unfortunate that nearly 50 million Americans have no medical insurance.[226] The problem is that medical insurance is very costly, to the point of being impossible to afford for people with serious pre-existing conditions. Without insurance, medical costs can quickly drain away emergency funds and even entire life savings.

If you are fortunate enough to work for an employer who provides (or at least supplements) health care insurance, take full advantage of the benefit. If your employer offers several plans, compare the trades between cost, co-pays, deductibles, and available network of doctors. If you are self-employed or unemployed, try teaming with groups of other individuals to get more attractive rates and coverage. Alternatively, look for plans through memberships or discount clubs. You can also comparison shop at websites like *www.ehealthinsurance.com*. Finally, if you are unemployed and simply cannot afford health care, seek treatment through lower cost clinics or county-run health offices.

It is often possible to negotiate pricing with providers and hospitals. Specifically, if you have a large outstanding balance with a health care provider, you can often get that balance significantly reduced if you are willing to pay a negotiated amount in full.

Personal aside: Several years back, my mother had a large hospital bill (over $6,000). I called and explained that she was elderly, had little income, and would only be able to pay about $50 per month. As her son, I offered to pay a lump sum if they could lower the bill. Within 24 hours, they cut the bill to $1,200, which I promptly paid. The hospital was content with the partial payment because it was far more than they would have received otherwise. Likewise, I was happy with having paid off the debt at a fraction of the original cost.

LIFE INSURANCE

Life insurance is often misunderstood. The point to keep in mind is that life insurance is not designed for the person who has died; rather, it acts as a safety net for those left behind. Because of this, it is not a question of whether someone needs life insurance, but only *how much* they need.

How much to get?

When a person is young or without dependents, their life insurance needs are very modest, perhaps only providing for burial expenses. Once a person gets married and has children, life insurance must provide for the family's continued well-being in case of the loss of the breadwinner. This can be quite substantial. As the family matures and children move away from home, life insurance needs will decrease, ultimately settling at the modest levels required to settle an estate.

Example: Jack, age 35, is a corporate salesman. His wife, Jill, is a homemaker who homeschools their three kids, ages 5, 7, and 10. Jack earns $100,000 a year. They live in a home with a mortgage of $300,000. Family debt (e.g., cars, credit cards) totals $40,000. The family's monthly expenses are $4,000 (half of which is a mortgage payment). Jack has a 401(k) retirement fund with a $100,000 balance. How much life insurance does he need and of what type? What about when he turns 70? Will he still have the same needs?

If Jack died today, his family would be without his significant income. His wife is a homemaker who has chosen to homeschool their children, meaning that her ability to replace his income is very limited. The amount of life insurance needed to act as a proper safety net could be calculated by summing up the following needs:

1. Pay off home mortgage.
2. Pay off other debt.
3. Establish savings large enough to provide an income stream to meet the family's long-term needs.

The mortgage payoff is easy to determine, $300,000 in this case. The additional debt adds another $40,000. The third part of the calculation is the hard part. How much money is needed to take care of future needs? Once the house is paid off, the monthly expenses will come down to $2,000. Let's assume that the remaining insurance money is put into a diversified investment account (stocks, bonds, money markets) with an average yield of 10%. Backing out the amount needed to earn $2,000 of interest each month gives a net deposit of $240,000. The insurance breakdown would be:

Mortgage:	$300,000
Debt:	$ 40,000
Savings:	$240,000

The grand total comes to $600,000 in life insurance needs. This example neglects many things (e.g., Social Security survivor benefits, future education needs, inflation, investment variations, etc.), but it serves as a basis for the recommendation of many financial advisors. Many experts recommend that your life insurance be equal to five to ten times your annual gross salary. In this simple example, it was shown that Jack needed about six times his annual income.

A person's life insurance needs change significantly during the course of a lifetime. The five to ten times rule does not generally hold early or very late in life. For example, if Jack lives to be 70, his children will all be grown, his house will already be paid off, and he will likely have additional retirement assets for his surviving wife. Therefore, a large insurance policy is neither necessary nor cost effective at that age.

What about Jill's insurance needs? Since she is a homemaker, her insurance needs are significantly different than his. Even with Jill's death, Jack would continue to work, so his family's income would remain unchanged. Normally this would indicate that very little life insurance is needed for Jill. In this case, however, the family may decide that private schooling should replace homeschooling, and the costs of that schooling would need to be included in her insurance policy. This demonstrates how life insurance needs are unique to every individual. Once again the thing to keep in mind is that life insurance is about meeting the needs of those left behind.

Write a Will

Regardless of your net worth, you should have a Last Will and Testament. A will should name an executor who you trust to handle your affairs when you die. It should clearly spell out how your assets are to be distributed. If you have children, your will should name who you would like to take custody of them should both parents die together. A will serves all these vital purposes, but it does one thing more. It gives you a sense of peace. Death may come suddenly, or it may come slowly. Either way, knowing that you have prepared your estate and named caregivers for your children will give you peace of mind.

For most people, writing a will is inexpensive and easy to do. You can use online services, such as *www.legalzoom.com*, or software packages, such as Quicken's Willmaker. Another option is to hire a local attorney to draft up a will for a modest fee.

DON'T GET SCAMMED!

The final step in becoming financially prepared is learning to hang onto your money. Scamming is big business. Billions of dollars are stolen from unsuspecting people each year. Being scammed and being robbed are tantamount to the same thing since both leave you with less cash in your pocket and the same violated and angry feelings. Below are a few basic observations about scamming:

1. At the root of nearly every scam is an exchange of money (or other valuable asset)—something leaving your hands and entering theirs.
2. In this electronic age, the most trusted form of payment is cash. Cashier's checks, money orders, bank checks, wire transfers, and personal checks are all readily forged.
3. Everyone is a target for scamming, from the single mother working two jobs to pay her rent, to the millionaire living in the Manhattan townhouse. Some scams amass money by targeting lots of "little people;" others shoot for the big payoffs.

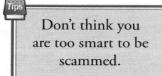
Don't think you are too smart to be scammed.

What can you do about it? First and foremost, don't fool yourself into thinking that you are too smart to be scammed. Everyone is vulnerable. With that said, there are many things you can do to minimize your vulnerability by making yourself a tougher target than the next person.

Start by familiarizing yourself with the most common scams (some of which are discussed below). Talk about them openly with your entire family, including your kids. A great resource for learning about scams is Clark Howard, heard on AM radio, seen on CNN, and found at *www.clarkhoward.com*.

Look at every financial transaction with skepticism. Ask yourself a few questions. What could go wrong? How will you handle problems? How can you protect yourself? Are you feeling rushed? Does something seem not quite right? Use a protected form of payment whenever possible (such as a credit card, PayPal, etc.). Most credit cards are protected against fraud if reported within 60 days. The maximum loss you will typically suffer is $50. For this reason, if you buy something and it hasn't arrived within 60 days, consider reporting it as a fraudulent transaction to your credit card company. You can always cancel that report should the item arrive later.

When selling something, don't release the item until you have a cleared payment from the purchaser. If payment is made through a personal check, have your banker contact the buyer's bank to verify funds. You should then wait until the check clears before releasing the item. Likewise, if payment is made with a bank-issued check, have your banker call the issuing bank to verify authenticity. For small transactions, you might consider using postal money orders with serial numbers that allow online verification. Don't accept wire transfers since they are often fraudulent.

Common Scams

A discussion of several common scams appear below:

- **Need Your Info:** Never give out personal or financial information to unknown callers or emailers. If someone calls or emails asking for your Social Security number or bank account information, it is a scam. There are many variations of this scam, including a court clerk claiming to be calling about jury duty, a bank representative claiming there is a problem with your account, or a prize company suggesting they want to deposit money into your bank account.

- **The Sure Thing:** Beware of any investment that claims to "beat the market." This is often a huckster trying to convince you to part with your life savings through a Ponzi scheme or by investing in very high risk activities. If anyone can reliably beat the stock market without added risk, they surely would not have to go looking for customers!

- **Money Up Front:** Don't pay for a service until the work is completed. If supplies are needed to perform the service, you may be required to pay enough to cover those costs. If the service is not done to your satisfaction, refuse final payment, and take up the issue with the parent company.

- **Online Hucksters:** Don't buy goods from questionable sellers. In the past, the shady salesperson might have been a street corner peddler with a coat full of cheap knockoff watches. That danger is largely replaced with online auction hucksters. If you decide to purchase something through an online auction, understand that it is inherently riskier than making retail purchases. Once again, use only protected forms of payment.

- **Return to Sender:** Never refund money back for accidental overpayment. If someone overpays you for an item and then asks for a partial refund of the overpayment, it is *always* a scam. Their original payment will ultimately prove to be worthless—often taking weeks to bounce.

- **Lottery:** You will never receive a legitimate limited-time-only, "act fast" lottery offer. Neither will you ever win the lottery through email. If you win any form of lottery, you will receive a certified letter as well as personal contact from the issuing company. You will never be required to pay any form of upfront payment, such as issuance, processing, registration, tax, or transfer fees. If it is truly your lucky day, there will be no doubt in your mind.

- **Nigerian Royalty:** If you receive a chain letter or email offering you a handsome commission for depositing a Nigerian prince's inheritance (or other such nonsense), it is a scam. If you reply, you will be led through a carefully crafted scheme that has you send money for one reason or another.

- **Hit Man:** If you receive a letter from a professional assassin claiming that he has a contract to kill you but will let you live if you pay him a fee, it is a scam. Report it to your local law enforcement agency or the FBI. But relax. No one is out to get you, only your money.

- **Gypsies and Curses:** If a gypsy, palm reader, tarot card reader, or anyone else tells you that your money is the source of a curse, it is a scam. They will ultimately try to convince you to give it to them to throw over a bridge, burn, or dispose of in some other way. Get away from them and seek better counseling.

NATIONAL ECONOMIC UNCERTAINTY

There is a growing sense of economic uncertainty in this country and throughout much of the industrial world. If asked why, some people will give personal anecdotal evidence such as "I see lots more people out begging on the street," while others will cite more widespread indicators, perhaps the recent unemployment numbers or massive national debt. Regardless of the scale of the argument, the consensus among many seems to be the same: we are headed for a financial calamity of epic proportions.

By the government's own admission, the nation's current levels of deficit and federal debt are unsustainable. In July of 2010, the CBO released a document, titled, "Federal Debt and the Risk of a Fiscal Crisis." Below are a few key quotes from that document:

> *"Over the past few years, U.S. government debt held by the public has grown rapidly—to the point that, compared with the total output of the economy, it is now higher than it has ever been except during the period around World War II."*

> *"Further increases in federal debt relative to the nation's output (gross domestic product, or GDP) almost certainly lie ahead if current policies remain in place."*

> *"Unless policymakers restrain the growth of spending, increase revenues significantly as a share of GDP, or adopt some combination of those two approaches, growing budget deficits will cause debt to rise to unsupportable levels."*

This talk of a possible economic collapse is not intended to scare anyone. With that said, it probably should. The facts are clear and undeniable, the concerns real and serious. It is only the final outcome that is open to debate. The uncertainty stems from the future actions our government will take (or fail to take) in the coming decade. These actions will undoubtedly shape our country for generations to come. What is not in debate is the fact that our country faces a grave situation, one that cannot be ignored.

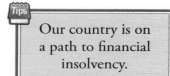

Our country is on a path to financial insolvency.

It is vital that every citizen understand the dangerous financial quicksand that the nation has fallen into. Avoiding a financial meltdown will require not only this understanding, but also the courage and steadfastness by policymakers and citizens alike to make very difficult choices. Without such grit, there is no hope of avoiding the coming financial catastrophe.

ORACLES AND DOOMSAYERS

Understand that *no one* can accurately predict the financial future; not the president, not the Federal Reserve Chairman, not even Warren Buffett. Most certainly not you or I. Everyone is guessing to one degree or another. If this weren't true, then the wealthiest people in the world would never lose money because they would simply hire those who could accurately make these predictions. In fact, the exact opposite is true. The richest people lose the most money during financial downturns. This is in part due to their net worth being tied to the investments that they own, but it is also because world economies are globally connected. If problems occur in one country (or sector), that problem is felt by people throughout the world.

Consider how the recession of 2008-2009 affected the net worth of some of the world's richest people (see Table 13-1). Certainly, these people are all still very rich, but obviously they had no specific insight that helped them avoid losing billions of dollars. While it is true that there were a dozen or so economists who predicted the financial crisis, there were thousands of other very respected economists who did not. Economics is a very complex field that is affected by countless factors, not all of which can be understood in real time. Nearly all financial happenings are easily explained after the fact, but many are completely invisible even a single day before the event unfolds. In effect, financial catastrophes write their own history.

Table 13-1 Changes in Net Worth from 2008-2009[255]

Person	2008 Net Worth ($ Billions)	2009 Net Worth ($ Billions)	Loss ($ Billions)
Warren Buffet	62.0	37.0	25.0
Bill Gates	58.0	40.0	18.0
Kirk Kerkorian	16.0	5.0	11.0
Paul Allen	16.0	10.5	5.5
Anil Ambani	42.0	10.1	31.9
Lakshmi Mittal	45.0	19.3	25.7
Carlos Slim	60.0	35.0	25.0

Recognize also that being right once or twice doesn't make someone an oracle. There have always been those who predict doom. When things turn down, these people will run around yelling "See I was right!" However, they won't bother to mention that they've been predicting doom in one form or another for decades, most of them making good money on books and speaking engagements. To some degree, everyone wants to believe doomsayers because if they do, and then preemptively act to avoid the downfall, they will have beaten the system. Likewise, there have always been people who see bunnies and rainbows on every horizon. When finances are growing, they will remind you of their optimistic predictions. However, when the skies cloud up, these self-proclaimed oracles are difficult to find.

LIVING AS A RATIONAL DOOMSAYER

Anyone who predicts an impending financial collapse stands the risk of being described as a doomsayer, or worse, a scaremonger. Such labels may not be fair but can proudly be worn as a badge of honor if considered in the right light. To do this, a distinction has to be made between two types of doomsday predictions. The first are predictions based solely on fear, hate, greed, or other emotions, and are often a means to some particular end (political or otherwise). Such predictions do not stand up to scrutiny and almost always prove to be inaccurate. Doomsayers who are driven by these motives are therefore not to be taken seriously.

The second type of doomsday predictions are based on fact along with a healthy dose of conjecture. Such predictions are typically supported by historical events and financial details and avoid political or religious underpinnings. If someone must be categorized as a doomsayer, this is the type to be. If there weren't such people in the world, no one would ever see catastrophes coming and would therefore never prepare for them.

> Tips
> Take financial actions based on rational analysis, not the rants of fear mongers.

Unfortunately, there are so many people crying wolf for the wrong reasons that few people are really listening to the warnings anymore.

THE STATE OF OUR COUNTRY

The most accurate data reflecting the country's financial wellbeing can arguably be found by consulting the U.S. Congressional Budget Office (CBO). The CBO publishes detailed annual budget reports as well as monthly updates to those reports. The budget reports provide insight into a host of important economic conditions, including unemployment, deficits, and national debt. They not only provide a point-in-time snapshot but also offer predictions of what the future might bring. A vast amount of information is publicly available at *www.cbo.gov*. A good place to start is by reviewing the annual CBO Budget and Economic Outlook, a report which captures the current financial state of the country and makes predictions for the coming decade.

Two useful websites that detail the country's financial health are:

- *www.usgovernmentspending.com*—Offers a wide assortment of up-to-date information on the nation's debt, deficit, and budget
- *www.treasurydirect.gov*—Provides information about the public debt and treasury securities

NATIONAL DEBT

The national debt is the total value of the outstanding bills, notes, bonds, and other debt instruments issued by the Treasury and other federal agencies. It is referred to by the CBO as the *federal debt* (or *gross debt*). Consider the following plot of the total federal debt as a percentage of *Gross Domestic Product* (GDP). The GDP is the total market value of all goods and services produced domestically in a given year. From the chart, it becomes quite clear that the nation's debt levels are historically high and rising rapidly, surpassed only by those during World War II.

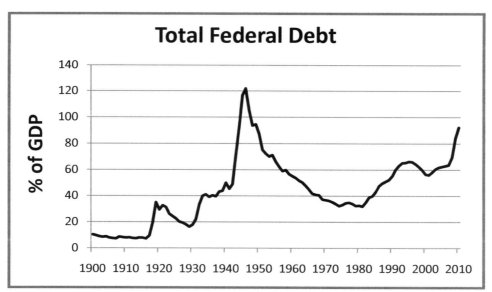

Gross debt as a percentage of GDP[256]

The national debt is divided into two components: *debt held by the public* and *debt held by the government*. The debt held by the government is the debt owed to intra-government institutions, such as the Social Security Trust Fund.

For several good reasons, the debt held by the public is the debt number most frequently reported. Obviously, it makes the debt appear smaller, which is good for the politicians. It also has the greatest economic significance because this represents the money that the government has to pay interest on. Unlike the debt held by the government, this is money owed to others, and outside investors demand reasonable returns on their money. The larger these interest payments become, the less money that is available for other government expenditures. Also, since Treasury securities must be sold to finance this debt, it is in direct competition with private investments. Buyers of Treasury securities elect to purchase government debt (usually for safety) rather than investing in businesses or other organizations. This leads to decreases in private investment, which in turn lead to slower economic growth, unemployment, lower wages, and reduced productivity.

The following plot shows the federal debt held by the public as a percentage of GDP from 1940 to 2010. CBO projections forecast for the next 25 years are also shown to the right of the vertical line. What becomes clear immediately is that the country is headed for uncharted financial territory. In our nation's entire history, the debt held by the public has exceeded 50% of the country's GDP only twice: once during World War II, and now. The current debt held by the public is over 60%, and that ratio is expected to grow rapidly over the next twenty-five years.

Current CBO projections show that the debt held by the public will exceed 200% of GDP by 2037.[257] Levels above 100% are almost universally viewed as unsustainable—meaning that our economy would be unable to meet the country's financial obligations.

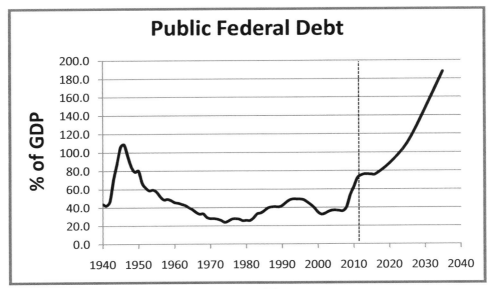

Public debt as a percentage of GDP[256]

Table 13-2 Breakdown of Federal Debt Held by Public, February 2011[258]

Country	Ownership ($ Billions)	% of Foreign Public Debt
China	1,154	25.8
Japan	890	19.9
United Kingdom	295	6.6
Oil Exporters	218	4.9
Brazil	194	4.3

As of February 2011, the federal debt held by the public was roughly $9.9 trillion. Of that debt, about 53% was held by domestic investors and 47% held by foreign investors. Table 13-2 shows the holdings of the top five foreign investors.

Two important points should be noted. First, the federal debt held by the public (in $) is the highest its ever been in the history of our country, and by all estimates, will continue to rise for the next twenty-five years. Second, nearly half of this public debt is held as Treasury securities by foreign investors. For many, this causes great concern. If foreign powers own our country's debt, it is believed that they can more easily influence policy. Also, by owning so much of our debt, these foreign investors have the ability to significantly affect the nation's economy. One way to do this is to demand higher interest rates for maturing debt. The debt level is too high for our country to repay, thus enabling those who hold the debt to have greater control over the terms of the loans.

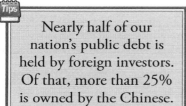

Nearly half of our nation's public debt is held by foreign investors. Of that, more than 25% is owned by the Chinese.

This is not to suggest that debt held by the government should be overlooked. This debt is essentially an IOU that the government has written to itself. When it comes due, the government will have to seek additional public debt to pay for the expenses. For example, the Social Security Trust Fund, which is the largest financer of this inter-government debt, has IOUs that must be repaid beginning in 2016. If money is not available to cover these costs, then a dollar-for-dollar shift must occur between government and public debt.

DEFICIT

The annual *deficit* is the amount that the federal government overspends. Said more technically, it is the amount that the federal government's financial outlays exceed its revenues for any given year. The opposite of a deficit is a *surplus*. According to the CBO's report, the fiscal year 2010 federal budget experienced a deficit of about $1.3 trillion. Just to put that into context, that means the government overspent by about $4,000 for every man, woman, and child currently living in the United States. This shortfall represents about 9% of the nation's $14.9 trillion GDP. The deficit-to-GDP ratio is a better metric of just how serious the overspending really is because it accounts for the change in the size of the economy and the value of the dollar.

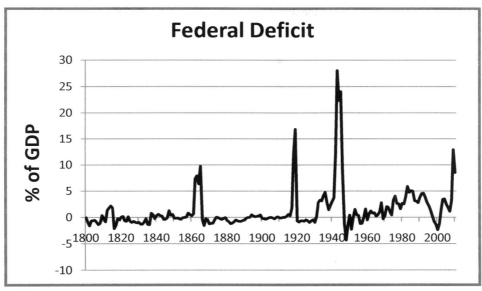

Deficit-to-GDP ratio[258]

Consider the plot above, which shows the deficit-to-GDP ratio all the way back to 1800. Up until about 1945, sharp increases in the ratio were a result of wars or the Great Depression of the 1930's. That trend of spiking up and then dropping back to near zero after the crisis is no longer holding true. Note that in over 200 years, the deficit-to-GDP ratio has exceeded 10% only four times: during the Civil War, World War I, World War II, and in the aftermath of the 2008 financial crisis.

WHERE DOES IT ALL GO?

With the gross overspending occurring almost daily in Washington, you might wonder how the money is being spent. The top most pie chart on the next page provides a high-level breakdown of the FY 2011 Budget. It's probably no surprise that the big expenditures are in defense, health care, pensions, and welfare programs. These four programs make up about 82% of the annual budget.

This makes clear the difficulty in reigning in costs. It is believed to be political suicide for leaders to propose reductions in any of these programs, let alone all four. This reluctance to face the music is driving our country toward a serious, yet inevitable, financial reckoning.

The problem becomes even more evident when considering the projected budget for future years. The second pie chart shows the projected budget for FY 2016. The projection shows that pensions, health care, and interest paid to our debtors will continue to grow. Of these three, the interest payments are the hardest to predict. If the holders of U.S. debt demand higher interest rates, which is likely given that they are currently at historic lows, the nation's budget will become unmanageable.

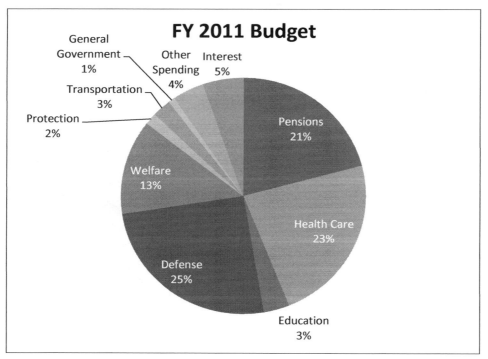

FY 2011 Budget[259]

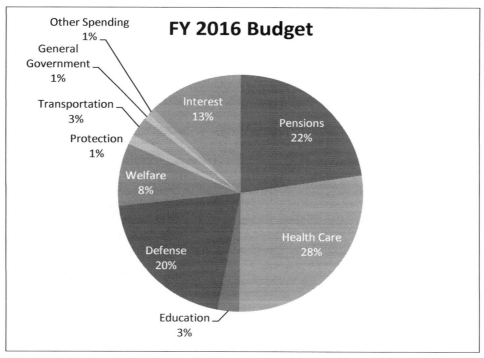

FY 2016 Estimated Budget[259]

ADDITIONAL ECONOMIC CONSIDERATIONS

There are many other economic bellwethers in addition to the national debt, deficit, and budget. A short list might include the trade deficit, value of our currency, unemployment rate, housing values, and inflation. Many of these are cyclic in nature, going up and down based on world conditions and events, making them difficult to draw clear conclusions from. The debt, deficit, and budget, however, continue their march onward with little deviation from their inevitable path to a national financial crisis.

THE COMING FINANCIAL MELTDOWN

Making the case for an imminent financial catastrophe is remarkably easy. The necessary groundwork has already been laid. By simply putting a few things together, a clear picture emerges. It is only the specifics of how and when the tragedy will unfold that are not entirely predictable.

Start with a few undeniable facts:

- The nation is heavily burdened with debt ($15.4 trillion in 2011).
- Nearly half of the nation's debt is held as Treasury securities by foreign investors.
- The deficit-to-GDP has grown to historically high levels (>10% in 2011). This ensures that the nation's debt will continue to grow year after year.
- The country's leaders have shown little resolve to seriously address these issues.
- Current interest rates charged on the nation's debt are at historically low levels and likely to rise.
- The unemployment rate is at unhealthy levels (9% in May of 2011) and is expected to remain elevated for several years.
- The housing crisis has left roughly one in four homeowners owing more on their home than it is worth.
- The value of the dollar against most major currencies has been steadily declining for the last decade.

Now consider a simple "what if" economic scenario:

1. The country's economy takes a hit. This could be due to a variety of reasons. Let's assume that it was at least partially due to bad lending practices that led to a high rate of foreclosures.
2. The foreclosures in turn lead to depreciated housing values as well as the banks tightening their purse strings, making home and business loans harder to receive.
3. The decreased availability of loans leads to fewer businesses expanding and hiring workers.
4. People become reluctant to spend as much because of the economic uncertainty.
5. This lack of money flowing forces businesses to close their doors and lay off workers.
6. Increasing unemployment keeps the cycle going by keeping spending low while increasing the number of home foreclosures—putting additional pressure on banks and retailers.
7. The government responds to pressure to "fix the problem" by printing more money and making it flow more freely. While the new money does help the banks lend more easily, it also leads to inflation and a growing deficit.
8. All of these factors combine to provide a negative sentiment to those who hold U.S. debt. They in turn demand higher interest rates from the government. The U.S. government has no choice but to comply because it is unable to finance its own debt.

9. Higher interest rates lead to even larger deficits and greater inflation, which in turn lead the debtors to demand even higher interest rates.

10. The cycle of destruction continues until things eventually come to a crashing halt with one of two possible outcomes: (1) the global economy steps in with a bailout built around debt forgiveness or other assistance, or (2) the nation defaults on its debt; the value of the dollar plummets; and the U.S. economy collapses.

The unfortunate truth is that these events are unfolding even as this book is being written. This is not to say that every step must lead to the next. If policymakers suddenly decide to become good stewards by better managing the nation's financial situation, a complete collapse may still be avoidable. The only way to improve the financial condition of the country is to establish a large annual surplus (compared to our current large annual deficit). But having a surplus is not enough. The surplus must to be put toward reducing the federal debt rather than paying for pork barrel projects or other spending activities. Establishing a surplus requires that revenues increase and expenditures decrease.

An increase in revenues implies that there is either a significant growth in the nation's GDP or an increase in taxation levels. In a healthy, growing economy, the GDP might rise a few percent annually, but this isn't nearly enough to offset the current deficit levels. That means that increased taxes will almost certainly be needed.

People scavenging from restaurant garbage during Argentina's 1999-2002 financial crisis *(Wikimedia Commons/Adam Jones)*

A decrease in spending will only be significant if it is made in the areas of our nation's major budget categories: defense, health care, pensions, and welfare programs. Voices of opposition already reach a fevered pitch any time there is talk of cutting even one of these programs. It would take tremendous courage from our elected leaders to make the necessary budgetary cuts to all four.

Even if such courage is found, it is important to understand that such an intervention would have serious consequences. Consider how our economy would react to tax hikes or worse yet, significant cuts in these budget categories. The country would almost certainly fall into a depression. Eventually, we would dig our way out healthier than before but not before widespread suffering was felt across every community. Internationally, there would probably be a huge sigh of relief because there is simply no entity large enough to bail out the United States. The truth is that we have to get our own house in order. The question is: Will we?

DP PLAN EXAMPLE

Table 13-3 Sample DP Plan Entry

Need: Financial Preparedness			
Danger	**Goals**	**Needs**	**Implementation**
A significant, unexpected financial expense	Pay unexpected expenses without getting into debt	An emergency fund	Establish an emergency fund with three month's of expenses.
Death in the family	Continue current lifestyle despite the loss of spouse	Life insurance	Insure breadwinner at six times annual salary. Insure non-working spouse at $200,000.
Property loss	Rebuild house and replace contents if destroyed Repair or replace car if damaged	House insurance with replacement cost coverage Car insurance	Insure home at 1.25 times value, and contents at $150,000. Insure car for comprehensive and collision coverage.
Loss of income due to disability or retirement	Have a reasonable quality of life if the breadwinner becomes disabled Have a comfortable quality of life during retirement	Disability insurance Retirement savings and income	Purchase disability insurance, covering 40% of salary. Save 10% of salary in 401(k) retirement account; take advantage of 5% employer matching. Contribute to employer's pension plan.

Quick Summary - Financial Preparedness

➢ Prioritizing can be done using the personalized DP Plan worksheets in the *Appendix*. Begin by identifying everything you need to reach a minimum level of self-sustenance. Once that level is reached, expand your capabilities to support your family through a prolonged or more severe disaster.

➢ Financial preparedness plays an important role in disaster preparedness, enabling you to stockpile supplies, cope with unexpected financial pressures, and minimize losses.

➢ Seven key steps to achieving financial preparedness are: staying employed, living below your means, limiting your debt, saving 10% of your gross income, getting the most from your money, being adequately insured, and not getting scammed.

➢ The ultimate goal of saving is to become financially prepared—out of debt, socking away money each month—and thus prepared to deal with life's unexpected challenges.

➢ A savings plan begins by establishing an emergency fund of at least three months of expenses. The emergency fund is only used to cope with hardships from unexpected events.

➢ Adequately insure your property, life, health, and ability to work. Remember, life insurance should be determined by the needs of those left behind.

➢ Leave behind a properly executed will to dictate how your possessions will be distributed and who will care for your surviving children.

➢ Protect your money by being ever vigilant against scams.

➢ Our country faces a grave threat from decades of overspending. It will require unprecedented courage and a willingness to sacrifice to avoid a financial catastrophe.

Recommended Items - Financial Preparedness

❑ A detailed budget

❑ Emergency fund (minimum of 3 months)

❑ Retirement savings

❑ Adequate life, health, home, and auto insurance

❑ Last Will and Testament

TRANSPORTATION

Challenge

While traveling through the Colorado Rockies, you take a wrong turn and get lost. Your car becomes stuck in deep snow on an infrequently traveled mountain pass. Do you know how to free your vehicle? If cell phone service is unavailable, how will you call for help? Should you leave your car and hike to safety? If rescuers are unable to find you, are you prepared to survive in your vehicle until weather conditions improve?

Reliable transportation, when coupled with accurate information and navigation, can save your life. Understand that, to be effective, they all have to work together. This is easiest to appreciate when you are trying to escape an immediate threat. For example, it would do you little good to hop in your car and race away from an approaching wildfire, only to discover that you have become hopelessly lost (navigation); that the winds had changed direction, putting the fire directly in your path (information); or worse yet, that your car has a leaking radiator (transportation).

Transportation, navigation, and information all offer valuable benefits, but when combined, they form a powerful defense.

- **Transportation**—keeps you mobile and able to get out of harm's way (the cardinal rule)
- **Navigation**—allows you to efficiently travel from point A to B as well as make detours without getting lost
- **Information**—helps you to avoid hazards that might interfere with your evacuation

More generally, transportation, navigation, and information serve two critical functions in disaster preparedness: (1) they help you to escape a dangerous area, should an evacuation be needed, and (2) they enable you to seek out supplies and assistance when you are unable or unwilling to evacuate.

TRANSPORTATION

Remember the cardinal rule that some disasters are only survived by getting out of their way. Evacuation is more fully addressed in *Chapter 5: Shelter* but warrants a little reiteration here. There are four basic steps to prepare for evacuation:

1. Fully fuel your vehicle and any spare gas cans. Store enough fuel to travel at least 500 miles—the specific number of gallons depends on your vehicle's fuel efficiency.
2. Identify multiple escape routes using maps or a GPS. Pick several possible retreat locations in case traffic flow prevents you from traveling to your preferred one.
3. Listen to TV or radio broadcasts to determine the best time to evacuate and the optimal escape routes.
4. Pack your vehicle with supplies, including things you might need for roadside emergencies (see *Roadside Emergency Kit* later in this chapter).

Once you make the decision to leave, take additional steps to prepare your home as outlined in *Chapter 5: Shelter*.

To successfully evacuate, you need a reliable vehicle capable of transporting your family, emergency supplies, and perhaps valuables away from the disaster. Of course, not all disasters will require (or even permit) evacuation, but there are many cases when getting out of harm's way is the first and best choice.

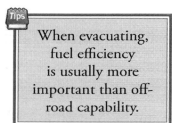

When evacuating, fuel efficiency is usually more important than off-road capability.

Contrary to some suggestions, it is not necessary to have a heavy-duty, off-road vehicle capable of climbing rocky ridges. Most of the time, evacuation will force you to vacate one city in favor of another that is less affected by the threat, meaning that a *dependable* car capable of traveling the roadways is all that is needed. For many threats, having a vehicle with good fuel economy is of primary importance because it allows you to travel further on a given quantity of gas—which is often in short supply.

With that said, if you have prepared a specific retreat that requires off-road access, such as a cabin up in the mountains, then of course you should only attempt to reach it with a suitable vehicle. Four-wheel drive vehicles may also provide some additional capabilities in wintry conditions, but they are by no means a guarantee of safe travel.

SECONDARY TRANSPORTATION

For situations in which you cannot (or choose not to) evacuate a disaster area, a highly fuel-efficient backup method of transportation may prove valuable. It may also serve you well in cases where a collapse in the country's infrastructure has led to a major fuel shortage—perhaps due to a solar storm or other significant disaster event. Backup transportation might be a bicycle, moped, motorcycle, golf cart, or other fuel-efficient vehicle. Spare parts and maintenance items, such as inner tubes, batteries, oil, and an air pump should all be stockpiled.

NAVIGATION

In this modern technological age, it is possible to pinpoint your exact location nearly anywhere on the planet. The Global Positioning System (GPS) uses overhead satellites to triangulate a user's position. Commercial GPS receivers are available for vehicle-mounting and handheld use. Vehicle-mounted GPS systems provide information more of interest to drivers than those lost in the wilderness. Handheld GPS units, on the other hand, are useful when traipsing around the woods since they frequently provide additional topographical information.

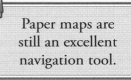

Using a GPS unit requires a fairly clear view of the sky. Systems don't work well indoors without an outdoor antenna and may have only marginal success during overcast conditions. It is worth noting that some disasters, such as a global war or large volcanic eruption, could leave the nation's GPS system inoperable. Finally, GPS systems are not particularly good at planning routes because they typically have very small screens. Old-fashioned paper maps are still the best choice, requiring neither batteries nor satellite signals. For this reason, it's a good idea to keep physical maps of your current location, any possible retreats, and the travel routes between them.

In the case of a widespread disaster, your evacuation routes may become extremely congested. What would normally take two hours to travel might now take a full day. To make matters worse, fuel is likely to be in short supply, leaving motorists stranded and further blocking the escape routes. If possible, before leaving home, carefully plan your escape route with consideration given to likely traffic flow and road conditions. Identify alternate routes in the event that your primary one doesn't work out. Finally, be prepared to violate conventional traffic rules in times of emergencies—but never risk endangering yourself or others.

INFORMATION

Important travel information might include: condition of the roadways, location of emergency shelters, recommended escape routes, coming weather events, curfews in effect, and availability of gasoline. This type of information may have to be collected from a variety of sources, including the Internet, public broadcasting (TV or radio), emergency services, short-wave radio, and friends or family outside the disaster area. Table 14-1 provides a few information resources relating to travel. Refer to *Chapter 12: Communication* for a more complete discussion of available information resources.

Table 14-1 Traveler Information Resources

Nationwide Traveler Information (only available in some areas)	511
National Traffic and Road Closure Information Maps and associated links to traffic conditions and road closures across the United States.	www.fhwa.dot.gov/trafficinfo
Google Maps Provides maps, street views, and driving directions.	www.maps.google.com

IMPORTANT PAPERS

One of the most important things you can do to prepare for a possible evacuation is to put copies of important documents onto a digital memory device, such as a USB flash drive, external hard drive, CD, or DVD.

A brief listing of some of the most important items to consider storing is given below:

- Addresses and phone numbers of points of contact (family, friends, insurer, doctor, etc.)
- Driver's licenses
- Social Security cards
- Birth and death certificates
- Adoption papers
- Insurance cards and policies
- Passports
- Recent photos of family members (suitable for missing person's posters)
- Property deeds
- Automobile titles
- Firearm serial numbers
- Weapon permits
- Marriage license
- Home inventory video or photos
- Bank and investment account information
- Tax records
- Computer account logins and passwords
- Last Will and Testament

A complete listing of all recommended information to store electronically is given in the *Appendix*. Having an easy to carry collection of important information can serve many purposes, including helping you replace original documents, provide proof of identity and ownership, and assist with insurance claims.

Store your information on password-protected or encrypted memory devices.

Given that the memory device will contain significant personal information, security becomes a major concern. To prevent the information from ever being compromised, you should use password protection or data encryption. At a minimum, you should password protect the individual files and place them in password protected zipped folders—instructions about how to do this are readily available on the web. Password protection will deter the casual criminal but not the sophisticated computer hacker. A vastly more secure method is to store the information on a hardware-encrypted USB flash drive (e.g., IRONKEY). These devices are more expensive than conventional memory sticks, but they are nearly impossible for anyone short of the National Security Agency to access. It is definitely worth investing a few dollars to ensure that your identity remains safe.

Roadside emergencies *(Wikimedia Commons/Wing-Chi Poon)*

BEING PREPARED ON THE ROAD

Some people might question the need to be prepared while on the road. After all, what could possibly happen while driving to work or the grocery store? Once again, this is a very rational argument for the typical day. However, remember that being prepared is not about the typical day, it is about the exceptionally bad day. Every year there are countless reports of people perishing in one way or another because they were unprepared while on the roadways.

Consider the following three reasons to be prepared when on the road:

1. People tend to spend a great deal of time in their vehicles, meaning that if you do find yourself in a dangerous situation, there's a reasonable chance that you will be in your car.
2. Some disasters force you to immediately evacuate the area, leaving you with only the supplies in your automobile.
3. When on the road, you are vulnerable to numerous hazards—especially when traveling in inclement weather. They might include:
   * Vehicle breakdown, such as a flat tire, dead battery, running out of gas, or mechanical failure
   * Dangerous weather events, such as tornados, hail, or heavy snowfall
   * Getting stuck in snow or mud
   * Auto accident
   * Flooded roads or submerged vehicle
   * Vehicle fire
   * Widespread disaster, such as a nuclear contaminant release, terrorist attack, or volcanic eruption

MAINTENANCE AND FUEL

The first step to being better prepared on the road is to keep your vehicle in good working order. Reliability, not curb appeal, is what's important when your life is depending on your car. Have your vehicle inspected regularly, and keep up on the routine maintenance items (e.g., tires, oil, radiator fluid, belts).

You should also endeavor to keep your vehicle's fuel tank at least half full at all times. Maintaining a minimum level of fuel gives you the ability to immediately evacuate an area without having to hunt for an open gas station. This obviously requires more frequent trips to the gas station but could give you a life-saving head start over other motorists. Certainly with an imminent threat, such as a dangerous weather event, you should fill your vehicle as one of your readiness steps. Having a tank full of gas also helps you to survive if you ever become stranded in the cold.

ROADSIDE EMERGENCY KIT

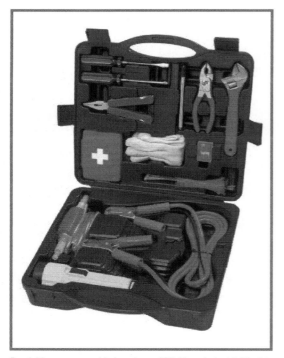

Roadside emergency kit *(courtesy of Michigan Industrial Tools)*

Every prepper should keep some basic supplies in their vehicle. The supplies should include items needed for unplanned evacuations and roadside emergencies, such as auto accidents or being stranded in your vehicle. By keeping a small backpack with the supplies, you can also quickly pull together a "grab-and-go" bag in the event that you must leave your vehicle behind. The recommended supplies are divided into two kits: the "Just the Basics" kit (things that everyone should keep in their car) and the "Kitchen Sink" kit (less critical items that might be useful when dangers are more likely, such as when traveling on a long trip).

The supplies can be stored in a large duffle bag, making them easier to quickly load and unload. You should review the two lists and tailor them to address the dangers that you see as most likely. Just as with the bucket survival kit described in *Chapter 2: Staying Alive*, most retail roadside emergency kits are grossly inadequate. It is much better to assemble your own kit with quality, hand-picked supplies that will adequately meet your needs.

Table 14-2 "Just the Basics" Roadside Emergency Kit

Item	Use
Cell phone with car charger	Call for help
Small gas can	Retrieve gas
Folding shovel	Dig out tires; make a fire pit
Jumper cables or Jumpstart battery booster	Jump a dead battery
Flashlight with spare batteries and car adapter	Safely navigate the dark; wave down assistance
Roadside triangle reflectors or flares	Warn others of a disabled vehicle, accident, or roadside hazard

Small tool kit (e.g., screwdrivers, adjustable wrench, pliers)	Perform basic repairs
First aid kit	Assist those with medical needs
Tow strap	Free a stuck vehicle; tow a disabled vehicle a short distance
Notepad and pen	Leave notes when you abandon your vehicle; write down tag numbers of a drunk driver
ResQme device	Cut seat belt if trapped; break out windows if vehicle becomes submerged
Roll of heavy-duty duct tape	Tape broken windows; fix tears; build shelters; secure enemy's hands; countless other uses
Windshield ice scraper	Clean snow and ice from window
Spare tire, jack, lug wrench, and small board to put under jack	Change flat tire
Warm blankets	Keep warm when stranded
Pair of comfortable walking shoes and socks	Walk to safety
Maps and/or GPS unit	Navigate to safety
A few bottles of water	Stay hydrated
Backpack	Use as a grab-and-go bag if forced to leave the vehicle
Essential personal medicine	Enough to get you to safety
Cash	Pay for gas, roadside assistance, food, water, or lodging when credit is unavailable
Heavyweight canvas bag	Store your emergency supplies; load and unload easier

Table 14-3 "Kitchen Sink" Roadside Emergency Kit

Item	Use
Everything from the "Just the Basics" kit	Takes care of most common needs
Class ABC or BC fire extinguisher	Extinguish small car fires
Leather work gloves	Protect your hands while changing tires, digging out car, etc.
Warm weather clothing (hats, gloves, coat)	Keep warm when stranded or hiking to safety
Large funnel	Fill radiator; add oil; funnel urine into plastic bag
Gallon-size freezer bags	Urinate in bag when stuck in traffic, or unable to go outdoors

Bulb-style siphon	Siphon fuel from a vehicle or gas container
Bag of coarse sand	Provide traction in mud or snow
Fix-a-Flat tire sealant	Quick temporary fix to a flat tire
Tire pump	Fill a leaky tire
Tire gauge	Check tire pressure
Hand-ratcheted winch (a.k.a. a "come-along")	Pull your vehicle out of the mud or snow
Permanent marker and plastic transparencies	Leave a weatherproof note on your windshield
Pack of wet wipes	Clean up after treating injury, or being contaminated
Bungee tie down cords	Strap down supplies to roof or truck bed
Plastic wire ties	Secure shelters; tie enemy's hands; make repairs
Disposable camera	Snap evidence at scene of accident
Oversized reflective emergency blanket	Use as lightweight, portable blanket
Emergency food, such as high calorie food bars	Eat when stranded, or when needing energy
Drinking water	Drink when stranded, or when unable to get to water source
Respirator; either low-cost Type N95 or gas mask	Protect from airborne threats
12-hour Cyalume chemlights	Provide night safety
Waterproof matches	Start a fire when stranded
TinderQuik (to start a fire even when wet)	Use as tinder for fire
Parachute cord	Make shelter; secure items
Lightweight rain poncho	Keep from getting wet; also doubles as a temporary shelter
Rescue whistle	Call for help
Quality fixed or folding-blade knife	Used for self defense, cutting supplies, shaving wood, cleaning animals, etc.
Position locator beacon	Signal for rescue from anywhere in the world
NukAlert	Detect high levels of radiation
Travel toiletries (toothbrush and paste, feminine hygiene supplies, comb, wash towel, etc.)	Keep yourself clean during an unexpected evacuation
Change of clothes	A fresh set of clothes can help to feel refreshed

FREEING A STUCK VEHICLE

Many travel-related tragedies begin when a vehicle becomes stuck in the mud or snow. If you travel with a well-stocked roadside emergency kit, you will have either a tow strap or what is known as a "come-along." In which case, freeing the vehicle requires that you either attach the come-along to a solid object, such as a tree or pole, or have another vehicle use the tow strap to pull you free. If you don't have either item, then you will have to free your automobile the hard way (i.e., pushing and rocking).

When trying to push your vehicle free, there are two cases to consider:

- **Help is available**—perhaps there is one or more adult passengers in your car, a good Samaritan stops to render aid, or you can easily ask for help from someone nearby
- **Help is not available**—perhaps you are alone on a road that is not well-traveled, and you have no way to call for assistance

In the first case, take the help. Freeing a stuck vehicle is much easier to do with a few extra hands to push. Try following the steps outlined below to free your vehicle:

1. Put your hazard lights on. This warns other drivers to be careful when passing you.
2. Use a shovel or other stiff object to remove snow, mud, or obstructions from in front of the wheels (or behind it if going in reverse).
3. Determine which wheel is slipping and put coarse sand, cat litter, an old piece of carpet, or a board in front of the wheel (or behind it if reversing). In a pinch, you can use your car's floor mats for traction. Regardless of what you use, warn everyone to be careful about the items possibly dislodging and flying out from under the wheels.
4. If your car is an automatic, put the transmission into Low gear or Reverse (depending on which way you are trying to pull out). If the car is a manual, put the transmission into 1st or Reverse. These low gears give you the most torque at the wheels. If you have a four-wheel drive vehicle, enable that feature.
5. Gently drive forward or backward, having everyone push in the direction of travel. Try not to spin the tires. Slow and easy wins the day.

If help isn't available, try the same procedure listed above on your own. If this doesn't work, try rocking the vehicle back and forth by shifting from Low to Reverse repeatedly. The secret to being successful is to time the oscillation of the vehicle with the gear changing, rocking the vehicle further and further out of the ruts with each cycle.

If you are unable to free your vehicle, don't hesitate to call for help. If you have a cell phone, contact family or friends, or call for a tow truck. If the situation is serious, such as in very cold weather, call 911 for emergency assistance. If you are on an interstate, you can also put on your emergency flashers and wait for a state trooper or roadside service vehicle to stop and assist you, although there is no guarantee as to how long it will take for you to get noticed—perhaps a minute, an hour, or longer.

If you can't call for help, consider waving down a passing motorist. But be aware of the risk in doing this. You are introducing an unknown person into a compromising situation, and that can be dangerous, especially if you are a woman or appear to be wealthy. Proceed with caution and follow your instincts. It

Freeing a stuck vehicle *(Wikimedia Commons/AgnosticPreachersKid)*

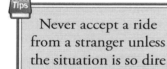

Never accept a ride from a stranger unless the situation is so dire that your survival depends on it.

is usually a better idea to ask a good Samaratin to drive up the road a bit (perhaps where cell coverage is better) and relay your request for help to family or friends. Do not accept a ride from someone you don't know unless the situation is so dire that you feel that your survival depends on it. If you do accept a ride, leave a detailed note on your vehicle's dashboard describing where you went, along with a description of the person, their car, and the tag number.

STRANDED

There are numerous ways that you can become stranded in your vehicle, such as your vehicle breaking down, becoming stuck in snow or mud, being trapped by impassible roadways (perhaps due to snow or floodwater), or losing control of your vehicle and sliding off the roadway. All of these scenarios (and many others) can leave you vulnerable, miles from home, and with limited options.

BENEFITS OF AN AUTOMOBILE

If you should ever become stranded, your automobile might save your life. A vehicle offers many benefits, and you should take full advantage of them. Consider the list of benefits that your car provides:

1. It functions as an excellent shelter, keeping you dry, out of the wind, and off the cold ground.
2. If your car still runs and has fuel, it can help you stay warm for several hours.
3. It is much more visible—generally, you are easier for rescue services to find if you remain with your vehicle.

Taking unnecessary risks can leave you stranded *(FEMA photo/Walter Jennings)*

4. It offers useful supplies, including:
   a) fuel for a fire (gasoline, oil, tires)
   b) a method of starting a fire (cigarette lighter, sparking across the battery, focusing with a headlamp reflector)
   c) a method of signaling for help (horn, headlights, emergency flashers)
   d) a radio for keeping up to date on emergency events
   e) a way to recharge your cell phone (cigarette lighter socket)
   f) a high-power two-way emergency radio (if equipped with Onstar)
   g) traction for getting a stuck vehicle free (floor mats placed under the wheels)
   h) lights to help you function at night (overhead light, headlights)

If things turn desperate, don't hesitate to scavenge from your vehicle. Use anything that will help you to survive, even if it means damaging your vehicle. For example, people who have been stranded in very cold weather have discovered that burning the tires not only provides warmth, but also acts as a smoky signal fire.

LESS SERIOUS SCENARIOS

If you should become stranded and can access a phone to call for help, there is a very good chance that the situation will be resolved within a couple of hours. For these less serious circumstances, your priorities are to stay comfortable and remain safe. The biggest threat you face is a collision with another motorist, especially during inclement weather with hazardous road conditions or limited visibility. If you have flares or warning triangles, space them along the roadway behind your vehicle. Beyond that precaution, put your hazard lights on and wait for assistance to arrive.

To conserve battery power, turn off everything that uses electrical power that isn't absolutely necessary, including headlights, overhead lights, heating/cooling blower motor, and the radio.

> **Tips**
> Establish a routine
> of starting, heating,
> and then shutting off
> your car.

If the weather is very cold, you may need to establish a routine that carefully balances conserving fuel, maintaining your battery power, and staying comfortable. Follow a regimen of starting the vehicle, warming the vehicle thoroughly (perhaps for 5-10 minutes with the heater blowing wide open), and then shutting it back off until it becomes cold again. Continue this routine until you determine that the battery is getting weak, at which time allow the vehicle to run 20 to 30 minutes to recharge the battery before returning to the regimen. Limit getting in and out of your vehicle to preserve the heat contained within. However, do open at least one window a small amount to introduce fresh air into the car. Also, be sure to check that the exhaust pipe is clear before starting your routine. If it becomes blocked, deadly carbon monoxide gas can back up into the vehicle's cabin.

In warmer weather, you will probably be fine shutting off your vehicle and opening all the doors, or simply finding shade under a nearby tree. Obviously, don't remain in a hot vehicle with the doors closed and windows rolled up. It is better to open the doors than roll down the windows because if your battery fails, it may be impossible to roll back up the windows—leaving you exposed should the weather turn colder. Opening the doors also signals to others that you are having difficulties and require assistance. In very hot weather, you may wish to follow a regimen of turning the engine on and running the air conditioner for a brief time (as described above), although it is usually better just to rest in a shady area.

MORE SERIOUS SCENARIOS

If you should become stranded in a remote region or in conditions that make your discovery unlikely, the situation can go from simple annoyance to something truly life-threatening. The biggest dangers are hypothermia and carbon monoxide poisoning in cold weather, and dehydration in hot weather. Make every effort to address these three dangers, and you will go a long way to ensuring your survival.

Hypothermia is a condition in which your body gets too cold to regulate its temperature. Symptoms may include gradual loss of motor skills and mental acuity, fatigue, mumbling, slowed breathing, slurred speech, and cold pale skin. Hypothermia is best avoided by prior planning—packing warm blankets and spare clothing as part of your roadside emergency kit. Pay special attention to keeping your hands and feet warm since they are often the first body parts affected by the cold.

Stay in your car to avoid exposure to cold winds as well as to help conserve your body heat. Depending on how much fuel you have in the vehicle, you may get several hours of heat by cycling the vehicle on and off, running the heater for a few minutes each time (as discussed previously).

> **Tips**
> Keep your exhaust
> pipe clear to prevent
> CO from backing up
> into your vehicle.

Carbon monoxide (CO) poisoning can occur when the vehicle's exhaust pipe becomes blocked or clogged, most likely from snow, causing CO gas to back up into the cabin. Symptoms of CO poisoning include headache, fatigue, lightheadedness, shortness of breath, nausea, and dizziness. Before running your vehicle, check that the exhaust pipe is clear. Also, if there is any possibility that melting or falling snow might cause a subsequent blockage, periodically re-check and clear the pipe.

Dehydration can occur at any temperature but happens most quickly during hot weather when you sweat. Once again, dehydration is best avoided by prior planning—packing supplies of water in your roadside emergency kit. In hot weather, your car acts as an oven (as many unfortunate pets have learned), so it is better to seek shade from a nearby tree.

Your primary goal when stranded in a remote location is to get noticed, ideally leading to your rescue. Open all your car doors and the hood to signal to any potential motorists that you are in distress. Also, if another car (or a rescue helicopter) approaches, do everything possible to get their attention—honking the horn, flashing your headlights, and waving your arms. Don't let vanity or pride kill you. If you are stranded in the vehicle with someone else, take turns sleeping so that one person is always on the lookout for potential rescuers.

If rescue is slow in coming, perhaps requiring you to spend one or more nights in your vehicle, you will have to weigh staying put and waiting for help against abandoning your vehicle and seeking help.

ABANDONING YOUR VEHICLE

All too often there are accounts of families stranded in their cars for days or even weeks. Many end tragically, usually because one or more family members leave the vehicle in search of help. Remember the order of needs discussed early in the book: shelter, water, and then food. Your stranded automobile can serve as a very good shelter, and leaving it should only be done when you determine that the danger of staying is greater than the danger of leaving.

How do you make this assessment? Ask yourself a few questions:

- Can you call for help using a cell phone, CB radio, position locator beacon, or OnStar?
- If you don't have cell coverage, is it possible to get to higher ground and call for help?
- How long will it take for someone to notice that you are missing? Do they know the route you are traveling?
- How likely is it that you will be discovered by a passerby?
- Can you signal for help?
- Are you properly equipped to weather it out for a few days?
- Do you know which direction to go for safety? Are you sure?
- Can you reach safety without getting lost or injured? Again, are you sure?

In most cases, the decision should be to stay with your vehicle. Of course there are exceptions—perhaps your vehicle has been submerged in water, or you have run so far off the roadway that it would be difficult for anyone to spot you, or you recall passing a gas station just a quarter of a mile back.

If you decide to leave your vehicle, consider taking the following precautions:

- Leave your vehicle only during the daytime. Wandering around at night is far too dangerous.
- Initially venture only to a distance such that you can still see your vehicle. This distance may be adequate for you to spot a gas station, house, or large road with passing traffic.
- If you expand your search further, take supplies with you. Assume that you *will* get lost and be unable to return to your vehicle. Be prepared to spend a cold night sleeping under the stars.

- Leave a clear, descriptive note on the vehicle's dashboard detailing who you are, where you went, as well as contact information for people who might be looking for you (see *Appendix* for note).
- Take the obvious path out, staying on the roadways if possible. If you have to make a turn at an intersection, clearly mark the way back to your vehicle.
- If the route is complicated, or you feel forced to explore, leave a trail of "breadcrumbs." One way to do this is to cut up strips of colorful cloth, string, or duct tape and hang them along your path.

DP PLAN EXAMPLE

Table 14-4 Sample DP Plan Entry

Need: Transportation			
Danger	**Goals**	**Needs**	**Implementation**
Roadside emergencies	Handle a medical emergency, vehicle breakdown, or engine fire	First aid supplies	Keep a trauma-based first aid kit in car.
		Tools and repair supplies	Keep a small tool kit along with a Jumpstart, car jack, spare tire, and gas can in the trunk.
		Fire extinguisher	Stock a 10BC fire extinguisher in trunk.
Immediate evacuation	Be ready to evacuate immediately	Reliable transportation	Maintain the car diligently.
		Adequate fuel	Keep gas tank at least half full. Keep ten additional gallons in two gas cans in garage.
	Map alternative routes out of an area as necessary	Maps or GPS	Keep maps of Virginia and surrounding states in glove box.
	Receive situation updates	Radio	Store talk radio stations in car radio memory.
Vehicle stuck in snow or mud	Free a stuck vehicle	Supplies for digging out tires and pulling vehicle free	Keep a folding shovel, tow strap, and a bag of sand in the trunk.
Stranded in vehicle	Call for help	Radio or phone capable of calling for help in nearly any conditions	Keep cell phone and spare battery in car. Equip car with 3-watt cell phone signal booster.
	Stay warm	Blankets and clothing to keep entire family warm in wintry conditions	Stock wool blankets in trunk, along with spare hats, gloves, and wool socks.
	Remain hydrated	Enough water for two days	Stock a case of 24 water bottles in trunk.

Quick Summary - Transportation

➢ Safely escaping a disaster area requires transportation, navigation, and information, all of which must work together.

➢ Having a reliable vehicle capable of taking your family out of harm's way is critical. For most situations, fuel efficiency is more important than off-road capability.

➢ Try to keep your vehicle at least half full of fuel at all times.

➢ Store enough fuel to travel a minimum of 500 miles.

➢ Stay abreast of emergency information so that you can properly time your escape and choose the correct evacuation route. Have multiple escape routes identified in case your primary one is congested or unavailable.

➢ Keep a roadside emergency kit in your car. More exhaustive supplies can be added to the kit when you are traveling long distances or through dangerous regions.

➢ Keep electronic copies of important papers on encrypted memory devices. Electronic copies can help when requesting replacement documents as well as prove ownership and identity.

➢ When stranded, try to stay comfortable and out of harm's way while awaiting assistance. For more serious situations, you may have to scavenge supplies from your car, take more drastic measures to get noticed, or evacuate to safety.

➢ Your vehicle can serve as an excellent shelter. Under most circumstances, you are more likely to survive if you stay with your vehicle than if you abandon it.

➢ If you must abandon your vehicle, take every precaution, including venturing only a short distance, taking supplies with you, leaving a clearly written note behind, and marking your path with improvised "breadcrumbs."

Recommended Items - Transportation

❑ Reliable vehicle capable of transporting your entire family plus emergency supplies

❑ Roadside emergency kit

❑ Enough fuel to travel a minimum of 500 miles

❑ Fuel-efficient backup transportation (e.g., bicycle, moped, motorcycle)

❑ Hardware-encrypted flash drive

CHAPTER 15

PROTECTION

Challenge

A dangerous riot is underway in your community. Small groups of violent criminals have taken to the streets, looting houses and stores as well as assaulting families. How will you secure your home? Do you have the weapons, expertise, and mindset necessary to protect your family?

There are numerous situations in which you might be required to protect yourself, your family, or your property. Most of these threats come from common criminals perpetrating violence on the innocent. Even under ideal conditions, police are unable to respond quickly enough to meet every threat. The reality is that sometimes you must take a stand and defend yourself.

During serious disaster situations, you may also be forced to protect your family when law and order breaks down due to widespread panic or chaos. Likewise, some events leave communities isolated with authorities unable to provide protection, perhaps due to high flood waters, impassable roadways, or widespread devastation. It is during these times of potential lawlessness that you and your fellow citizens must band together to establish order and security for your community.

TYPES OF VIOLENT THREATS

Violent threats can generally be categorized into three groups:

- Good people who are driven to do bad things because of need
- Career criminals who use disasters to commit crimes of opportunity
- Terrorists or other extremists whose intentions are to cause harm

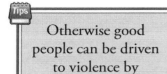

Otherwise good people can be driven to violence by desperation.

The first group could include your neighbors, people you might otherwise wave to while fetching the newspaper from your front yard. Under normal circumstances, they pose no threat and might even render assistance. However, when a serious situation arises, it may be these same people who find themselves hopelessly unprepared. Consider the actions *you* would be willing to take if your family was dying of dehydration and you believed your neighbor was hoarding a large stockpile of water. Desperate times can force people to take desperate measures. These people might convince themselves that you are being unfair and even rationalize that it is your duty to provide for them. If you fail to offer your supplies, they might resort to stealing or even taking your supplies by force. The best way to avoid this potential confrontation is to be discreet and not broadcast that you have stockpiles of supplies.

The second group of people are criminals, either traditional career offenders or those exploiting the chaos of the situation. Sensing that they are less likely to be caught by the authorities, they perpetrate crimes of opportunity: robbery, rape, and even murder. While these individuals invariably emerge during city riots, gangs of violent criminals are also prevalent after major natural disasters. A good example of this is the small bands of marauders who roamed the streets following Hurricane Katrina.

The third and final group is made up of people whose only intentions are to cause death and destruction, perhaps due to their political or religious agenda. These people are the most dangerous because there is no room for negotiation or reason. There is no appeasing them, except through suffering and death. Terrorists often view human life as expendable for purposes of making a statement or progressing their cause. Many are willing to sacrifice their own life for these purposes.

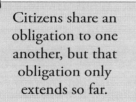

Citizens share an obligation to one another, but that obligation only extends so far.

Each of the three types of threats must be handled differently, but all should be recognized as very real dangers. Perhaps in the first scenario, you might assist those in need by providing training or helping to meet their most pressing needs. It is easy to argue that every citizen shares an obligation to help one another in times of need. However, that obligation should not extend to the point of endangering your own family.

In the case of criminals, your goal is to avoid conflict while removing your family from harm's way. This may require nothing more than a show of force to indicate that you are not an easy target, or a little extra caution about where you travel. It is not your responsibility, neither is it particularly wise, to adopt the role of police officer. It is better to focus on taking care of your own backyard by pooling with others to establish a sense of order to your neighborhood and immediate community.

When dealing with terrorists or other extremists, there is little choice but to step outside your normal civilized behavior and operate with the mentality of a ruthless soldier fighting his enemy. There is no room for negotiation or restraint. Forgiving yourself later is easier than living with a lifetime of regret and sorrow. Terrorists' actions are often designed to be widespread and horrific, so stopping them is in everyone's interest.

DEPENDING ON THE POLICE

During large-scale disasters, local police may be unable to maintain law and order, leaving you to defend your family and property. This is in no way meant to suggest that you should give anything but

Protecting your family when no one else will

your complete respect to those in uniform; only that you should be prepared for the authorities to be slow, or even completely unable, to respond to your calls for help.

During widespread disasters, policemen have also been known to abandon their posts to care for their own families—certainly an understandable course of action. Those who remain are often tasked first with evacuating people in need. Crimes, both property and violent, are reluctantly tolerated by authorities until conditions improve. When faced with a breakdown in conventional law enforcement, citizens have been known to band together into small neighborhood militias to protect themselves from looters and gangs of violent criminals.

YOUR RIGHTS AND RESPONSIBILITIES

The United States has a proud history of protecting the rights of its law-abiding citizens to own weapons for the stated purposes of self-protection and maintaining a free state. The Second Amendment to the Constitution states:

> "A well regulated Militia, being necessary to the security of a free State, the right of the people to keep and bear Arms, shall not be infringed."

Photo by Robert H. Boatman, *Living with Glocks*

Exactly what this right guarantees is the subject of countless arguments and court decisions. What is clear, however, is that with only a few restrictions, Americans can legally own numerous types of firearms, including rifles, shotguns, and handguns. Militarized weaponry, such as automatic assault rifles and explosives, are more controlled, requiring special licenses for ownership.

As the head of a household, it is your responsibility to provide some measure of protection for your family. This is not to suggest that you must own a closet full of firearms. Protection comes in many forms, including having a strong support group willing to come to your aid should the need arise. Many people do, however, choose to own firearms for the purposes of personal and home defense.

When taking protective measures, there is a temptation to go overboard and adopt the position that if a little is good, then a lot is better. Why stop at a handgun, when you can have an automatic assault rifle with armor-piercing shells? This view is not only overkill, but it also transforms people who claim to be preparing for possible disasters into "paramilitary nuts" in the eyes of neighbors, friends, and the authorities. Once again, it is best to take reasonable, judicious measures when preparing.

In this country of personal freedoms, gun buffs and pacifists are to be equally respected.

Regardless of the moderation with which this topic is approached, there will undoubtedly be some people who are unwilling to consider owning weapons due to their strong aversion to violence. Likewise, there are adamant gun buffs who cherish their right to own a wide assortment of firearms. In our country of personal freedoms, both viewpoints are to be equally respected.

Table 15-1 Firearms Comparison

Weapon Type	Advantages	Disadvantages
Rifles	Excellent range and accuracy; small calibers can be easily handled by women; larger calibers have excellent stopping power	Not concealable; small calibers have limited stopping power; high penetration; prone to be taken away
Shotguns	Excellent stopping power; less accuracy required; limited penetration; intimidating	Not concealable; heavy ammunition; limited capacity; somewhat limited range; prone to be taken away
Handguns	Concealable; lightweight; easily handled by women; good for close range protection; good stopping power; large capacities available; not prone to be taken away	Very limited range; requires skill to be accurate; low calibers have limited stopping power; automatics can jam

FIREARMS

Irrespective of your position regarding firearms, be honest enough to accept that we live in a world of guns—with roughly 200 million firearms in the United States alone.[227] If you don't own a firearm, you will be at a distinct disadvantage if violence comes your way. You don't have to like guns, but you should at least recognize that they are highly effective weapons. Dating back to the 19th century, Samuel Colt's revolvers inspired the old saying: "God created man, Sam Colt made them equal." The underlying message being that a firearm gives even the smallest person the ability to defend himself against the largest.

The three types of firearms to consider are: rifles, shotguns, and handguns. As outlined in Table 15-1, each has distinct advantages and disadvantages when it comes to personal protection. For most people, a handgun or shotgun is the best option for home defense. Handguns are easier to handle and good for close range combat, whereas shotguns have considerably more stopping power and can be very intimidating. When loaded with the correct ammunition, both weapons have limited penetration, making it less likely that the bullets will pass through the walls of your house to hit a family member or neighbor.

AUTOMATICS VERSUS REVOLVERS

People tend to have strong, differing opinions about which is better: an automatic or a revolver. The truth is that either type of gun is adequate in the hands of someone who knows how to use it. Automatics offer greater capacity and shooting speed, but revolvers are generally considered more reliable and easier to use. If you are an experienced handgunner, then an automatic is a reasonable choice. Otherwise, a revolver would probably serve you better.

In addition to choosing between a revolver and an automatic, you must also select the right caliber of handgun. Don't fall into the trap of thinking that bigger is better. A firearm is not only useless but also dangerous to innocent bystanders if you can't use it accurately. A .38 caliber handgun will serve you better than a supersized .44 magnum if the kick and weight of the larger gun prevent you from hitting your target.

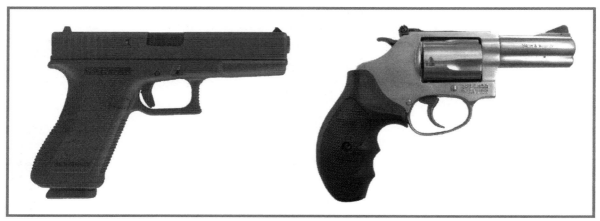

Automatic and revolver

GUN SAFETY

There's an old saying among gun owners: "Guns don't kill people; people kill people." The point they're making is that guns are not inherently dangerous; rather, it is the violent intent or reckless handling that causes injury. Whether you prescribe to that line of thinking or not, gun safety is paramount if you own a firearm. Gun safety can be divided into two topics: gun handling and gun storage.

Gun handling refers to the manner in which you hold and operate a firearm. Gun handling rules must be practiced each and every time you pick up a weapon. It takes only a single instant of negligence to cause a lifetime of regret. Consider the following gun handling rules.

Gun handling rules to *live* by:
1. Treat every gun as if it is loaded until you know differently.
2. Never point a gun at anything you don't intend to shoot (i.e., keep it pointed in a safe direction).
3. Keep your finger off the trigger until you're ready to fire.
4. Know your target and what is behind it.

GUNS IN THE HOME

To be a responsible gun owner, you *must* be diligent in practicing gun safety within your home. At a minimum, this means keeping trigger or cable locks on every firearm, with the keys kept in a location that is inaccessible by children. Storing your firearms unloaded, with the ammunition in a separate location, adds another degree of safety. Additional safety can be had by locking the firearms and ammunition in a quality gun safe.

The argument is often made that if a gun is kept secured, then it will not be readily accessible, and thus be of limited value for fast-acting situations. This may well be true, but in households where children are present, *safety should always trump readiness*. On average, about 200 children die in the U.S. every year from accidental shootings, with about 20 times that number injured.[227] If proper safety precautions are not taken, the chances are far greater that a child will be injured or killed in your home playing with your firearm than at the hands of an intruder. When it comes to guns, be safe above all else.

Trigger lock by Homak Security

THE "RULE OF 43"

An often-cited statistic claims that having a firearm in your home leaves you 43 times more likely to have it used for suicide, criminal homicide, or accidental shooting than for killing someone in self-defense.[228] At first glance, this would lead any rational person to seriously question ever having a gun in their home. However, as pointed out by experts who analyzed the report, the data was not properly presented and the conclusions not fairly

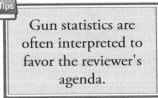

Gun statistics are often interpreted to favor the reviewer's agenda.

drawn.[229] Most people would agree that it is not the number of dead intruders that should be tabulated when determining the effectiveness of having a gun in your home but rather the comprehensive protective benefits, including how many lives were saved and injuries prevented.

This point is best made through a hypothetical example. Assume there are one thousand homeowners who *don't* have guns, and one thousand homeowners who *do* have guns. Of the homeowners who don't have guns, ten are murdered by intruders. Likewise, of the homeowners who do have guns, only one is murdered by intruders. One intruder is also killed by the homeowner, with the other intruders being scared away. Also, assume that there are four fatal gun accidents in the second group of homes.

The data could be presented with two very different, but equally accurate, statements:

- **Anti-gun group:** In homes where guns were present, four times as many innocent people were killed by guns than were intruders.
- **Pro-gun group:** Guns reduced the number of murders caused by home intruders by 90%.

You get the point. Data is often about how you present it. With the Rule of 43, one must consider what the data actually means. There may indeed be far more unintentional than intentional deaths in homes due to guns, but considering that less than 0.2% of defensive gun usages actually result in the death of a criminal, the Rule of 43 may not tell the entire story.[230] Gun ownership is a family decision that requires careful consideration of much more than simple statistics.

Firearm training *(courtesy of FBI)*

DEADLY FORCE

It should go without saying that the use of deadly force must be taken very seriously. Understand that if you kill someone, the burden will be on you to convince the authorities (or a jury of your peers) that your actions were reasonable *and* necessary. Laws for every community are different, but generally this means that your life, or that of another, must have been in imminent danger. There are many cases of people who used deadly force when other options existed and in turn were prosecuted and jailed. Should you ever be forced to kill someone, your fate will depend on the situation, the efforts you made to avoid confrontation, and the disposition of the district attorney. If someone breaks into your home and attacks your family in the dead of night, then the use of deadly force is not likely to be questioned. On the other hand, if you have a fight with another patron at a bar and end up shooting him in the parking lot, you will likely have a much harder time convincing authorities that it was necessary.

In a disaster situation, you may find it necessary to protect your family or property. The threat may come from lifelong criminals pursuing crimes of opportunity, looters looking for sellable goods, or neighbors just desperate for supplies. Under such circumstances, you must assess the situation and decide upon a reasonable course of action. Each of these three threats would likely be handled very differently. Criminals might have to be dealt with forcefully, looters scared away, and neighbors negotiated with. Be careful not to let the feel of a gun in your hand cloud your judgment.

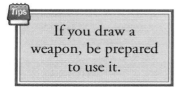

If you draw a weapon, be prepared to use it.

If you draw a weapon, be prepared to use it. The introduction of a gun into an altercation can cause things to quickly spin out of control. Weapons can be taken away and used against you. To make matters worse, the culprit could easily claim that his life was threatened, and therefore, he had to use deadly force against you.

Personal aside: Many years ago, I was accosted by three men as I returned to my vehicle at a shopping mall. I had never met them before and could only assume that they planned to rob or carjack me. They quickly approached my vehicle from all sides, one of them carrying a small fixed-blade knife. At the time, I carried a 9mm Beretta for personal protection. Assessing the situation as one requiring the use of a firearm, I drew it and took aim at the closest man. When they saw the pistol, all three immediately stopped. It was as if they had suddenly hit an invisible wall. I remember feeling both fear and determination. I absolutely did not want to shoot them. On the other hand, I was fully prepared to do so. After shouting a few threatening words, the three men slowly retreated to a nearby vehicle and sped away. I immediately went to the nearest police station and reported the incident. The investigating officer took my statement and said that my actions appeared to be justified.

GAINING PROFICIENCY

If you own a firearm, take the time and training necessary to learn how to use it safely and effectively. Gun safety classes are offered from local firearm clubs as well as the National Rifle Association (NRA) and Civilian Marksmanship Program (CMP). In these classes you will learn about many gun-related topics, including:

- Handling a firearm
- Understanding how a gun operates

- Selecting the right ammunition
- Shooting safely and effectively
- Selecting, cleaning, and storing a firearm

Everyone in your family who may be expected to handle a firearm should become proficient with essential gun-handling skills before you bring one into your home.

Below are a set of progressive steps that will help you gain proficiency with your weapon:

Step 1: Learn and practice the four gun safety rules. If you cannot safely handle a firearm, don't touch it.

Step 2: Become familiar with how your firearm operates. There are many types of handguns, shotguns, and rifles, so don't assume that just because you've shot in the past, that you know how the weapon operates. Learn to activate the safety; load, unload, and clear the weapon; and disassemble, clean, and reassemble the weapon.

Step 3: Learn to quickly and safely clear malfunctions. Firearms routinely malfunction (especially automatics), so you must understand the procedures to clear misfires (a.k.a. failures to fire); double feeds and stove pipes (a.k.a. failures to eject); and failures to feed.

Step 4: Practice dry firing. Before you fire a single round through the weapon, perform some basic dry fire exercises. Practice drawing and holstering your weapon. Then focus on the fundamentals: stance, grip, sight picture, breathing, and trigger control—see *Shooting Fundamentals*. Once you're comfortable with the basics, simulate using the weapon in various real-life situations. These might include readying your weapon from the nightstand drawer, drawing and firing it at an aggressor who is grappling with you, and firing from a prone position while lying on the floor beside your bed.

Step 5: Head to the range. If you've never shot a firearm before, seek professional instruction. Many gun ranges offer one-on-one instruction for minimal cost. Leave your ego at the door. Start with a man-sized paper target placed at three to five yards away. Begin by practicing your shooting fundamentals. Once you can control where your bullet hits to within a couple of inches, move the target back a few yards and re-work the fundamentals. Eventually, you may wish to work up to more advanced exercises, such as draw-and-firing, timed firing, firing on the move, one-hand shooting, weak-side firing, double tapping, etc.

Step 6: Practice in more real-world environments. The International Practical Shooting Confederation (IPSC) holds competitions that test your ability to shoot quickly and accurately in dynamic settings. Find a local gun club that hosts the competitions and learn the necessary skills. You may also discover that firearm enthusiasts, security specialists, or ex-military personnel have set up similar real-world gun courses for competition and training.

IPSC competition *(Wikimedia Commons/Damir Colak)*

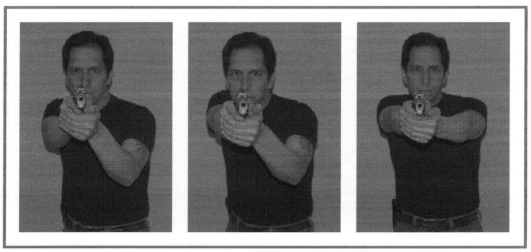

Weaver, Modified Weaver, and Isosceles shooting stances

Shooting Fundamentals

Whether you're shooting a single-action revolver, 1911 automatic, 12-gauge shotgun, or AR-15 assault rifle, the shooting fundamentals largely remain the same. You must take a stable stance, hold the weapon firmly, line up your sights, control your breathing, and squeeze the trigger. Much of what follows focuses on handgun fundamentals, but it could easily be adapted to rifle or shotgun use.

Stance—There are three common stances used with handguns: Weaver, Modified Weaver, and Isosceles (see photos above). Any of the three can work well, so try not to adopt a philosophy that one is inherently superior to the others. Try them all and decide which works best for you. For the Weaver stances, the feet are placed like those of a boxer, approximately shoulder width apart with one foot forward of the other. The primary elbow is either bent or extended. With the Isosceles stance, the feet are placed out to the sides or with one foot just slightly forward of the other. Both hands are extended. With all three stances, practice taking an aggressive forward lean. This helps to stabilize your body, but more important, it helps to set the "attack the target" mindset that you will need in a real-world confrontation.

Grip—Hold your firearm tightly enough to confidently control it while firing. In the case of a rifle or shotgun, this means that you must pull it tightly into the shoulder. Pistols are almost always fired with two hands unless an injury or specific circumstances prevent this. The strong hand grabs high on the grip much like a firm handshake. The weak hand wraps around the front of the firing hand with the thumbs pointing forward along the side of the gun (see photos). Both hands work together to control and steady the weapon. The idea is to have as much of your hands in contact with the pistol as possible, making it easier to absorb the recoil.

Two-handed grip for revolver and automatic

Sight Picture—The correct sight picture is with the top of the front sight blade aligned with the notches on the rear sight (see illustration). Use your dominant eye, which is not always the same as your handedness, to perform the alignment. Many people find that it helps to close their other eye. Some firearms come with their sights aligned and fixed at the factory, while others have small screws or knobs that allow the shooter to adjust the alignment. There are also laser, red dot, holographic, and other sight enhancement devices that can greatly improve the speed and accuracy of the average shooter.

Breathing—Most people find that their hand becomes steadier when they reach the rest phase of their breathing cycle. For this reason, try the following sequence when shooting: breath in, exhale half or all of your breath, then hold your breath momentarily while squeezing the trigger. Recognize that in a real combat situation, you will probably not have time to follow this routine but will instead resort to just forcing yourself to breathe.

Trigger Control—Incorrect trigger control causes more problems than all the other shooting fundamentals. The key is to pull the trigger straight toward the rear of the gun without affecting the sight picture. It is often said that the gun should surprise the shooter when it finally goes off. This is to emphasize the importance of not antici-

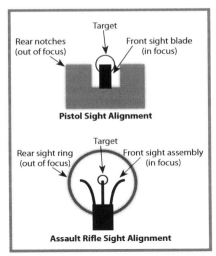

Correct sight picture for pistol and assault rifle

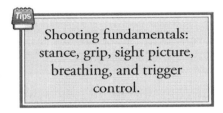

Shooting fundamentals: stance, grip, sight picture, breathing, and trigger control.

pating the recoil since such anticipation causes unintentional jerking of the firearm. An excellent way to see if you anticipate recoil is to have someone load the firearm, inserting a single dummy training round somewhere in the load. When the dummy round is finally "fired," you may find that your hand jerks in anticipation of the bang. Similarly, in the military, soldiers learn by balancing a dime on the end of their barrel when doing dry fire exercises. If the dime falls off, they are anticipating the shot. It takes time and concentration to overcome recoil anticipation, but it is critical to becoming more accurate.

CONCEALED CARRY

It is an unfortunate reality that violent crime permeates through nearly every community in America. Many citizens have chosen to carry concealed firearms as their first-line of defense against this violence. When asked why they feel the need to carry a firearm, concealed-carry weapon holders often answer "Because I can't fit a policeman in my pocket." Their point is that police are a reactionary force. They are not present when the crime unfolds but rather called in to make sense of the aftermath. Sometimes this puts law enforcement on the scene in time to resolve matters favorably for the victim, but many times they are simply too late.

Millions of people across the U.S. are registered to legally carry a concealed weapon (CCW). Countless others choose to carry weapons without registering. The requirements necessary to receive a CCW permit vary from state to state and may even vary between municipalities within the state. The four general categories of state CCW policies are:

Shall-issue—Carrying a concealed firearm requires a permit (except in the states of Alaska and Arizona). Granting of the permit is based solely on legal requirements, which typically include: minimum age, criminal convictions, residency, safety class participation, fee payment, and fingerprinting. The granting authority has no discretion in the awarding of permits beyond these requirements.

May-issue—Carrying a concealed firearm requires a permit. Granting of the permit is based on requirements and the discretion of the granting authority (usually the sheriff's office). A specific justification or additional actions may be required. Some states issue permits to nearly all applicants (e.g., Alabama), while others are more stringent in their issuing process (e.g., Delaware).

Unrestricted—A permit is not required to carry a concealed firearm.

No-issue—Carrying a concealed firearm is not allowed.

Some states, such as Virginia, also allow "Open-Carry." This means that firearms may be carried openly in public without a permit. However, if they are carried concealed, a permit may still be required. Table 15-2 shows the breakdown of current concealed carry laws across the USA. These are subject to change, and you should consult with local authorities to confirm the laws of your state and municipality.

Carrying a firearm in public provides a unique sense of security and responsibility. CCW holders often view society as consisting of sheep and sheepdogs. At first read, this might come off as derogatory, and perhaps it is meant that way in some circles, but for most CCW holders, it really refers to the fact that the vast majority of people (the sheep) are willing to live in a world where they assume others will keep them safe. Others (the sheepdogs) feel a responsibility to protect themselves and society as a whole from violent threats.

The decision to carry a concealed firearm should not be taken lightly. There are many people who hate firearms and see them as *the* source of violence in our country. You must be prepared to defend your position clearly and calmly, but also recognize that some people will continue to view you as being dangerous. In the eyes of the law, CCW holders are expected to behave to a "higher standard." This implies that you must do everything possible to avoid confrontation with others, even beyond the efforts that you would normally take.

If you decide to become a concealed carry holder, perform several preparatory actions:

- Learn and practice gun safety (see *Gun Safety*)
- Study local laws regarding where you can and cannot carry a concealed weapon (consult *www.concealedcarry.net* and *www.opencarry.org*)
- Understand laws regarding the use of deadly force (see *Deadly Force*)
- Become proficient with your weapon (see *Gaining Proficiency*)

Even if you ultimately decide not to carry a concealed firearm, you may still want to get a CCW permit. Having a permit enables you to legally carry a firearm when a disaster occurs (perhaps during an evacuation or when scavenging for food or other supplies). Note that a permit for one state does not necessarily allow you to carry a concealed firearm into another state. Some reciprocity does exist, but it is your responsibility to understand where you are legally allowed to carry. If you plan to travel across state lines with a firearm, you may need to get multiple permits in the respective states. Reciprocity maps are available at *www.concealedcarry.net*.

Table 15-2 State CCW Laws[260]

Alabama	MI	Kansas	SI	Oklahoma	SI
Alaska	SI, U	Kentucky	SI	Oregon	SI
Arizona	SI, U	Louisiana	SI	Pennsylvania	SI
Arkansas	SI	Maine	SI	Rhode Island	MI
California	MI	Maryland	MI	South Carolina	SI
Colorado	SI	Massachusetts	MI	South Dakota	SI
Connecticut	SI	Montana	SI	Tennessee	SI
Delaware	MI	Nebraska	SI	Texas	SI
District of Columbia	NI	Nevada	SI	Utah	SI
Florida	SI	New Hampshire	SI	Vermont	U
Georgia	SI	New Jersey	MI	Virginia	SI
Hawaii	MI	New Mexico	SI	Washington	SI
Idaho	SI	New York	MI	West Virginia	SI
Illinois	NI	North Carolina	SI	Wisconsin	NI
Indiana	SI	North Dakota	SI	Wyoming	SI, U
Iowa	SI	Ohio	SI		

Key: SI = Shall-issue, MI=May-issue, NI=No-issue, U=Unrestricted

Concealed Firearm Options

One of the most important decisions that a CCW holder must make is what type of firearm to carry. Certainly, there are many different handguns available that would serve as an effective concealed carry weapon. Consider the following four metrics when making your selection:

Size—A concealed carry weapon should be small enough to carry safely and comfortably. There's no point in trying to hide a .50 caliber Desert Eagle in your waistband. Someone is bound to eventually see the bulge under your clothes, and you will probably end up in front of the police trying to explain why you didn't do a better job of concealing your weapon. Pick a weapon that fits your lifestyle and wardrobe. If you normally wear jeans and a t-shirt, find a weapon and holster combination that works with those clothes. If you wear a suit coat or other outer jacket, then you have more options. Carefully consider not only the size of the gun, but also the weight. Some guns are made of steel, while others are made in part from lightweight metals or plastics. Lighter firearms are easier and more comfortable to carry concealed.

Caliber—There is a tendency for those new to firearms to assume that they need a gun that fires a very large bullet. While it is generally true that larger calibers correlate to greater stopping power, it is

A Kimber .45 automatic in an inside-the-waistband holster

important to understand that any gun can kill. Larger calibers are generally not ideal for new shooters because they are more difficult to control. If you must choose between control (which translates into accuracy) and stopping power, choose control every time. You must reliably hit your target, or you will pose a deadly risk to others. Reasonable calibers for most people are the .38 Special, 9 mm, .40 S&W, .45 Auto, and .45 GAP.

Capacity—Most revolvers hold five or six rounds. Automatics frequently hold seven to fifteen rounds, with extended magazines offering even greater capacity. Deciding on firearm capacity is a personal choice. Those who select a firearm with lower capacity will argue that there are very few situations that would require them to shoot more than a couple of rounds. Other shooters prefer to carry high-capacity guns because they feel it's better to be prepared for a prolonged gunfight with multiple targets. The choice is yours to make, but appreciate that lower capacity guns are usually smaller, lighter, and easier to carry. Spare magazines or speedloaders can also be carried to supplement the load.

Complexity—It is very important to select a gun that you are confident operating. When faced with a deadly conflict, you will likely experience a physiological reaction to the fear and adrenaline. This may manifest as sweating, trembling hands, and tunnel vision. At times like these, simplicity may save your life. This is one reason many people prefer revolvers or point-and-shoot automatics, such as the Glock, for personal defense. If you are an experienced shooter, training can help limit the physiological reactions and allow you to use a more complex firearm. Whatever weapon you choose, you must be able to safely draw, ready, and fire it under any circumstances.

LESS LETHAL WEAPONS

There are situations when you might want to defend yourself without resorting to using deadly force. Fortunately, there are many weapons that can deter an attacker without killing him. However, when dealing with "less lethal weapons," it is important to realize that they can, and frequently do, kill people accidentally. Also, understand that the weapons mentioned in this section are not a replacement for firearms.

A list of less lethal weapons is given below:

- **Taser**—very effective; works through most clothing; range up to about 15 feet; commercial devices aren't as capable as those used by law enforcement; outlawed in some states
- **Chemical spray (tear gas, pepper spray)**—safe to use on an attacker without risk of serious injury; not very effective; can help to escape from an attacker
- **Bludgeoning weapon (baton, bat, pipe)**—effective; requires you to get fairly close; has the risk of causing serious injury or death to your opponent

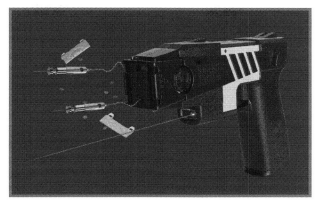

Taser X26

- **Knife**—often deadly; requires getting close; has a high risk of injury to self
- **Contact stun gun**—moderately effective; requires sustained direct contact; works through thin clothing; can easily injure self; requires you to be very close
- **Non-lethal bullets**—large rubber, bean bag, or plastic projectiles can be effective but are generally only available to law enforcement personnel
- **Bare hands**—can be effective with proper training, such as boxing or martial arts

HAND-TO-HAND FIGHTING

There will undoubtedly be some time in your life when you find yourself in a violent, physical altercation. To effectively handle the situation, it will be to your advantage to have three things: skill, courage, and physical prowess. These desired attributes are listed in order of importance. Countless competitions have demonstrated that skill can overcome physical disadvantages. Consider that no-holds-barred competitions are frequently won by smaller men (e.g., the Gracie family).

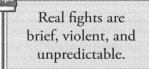

Real fights are brief, violent, and unpredictable.

Fights should be avoided for numerous reasons, not the least of which is that you might be injured or arrested. However, if you do have to fight, fight to win. More fights are lost by being hesitant than for any other reason. Fights are usually brief, violent, and unpredictable. They do not resemble what you see in the movies. Nothing is choreographed, meaning that you must adapt and focus on winning rather than putting on a show.

If you want to become skilled at unarmed combat, you must train. Enroll in a good martial arts or boxing club to get some hands-on experience. Recognize that many martial arts studios are designed more for sport than for practical defense, so choose your school carefully.

Martial arts training *(courtesy of U.S. Army)*

DP PLAN EXAMPLE

Table 15-3 Sample DP Plan Entry

Need: Protection			
Danger	**Goals**	**Needs**	**Implementation**
Home invasion, robbery, looting, or other violent crime	Defend family and property	Firearms capable of being used for home defense or riot crowd deterrence	Use a Beretta 92FS for home defense. Use a Remington 870 Police Magnum for riots or other external threats.
Accidental shooting	Keep firearms safe and away from children	Safety measures for all firearms	Keep firearms unloaded with trigger locks. Store ammunition in a locked fire safe. Teach everyone in the family about the dangers of firearms and proper handling techniques.
Robbery, carjacking, or armed altercation when away from home	Protect self and others	Firearm capable of being concealed	Apply for CCW permit from local Sheriff's office. Carry a Kimber.45 auto in an IWB holster along with a spare 7-round magazine.
Assault by unarmed assailant	Protect self and family	Skill in unarmed combat	Attend weekly mixed-martial arts training classes at local studio.

Quick Summary - Protection

➢ Disasters can force law-abiding individuals to resort to violence in an effort to survive.

➢ When the authorities are unable to keep order, criminals are more likely to commit crimes of opportunity, including looting, rape, theft, robbery, and vandalism.

➢ Be discreet by not broadcasting your level of preparedness.

➢ There are approximately 200 million guns in the United States. Without a firearm, you are at a distinct disadvantage when defending yourself.

➢ Three types of firearms to consider are shotguns, handguns, and rifles. Each has its respective advantages and disadvantages. Handguns and shotguns are particularly suited to home defense.

➢ If you're going to become a CCW holder, learn gun safety, understand local carry laws, and become highly proficient with your weapon.

➢ When selecting a firearm, consider weapon size, caliber, capacity, and ease of use.

➢ With children in the house, safety should always precede readiness. Safety methods include using trigger guards, separating weapons from ammunition, and storing weapons in a gun safe.

➢ Gun safety statistics are open to interpretation. The bottom line is to be safe above all else.

➢ If you own a firearm, take professional training in safety and proper handling techniques.

➢ As an alternative to firearms, consider less lethal weapons, including tasers, chemical sprays, knives, stun guns, and bludgeoning instruments.

Recommended Items - Protection

❑ Lethal weapon
   a. Shotgun with an 18- to 22-inch barrel (e.g., Remington 870, Mossberg 500), *or*
   b. Handgun (e.g., Beretta 92, Colt Python, Glock 17, Ruger GP100)

❑ Less-lethal weapon
   a. Taser, *or*
   b. Fixed-blade knife, *or*
   c. Baseball bat, baton, or other bludgeoning instrument

CHAPTER 16

SPECIAL NEEDS

Challenge

A major earthquake has struck your community. Roadways are so badly damaged that you are unable to evacuate. You have three children and an infirmed, elderly neighbor who are depending on you. What special needs do they have, and how will you meet those needs?

From the very beginning, this handbook has focused on addressing *family* preparedness needs. Most of the topics presented thus far affect everyone equally—we all have to eat, drink, and stay warm. The only exceptions are the two chapters targeted at adult concerns, specifically *Chapter 13: Financial Preparedness* and *Chapter 15: Protection*. This chapter addresses the special needs of the elderly, people with disabilities, children, pregnant women, and pets.

Many people have family members, neighbors, or friends with special needs. This could range from the elderly neighbor who requires nightly help preparing his meals, to the child born with a hearing disability. For purposes of identifying unique concerns, those with special needs have been grouped here into four categories: the elderly; people with disabilities or other medical needs; children; and pregnant women. This is not to suggest that these people are any less capable of caring for themselves than anyone else. Rather, it is simply a method of sorting a collection of preparations that may be useful in assisting others. The reader is strongly cautioned to avoid stereotyping any group of people.

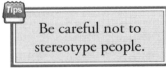

Be careful not to stereotype people.

THE ELDERLY

The definition of "elderly" is neither exact nor uniform in this country or anywhere else in the world. Certainly there are eighty year-old people who run marathons or hike mountains and suffer no physical limitations. For purposes of this discussion, however, an elderly person will be defined simply as someone whose age introduces limitations in their ability to handle disaster situations.

This section is divided into two parts. The first is targeted to elderly readers; the second is for those wishing to assist an elderly person during a time of crisis.

Special needs of the elderly *(Wikimedia Commons/Chalmer Butterfield)*

IF YOU ARE ELDERLY

Every person's capabilities are unique, and those in their "golden years" are no exception. Whether you climb mountains or are confined to a wheelchair, there are steps that you can take to better cope with a disaster. Begin by recognizing that regardless of age, nearly every topic in this book applies directly to you. You still need food, water, clean air, and life's other basic necessities. In order to be properly prepared, you *must* make the preparations outlined in the preceding chapters. However, as a senior, you should also recognize that you may have a few additional unique considerations.[231,232]

Start by answering a few questions to help you identify any limitations that you or your spouse may have.

PERSONAL ASSESSMENT QUESTIONS[233]

- **Medical Conditions**—Do you have any existing medical conditions that might require immediate attention during a disaster? Do you have an adequate supply of medications? How will you get to the hospital in an emergency? Are others aware of your medical conditions?

- **Personal Care**—Do you need assistance with bathing and grooming? Do you use any personal care equipment to help you bathe or get dressed? What will you do if your care provider is no longer available?
- **Water Service**—Do you have access to purified water, or if not, are you able to safely boil water?
- **Food Service**—Can you prepare your own meals, or do you rely on services such as Meals on Wheels? How will you eat if your food service is interrupted?
- **Feeding Devices**—Do you use any special devices or utensils to eat? What will you do if one breaks and you cannot quickly get a replacement?
- **Electricity-Dependent Equipment**—Do you rely upon medical equipment that requires electricity? Do you have a backup power supply, and if so, how long will it last? Is the electric company aware of your dependency?
- **Transportation**—Do you require a specially equipped vehicle to travel? Can you safely drive at night or in bad weather? Do you need assistance entering or exiting your vehicle? Can you travel alone?
- **Mobility**—Do you use a wheelchair or walker? Can you navigate your house if it becomes cluttered with debris? If your wheelchair or walker is damaged and inoperable, how will you evacuate your home?
- **Errands**—Do you need others to run errands, such as obtaining groceries and medications? What will you do if the person you rely on is unable to reach you because roads are impassible?
- **Building Evacuation**—Can you evacuate your home or office independently and without assistance? Are you able to cope with debris that might block your exit? Can you reach and activate an alarm? Do you need an elevator to evacuate?
- **Calling for Help**—Can you call for help by shouting as well as using a telephone? Do you require hearing aids to communicate? If your hearing aid becomes lost or inoperable, how will you communicate?
- **Service Animals/Pets**—Will you be able to care for your service animal or pets during and after a disaster? Do you have appropriate licenses and certificates for your service animal such that you can keep it with you at an emergency shelter?

Once you have identified any personal limitations or concerns, take the steps necessary to minimize their impact. Below is a list of recommended actions that might be applicable to an elderly person.[234] Many of these suggestions are also applicable to a person with a disability or serious medical condition.

RECOMMENDED ACTIONS

- Identify and stock any special backup supplies that you may need, such as spare oxygen bottles, wheelchair or hearing aid batteries, medications, and eyeglasses.
- If you are dependent on electricity for medical needs, notify your local power company. Also, equip your home with a backup power supply.
- If you receive periodic medical treatments or services such as home health care or food delivery, talk to your providers about their ability to continue services during an emergency.
- Wear medical alert tags or bracelets to help others quickly identify your disability or medical needs. Alert bracelets may be found online at *www.medids.com*.

- Consider getting an emergency medical alert system that allows you to call for assistance should you become immobilized.
- Compile a folder of your medical information. Give a copy of this information to a DP network member or trusted friend who lives outside your home. The folder should include:
  o A list of prescription medicines, including dosages
  o Descriptions of your medical conditions
  o Allergy information
  o Copies of your medical insurance and Medicare cards
  o Descriptions of your medical devices, including models and serial numbers
  o A list of contact information for your doctors and care providers
- Teach others in your home to operate your wheelchair. This will allow them to more easily transport you, should you become incapacitated.
- When making an emergency call, try to use a conventional landline rather than a cell phone. The landline will enable the dispatcher to more quickly identify your location. Be prepared to provide clear, specific instructions to rescue personnel. Equip your home with several phones, or keep one with you.
- Be careful about getting too hot or cold. Elderly people are more susceptible to temperature-related injuries.
- Millions of children live in households headed by grandparents. If you are raising grandchildren, be sure to consider their special needs as well (see *Children* later in this chapter).
- Don't forget the needs of your pets (see *Pets* later in this chapter).
- Establish a support network with your family, friends, neighbors, doctors, etc. Keep their contact information handy, and don't hesitate to call on them.
- Give a spare house key to someone in your local network. This will allow them to better check on you, as well as to keep an eye on your house in the event you have evacuated.
- If you decide to evacuate, let numerous people in your network know where you are going. Contact them once you arrive safely at your destination, and continue to stay in contact with them throughout the disaster.
- If you receive federal benefits, sign up for electronic payments, which are paid directly from the government to your bank account or a debit card. Disasters can disrupt the mail service for days or even weeks. Direct deposit also eliminates the risk of anyone stealing your benefit checks. You can sign up for direct deposit of federal benefits by calling 1-800-333-1795, or visiting *www.godirect. org*. Electronic payment will be required for all recipients beginning March 1, 2013.[235]
- Following a disaster, beware of fraudulent home repair contractors. Don't hire a contractor who comes to your door and solicits. Rather, find a reputable, insured contractor through the Yellow Pages or other conventional methods. Get more than one estimate for the repairs. Only deal with a contractor who is willing to put everything in writing, including firm repair costs. Ideally, pay for the work only after it has been completed to your satisfaction.

ASSISTING THE ELDERLY

The elderly are particularly susceptible in the event of a disaster because many have medical conditions or physical limitations that make it more difficult for them to evacuate or perform rigorous activities.

Some elderly people may be unwilling to leave their home for fear of having their belongings stolen or destroyed. They may also be reluctant to reach out to others because they don't want to be seen as a burden. As a good citizen, you are encouraged to assume some responsibility for caring for the elderly in your community. Below is a list of things you can do to help the elderly during a crisis:[231,232]

- Get to know the elderly in your neighborhood. Don't be offended if they seem suspicious of your intentions. The elderly are often the victims of scams (see *Chapter 13: Financial Preparedness*). Be patient and understanding of their caution.
- If you see a disaster coming, approach your elderly friends and neighbors beforehand, and let them know that you will be checking in on them. Ask if there are any preparations that you can help with, such as refilling prescriptions, picking up a few extra groceries, and removing any home hazards. Make it clear that you don't want payment for your assistance. Leave your phone number, and ask them to call you if they need anything.
- When a disaster does occur, periodically check on the elderly. Make sure that they have the necessary supplies to meet their needs, including food, water, medication, and heating/cooling.
- Continue checking on them after the disaster passes. Studies have shown that the elderly often have more economic and health-related issues after a disaster than younger people.
- If it is clear that an elderly person is suffering from a health problem, help him or her to seek medical attention.
- Help them to register for disaster recovery assistance should they require it.
- Keep an eye out to make sure they don't become targets of a fraudulent contractor. Intervene on the elderly person's behalf to let the perpetrator know that you are scrutinizing his activity.
- Don't treat the elderly as helpless. Many are perfectly capable of caring for themselves, so be careful not to stereotype anyone.
- Encourage elderly people to receive electronic deposits of any federal benefits they may receive.

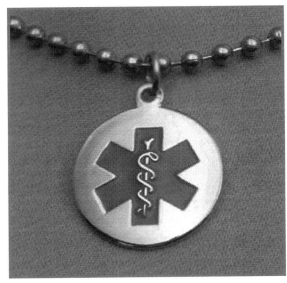

Medical ID tag *(courtesy of MedIDs.com)*

PEOPLE WITH DISABILITIES

The term "people with disabilities" is very broad in scope. The Americans with Disabilities Act defines a person with a disability as:[236]

- anyone with a physical or mental impairment that substantially limits one or more major life activities; or
- someone with a record of a physical or mental impairment; or
- any person who is regarded as having such an impairment.

Evacuating people with disabilities *(FEMA photo/Win Henderson)*

People may suffer from physical disabilities, requiring a wheelchair, cane, prosthetic, or other device. Or they may suffer from mental or cognitive impairments, including autism, psychiatric disabilities, speech impediments, and learning disabilities. As a result, providing a single set of general disaster preparedness recommendations is simply not possible.

From a preparedness point of view, the challenges facing people with disabilities are often comparable to those facing the elderly, since many of these challenges result from limited mobility or impaired senses. For this reason, much of what has been written earlier, including the *Personal Assessment Questions, Recommended Actions,* and *Assisting the Elderly,* directly applies to people with disabilities or other serious medical conditions. The extent to which the particular recommendations apply depend on the type and severity of a person's disabilities.

For a person with a disability, being ready for a disaster is another aspect of maintaining personal independence. They face the same needs as every other citizen, including food, water, shelter, protection, and so on. Each individual must answer, and subsequently address, how their disability might introduce special needs in the event of a catastrophe.

CHILDREN

Having children in your home requires you to make some special preparations. There are differing points of view regarding the best ways to handle dangers faced by younger family members. Many experienced parents would probably agree that children respond better to honest, clear discussions about threats—whether it is a neighborhood sex offender, the school bully, or more suited to this discussion, the approaching tornado.

Openly discuss the disasters that can affect your area. Explain the likelihood and expected impact of each. Encourage your children to think about how your family should prepare. Actively solicit their inputs, giving each suggestion the appropriate level of consideration. At no time should you make the discussion frightening. Instead, try to make this a positive, bonding experience in which everyone feels that they have contributed part of the solution. Explain to your kids that not everyone prepares adequately, but by doing so, you can not only help yourself but also those who are less prepared.

Teach your children not to panic. Kids mimic what they see. If you are calm and in control, they will draw strength from you. Practice emergency situations, including such things as sheltering in place, evacuating your home, and functioning without electrical power—see *Chapter 19: Trial by Fire*. Make sure that your children know how to call 911. Practice how they would interact with the dispatcher. Role-play to make this both informative and fun.

Teach your children how to evacuate from every room in the house. Most important, show them how to escape from their own bedrooms in case they should ever become trapped by a house fire. If old enough, teach them to use a fire extinguisher (see *Using an Extinguisher* in *Chapter 5: Shelter*). Finally, if your children will ever be exposed to firearms, teach them to handle them safely—see *Chapter 15: Protection*.

> **Tips**
> Children will mimic what they see, so act accordingly.

For younger children, consider providing them with comics, games, and coloring books that relate to disaster preparedness (found on FEMA's Ready Kids website at *www.ready.gov/kids*).[237] These materials strive to make learning fun while still providing accurate and useful information.

Older children might find the discussion of disaster types found at *www.fema.gov/hazard/types* to be interesting. Consider assigning a project, such as assembling a small disaster kit or writing a report on the dangers of a specific threat relevant to your area.

COPING WITH A DISASTER

When a disaster occurs, children may feel even greater stress than their parents. At the root of their anxiety is a loss of control and normalcy. Daily routines are disturbed. Friends may not be accessible. Sleeping conditions may have changed. Food and water may be in short supply. Worst of all, they may have witnessed devastation, physical injury, and emotional trauma. All this can add up to terrible anxiety. As an adult, you are able to put logical bounds on how bad things can get. You can see beyond the immediate chaos to a longer term solution. Children, on the

Children coping with disaster *(FEMA photo/Andrea Boohers)*

other hand, may feel their world has been turned upside down, uncertain that it will ever right itself.

Every child reacts to stress differently. Some may grow irritable, while others may become clingy, emotionally upset, or withdrawn. The best way to help them is to be calm and reassuring. Talk to your children about what they are feeling, and work to address their concerns. But don't lie to them. Trust is something that is difficult to regain once it is lost. Below are a few additional suggestions regarding helping children to cope with stress:[238,239,240]

- Be mindful of *your* behavior. Children watch their parents for clues on how they should act. If you begin shouting and acting frantic, they may become frightened. Put on a brave face, and let your children know that you will keep them safe.
- Try to understand your children's fears, and provide the necessary reassurance. These fears often center around three concerns:
    o The disaster will never end.
    o Someone they love will be killed or injured.
    o They will be separated from their family.
- Give each child a specific job, and praise their efforts when they rise to the challenge.
- Establish a routine. Children and adults alike are able to adapt to new conditions, and routines help to relieve the stress that uncertainty brings.
- Maintain discipline. Allowing your children to turn their anxiety into disrespect or argument can make the entire family less functional and heighten the level of stress.
- If possible, allow them to have contact with family or friends.
- Demonstrate your love. Give your children plenty of pats, hugs, and kisses. Physical contact can be very reassuring. Spend extra time with them, perhaps reading a book or playing a game. It is during times of greatest stress that love can shine the brightest. How you treat your children in these times of crisis is what they will remember their entire lives.

Once the disaster has passed, children may display very different reactions. FEMA provides the following descriptions of common reactions based on a child's age:[240]

- **Birth to 2 years**—Children who are not yet able to effectively communicate may become irritable, upset, or clingy. When they grow older (perhaps years later), their games may involve acting out elements of the traumatic event.
- **Ages 3 to 6 years**—Preschool children often feel helpless, small, and unable to protect themselves or others. This may cause intense fear and insecurity when separated from their parents. Preschool children generally cannot grasp the concept of permanent loss. In the weeks following the traumatic event, their play may involve reenacting the incident.
- **Ages 7 to 10 years**—By age seven, children have the ability to understand permanent loss. Some kids may become preoccupied with the details of the disaster. This preoccupation can interfere with their school or other activities. They may also display a wide range of reactions, including sadness, fear, anger, and guilt.
- **Ages 11+**—Pre-adolescent and adolescent children will have a more sophisticated understanding of the disaster event. Some may become involved in risk-taking behaviors, such as alcohol or drug

use. Others may become fearful and avoid activities they were previously comfortable doing. They may also have difficulty expressing their emotions and worries.

SCHOOL EMERGENCY PLANS

If your children attend a public or private school, contact the school administrator and ask a few basic questions about their emergency plans:

- How will school officials communicate with families during an emergency? Do they have your emergency contact information, including work, home, and cell phone numbers? What will they do if you are unreachable?
- Does the school store food, water, and other basic supplies?
- Is the school prepared to shelter-in-place?
- What are their evacuation plans?

In cases where schools institute procedures to shelter-in-place, be aware that you may not be permitted to pick up your children. Even if you go to the school, the doors may be locked or inaccessible for safety reasons. Also, if you want someone else to have the authority to pick up your children in case of an emergency (such as a grandmother or close friend), sign the necessary release paperwork with the school ahead of time.

PREGNANT WOMEN

The needs of pregnant women (or people with other medical conditions) during a disaster can be very unique. The suggestions given below are tailored to pregnant women, but most can be easily modified to fit people with other medical conditions.[241]

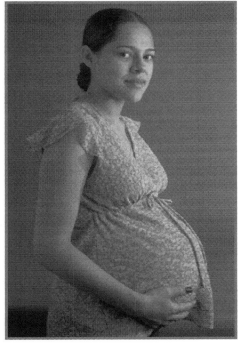

If you think you might need to evacuate, contact your healthcare provider ahead of time and get copies of your prenatal records. Also, let your doctor know the specifics of your new location. If you are close to your delivery date, or have a high-risk pregnancy, ask your caregiver for guidance as well as a referral for a doctor in the evacuation area. Also, compile a list of your medications, both prescription and non-prescription.

If you decide to evacuate, take an adequate supply of medications and prenatal vitamins, as well as your prenatal records and healthcare provider's contact information. If you are traveling by car, be sure to stop periodically to stretch and walk around. Wear and pack comfortable clothing and shoes. Also, take along some healthy snacks and water, so you won't have to survive on roadside junk food. Be extra careful about not consuming food that might be contaminated or spoiled.

A time for special care *(photo by USDA/Ken Hammond)*

Inform emergency personnel of your pregnancy before receiving any medical treatment.

Try to get adequate rest and relaxation. Avoid overexerting yourself or getting too hot. The bottom line is to give yourself permission to take it easy. Even though it might be difficult given the circumstances, try to keep your stress level down. If you feel anxiety, consult a trusted friend, family member, or clergy member.

Inform emergency personnel or relief services that you are pregnant before receiving any immunizations, medications, or x-rays. If your health care provider closes or becomes inaccessible, contact a local hospital or health department about receiving prenatal, delivery, and general medical care.

PETS

Americans have over 360 million pets, and that does not include exotic or wild animals. The simple truth is that people all over the world love their pets. A dog, cat, bird, horse, or even a turtle can feel like part of the family. Unfortunately, some disasters leave pet owners with very difficult decisions to make if prior planning was not done.

#1 Rule: Don't leave your pets behind.

The starting point to pet planning is to *do everything possible not to leave your animals behind.* History tells us that animals left behind will suffer terribly, with many perishing. You certainly don't want to be responsible for the death of your animal companion. Fortunately, proper planning can help.

Evacuating with pets *(FEMA photo/Leif Skoogfors)*

A useful video that discusses many important disaster preparations for pets is provided online by the Oak Ridge National Laboratory.[242] Every pet and livestock owner is encouraged to watch this important primer—see references at *http://disasterpreparer.com*.

The two basic disaster scenarios to consider are: (1) you must immediately shelter in place—perhaps due to a biochemical threat, or (2) you must evacuate—perhaps due to an approaching wildfire or hurricane. In the case of sheltering in place, you would normally just take your animals into your home's makeshift shelter (see *Chapter 5: Shelter*). If the animals are too large, you will need to make other accommodations, perhaps a barn, basement, or garage. The important thing is to get them out of harm's way.

If you decide to evacuate, take your animals with you. If that is not possible, shelter them in a safe location, such as a friend's house or boarding kennel. In 2006, the Pets Evacuation and Transportation Standards Act (PETS) was passed, mandating that communities have plans for evacuating, transporting, and sheltering animals. However, don't expect extensive assistance from government officials. They will have their hands full with many other pressing needs. The responsibility ultimately falls on you as the pet owner to ensure your pet's well-being.

This section is divided into three time frames: before a disaster arrives, when the event is occurring, and after it has passed. The material draws heavily from FEMA's guide, *Information for Pet Owners* and the American Veterinarian's Medical Association's guide, *Saving the Whole Family*.[243,244]

BEFORE A DISASTER

The time to take action to ensure that your pets will be properly cared for is *before* a disaster strikes. There are numerous preparatory steps to consider:

- Understand your local risks. Every region faces different hazards, ranging from flooding, to extreme weather events, to wildfires. Consider how these threats will affect your pets.
- Take clear photos of each pet. If your animal ever becomes lost, you can use the photos to create posters or show them to shelter workers.
- Make certain that your pets have up-to-date identification tags. If you know where you will be evacuating to, put that location and a phone number on the back of your pet's tag or on the collar using duct tape. Microchips placed under the skin can also greatly assist veterinarians or shelter workers when trying to reunite lost pets with owners.
- Locate a place to board your pet. This could be a professional boarder, veterinarian, or friend not affected by the disaster. For health and safety reasons, most emergency shelters *will not* accept pets. The one exception to this is service animals. By law, service animals are allowed into emergency shelters if the owner has proof of a medical need and the pet's current vaccinations. Some hotels and motels will also allow pets, but you should confirm this ahead of time. Your local animal shelter may also be able provide information regarding available shelter and boarding locations.
- If you decide to evacuate your pets with you, take adequate pet food, water, medications, veterinary records, vaccination certificates, litter, clean up material, muzzles, tie out materials, and other supplies that might not be available during transit.
- Take pet carriers, leashes, and harnesses as appropriate. Pet carriers should be large enough to have a small water bowl and a litter pan (for cats), and must have your contact information on them.

- Supplement your first aid kit with any additional specialty supplies needed for your pets. See FEMA's *Information for Pet Owners* for a complete listing of first aid supplies for various pet types and sizes.[243]
- Don't leave your pets unattended in your vehicle in hot weather. Temperatures can become far hotter in the car than outside and can quickly kill your animals.
- If you have livestock or other large animals, work with local and state authorities to evacuate them. During some emergencies, state authorities may not allow large animal trailers on the roadways.
- Identify a trusted neighbor or friend who will serve as the designated caregiver for your animals in case the emergency happens while you are away from home. Give him or her a spare key and simple instructions on how to care for your pets.
- Understand that it may be impossible to take any animal (except service animals) on a plane, bus, or train. Plan accordingly.

If you absolutely have no choice but to leave your pets behind, there are a few precautions you can take to improve their chances for survival:

1. Confine your pets to a safe area inside the home. Do not leave them chained or tied up.
2. Leave plenty of food and water.
3. Remove the toilet tank lids and raise the seats so they can drink from the toilets as needed. Also, fill the bathtubs with water, and brace the bathroom doors open.
4. Leave a notice on your front door that details what types of pets are inside, their names, where they are located, and when you left them. Also, provide your contact information and that of your veterinarian.
5. Return to your pets as promptly as possible. If you will be delayed, have someone else check on them.

Understand that these steps will not ensure your pets' survival, only help to give them a fighting chance. Leaving a pet behind should always be considered a last resort.

Some people mistakenly think that setting their animals free in their neighborhood gives them a better chance of surviving. On the contrary, releasing domesticated animals leads to horrible animal suffering because they are not experienced at foraging for food and are accustomed to companionship and love. Fear, hunger, and loneliness frequently drive what were once loving pets to attack other animals as well as people. Most pets released like this ultimately perish or must be put down by animal control.

DURING A DISASTER

If a disaster is imminent or underway, you need to take immediate action to protect your pets:

- Bring them inside to a safe, manageable area of your home. If you have an indoor kennel, you may wish to keep them in it for their protection and comfort. If the animals are too large to be brought indoors, accommodate them in a barn, basement, garage, or other structure.
- If you have different types of animals (such as a cat and dog), keep them separated. The anxiety of an emergency situation can cause them to act dominant and aggressive.

- Provide reassurance to your pets. If you have children, ask them to help comfort the animals. This will help your pets relax and give your kids a sense of participation.
- Be sure to have adequate food, water, and cleanup supplies available. Consider feeding moist food to reduce your pet's additional water needs. Don't overfeed your pets in an effort to keep them happy.
- Clean up any animal urine or feces immediately and thoroughly.

AFTER A DISASTER

Once the disaster passes, take a few final precautions to ensure your pets' safety:

- Begin by assessing if there are any immediate threats to your pets. These could include rising waters, downed power lines, roaming animals, and debris. If threats exist, keep your pets on a leash or indoors until the dangers subside.
- Recognize that your outdoor animals may act strangely or try to wander off because of new smells and sounds. Keep a close eye on them until things return to normal.
- If pets were left behind, they may require additional care. This may include professional decontamination, especially if they were exposed to high waters. Discuss your pets' needs with local emergency management officials or your veterinarian.
- Discard any contaminated pet supplies, including food and water bowls, bedding, and toys.
- If your pet is lost, check with animal control, local shelters, and veterinarians daily to see if it turns up. Also, put up missing posters, and ask neighbors to be on the lookout for your pets.
- Animals left unattended during a disaster may become violent and dangerous. If you see signs of personality changes in your pets, seek counsel from your veterinarian.

Victims of Hurricane Katrina *(FEMA photo/Jocelyn Augustino)*

Quick Summary - Special Needs

➢ Those with special needs may include the elderly, people with disabilities or medical conditions, pregnant women, and children.

➢ Elderly preppers should assess their physical limitations, if any, and take steps to minimize the impact of those limitations. Such steps may include stocking medicines or medical equipment, compiling medical information, and establishing a trusted network of supporters.

➢ From a preparedness point of view, the challenges facing people with disabilities are in many cases similar to those facing the elderly since challenges often result from mobility limitations or impaired senses. However, every case is unique, and needs must be addressed with individual consideration.

➢ Pregnant women are particularly susceptible during a crisis. They should be careful not to get too hot or cold, and to limit their physical activity. If forced to evacuate, anyone with a medical condition (such as being pregnant) should take their medical records and any appropriate supplies (e.g., vitamins, medications). They should also inform emergency personnel of their condition prior to receiving medical care.

➢ Children respond better to honest, clear discussions about threats. Don't frighten them, but don't lose credibility either.

➢ When around children, be mindful of your behavior. Kids watch their parents for clues as to how they should act. Put on a brave face, and let your children know that you will keep them safe. Maintain discipline and work to establish a daily routine.

➢ Contact your children's school administrator to ask about their emergency plans. Be certain that they have your current emergency contact information.

➢ The time to make emergency preparations for your pets is before a disaster strikes. Do everything possible *not* to leave your animals behind. If you must evacuate with your pets, take adequate supplies to care for them.

CHAPTER 17

CREATING A DP NETWORK

Challenge

A serious pandemic has swept across the country. Unfortunately, your family has been infected. Medical services have treated your family's symptoms and advised them to stay away from other people for at least 10 days to prevent spreading the disease. Who will you call on to help meet your family's needs?

One of the most important things you can do to prepare for a disaster is to form a support network. To do this, you must first recognize that you are stronger in a group of like-minded individuals than you are alone. Sometimes this is a difficult idea to accept because most people see themselves as independent and able to handle anything. However, if history has taught us anything about surviving widespread crises, it is that those affected find increased support and strength in numbers.

Everyone has unique capabilities and skills. Individually, these skills may prove valuable, but when combined with others, a much more cohesive solution can be found. Consider the contributions of a doctor, policeman, elementary school teacher, fisherman, lawyer, and cook. Any one of these people's skills might be valuable during a disaster, but when combined, they form an effective *community*.

You might believe that people will think you are bonkers if you talk about the need to establish a disaster preparedness support network. Perhaps ten years ago this would have been true, but today there is a growing sense that being prepared is vitally important. Simply put, you are not alone! People in large numbers around the world are beginning to consider and prepare for worst-case, "what if" scenarios.

Even if the people you approach have not previously considered the importance of an emergency preparedness group, it is not a hard sell. Nearly everyone has witnessed the widespread catastrophes of the past few years, including the tsunami in Japan, the tornadoes in Alabama, the oil spill in the Gulf of Mexico, the earthquake in Haiti, and the stock market collapse of 2008 to name but a handful. Events of this magnitude affect vast numbers of people, from relatives of victims, to business owners, to medical care

> **Tips**
> Invite others to be part of this grass roots movement to become better prepared.

providers. When inviting others to participate in this grass roots movement, you may find that many people will have a story to share about how their lives have already been affected by a recent disaster event.

PURPOSE OF A DP NETWORK

The purpose of a DP network is to more effectively meet life's basic needs during a disaster. Your group must decide how to meet those needs as a collective rather than as individuals. This is generally accomplished by pooling resources, sharing supplies, and leveraging skills. A few examples of each are given below:

- **Pooling resources**—A member of your network has a large pickup truck that can be used to transport debris to the local dump. Another member has a chainsaw to help cut fallen trees. A third person has a portable electric generator to help recharge batteries. Resources such as these can be shared to the benefit of everyone in the group.
- **Sharing supplies**—A member of your network has a large vegetable garden with more food than his family can consume. Another person has a 250-gallon Super Tanker that can be used to store extra water. A third member has several cords of cut wood in his backyard that can be used as fuel for wood-burning stoves. Once again, these supplies can be shared to the benefit of the entire team.
- **Leveraging skills**—A member of your network is a shortwave radio operator and can receive and relay information about your situation. Another member is an insurance salesman who can offer advice on how best to document and file claims. A third person works as a taxi cab driver and can provide details on alternate evacuation routes out of the city. Leveraging these skills can benefit everyone in the network.

NETWORK MEMBERS

It is important to distinguish between contacts and network members so as to correctly set your expectations. Contacts are people who exchange information with you. A neighbor who keeps you informed of inclement weather events is an example of a contact. When it comes to contacts, nothing is expected, and no agreements or understandings are reached. It is a simple exchange of information, which sometimes can be extremely valuable. Network members are different in that they have pledged to work as part of a team and have therefore accepted specific responsibilities.

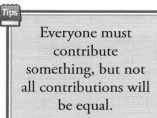

Everyone must contribute something, but not all contributions will be equal.

The fundamental rule for establishing an effective network is that everyone must bring something to the table. For some members, this might be tangible contributions, such as stored food or water. Others may elect to contribute a valuable service, such as emergency medical care or home repair. Contributions will never be exactly equal, and that's okay. However, everyone should adopt the mindset that they are part of a group and be fully committed to share and contribute to the welfare of the team.

Your network should strive to have a few members outside your immediate community. Their contribution to the group may simply be that of information conduits, relaying information to and from emergency services or concerned family members. This is particularly important for situations where local communication services have failed.

Disaster preparedness networks can include anyone but most often consist of people from four distinct groups:

- Family and friends
- Neighbors
- Service providers
- Church or civic groups

FRIENDS AND FAMILY

You will likely have the greatest success if you start your DP network with close friends and family members. These are the people with whom you feel most comfortable and who are most likely to share similar values and concerns. They may live next door, across town, or halfway around the globe. Friends and family are typically the easiest to enlist because there already exists a longstanding trust.

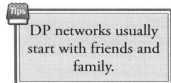

DP networks usually start with friends and family.

One way to broach the subject is to begin by discussing the tragedies currently affecting others, and then move on to express your concerns about how threats might affect your respective families. From there, you can suggest ways to work together to prepare for the most worrisome events. A DP network can start very informally—a few friends sitting around the kitchen table discussing how to keep their families safe, or it can be a more formal meeting with a detailed agenda.

If your friends or family members don't initially respond to your invitation with enthusiasm, be patient and give them time for the idea to settle in. Certainly, be careful not to come off as an overzealous alarmist, which might endanger your relationships. Starting a network is about building relationships, not destroying them.

NEIGHBORS

Inviting neighbors to be part of a disaster preparedness network is very much like inviting them to be part of a neighborhood watch program. In fact, the two can often be combined. Start with a discussion of the benefits of a DP network, much like a friendly business proposition. Then progress to discussing individual contributions. Be patient and give everyone a

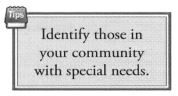

Identify those in your community with special needs.

chance to come to terms with your proposition. You don't want to be seen as a pushy neighbor. The good news is that neighbors tend to band together when times get tough. Everyone enjoys knowing that the people living next door are watching their back.

An important task for a neighborhood-based DP network is to identify people in the immediate community who might have special needs, such as the elderly or someone with a disability. Certainly, these people should be invited into your network as well, but even if they decline to participate, you should consider how best to help them through challenging times (see *Chapter 16: Special Needs*).

SERVICE PROVIDERS

Your doctor, accountant, hair dresser, pastor, veterinarian, and day-care provider are all examples of service providers. A person possessing such skills can be extremely valuable during a time when normal services are unavailable. Understandably, your relationship with these people might not be as strong as with family or friends, but these people may still be interested in being part of a network of fellow preparers.

A good way to begin is to discuss relevant concerns and preparations with them—perhaps asking your doctor about prescribing additional medications or your pastor about providing shelter in the case of a family that suddenly becomes homeless. The conversation can then easily migrate into a more personal invitation if they express an interest.

CHURCH OR CIVIC GROUPS

Churches and civic organizations (e.g., Lion's Club, Masonic Lodge, Kiwanis club) are excellent places to find individuals who might be interested in being part of a preparedness network. Both churches and civic groups are well versed in providing food and other relief, and the importance of community care and kinship is inherent in their structures.

A reasonable approach is to start by explaining what you are trying to accomplish to the church or civic organization's leaders. Encourage them to make an announcement to their broader audience. From there, hold a small question and answer session to work out the details of how the members will work together. Many local preparedness groups even hold meetings in their local church.

CREATING AN EFFECTIVE NETWORK

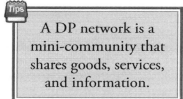

A DP network is a mini-community that shares goods, services, and information.

Neighborhood Ready

How do you establish an effective network? First and foremost, find people who share similar concerns. Everyone wants to survive, and many people are coming to realize that the world presents very real dangers. Your success in recruiting network members is more a matter of timing—connecting with people who are at the point in their life where they recognize and wish to prepare for these dangers.

Once you have identified interested parties, determine the best way for everyone to work well together. People tend to get in the right frame of mind when reminded that the activity is motivated by love and concern for one another. Being part of a disaster preparedness network requires you to be part Scout and part charity worker.

At its core, a DP network is simply a mini-community that looks out for one another. Its members share goods, services, and information; more important, they watch out for one another. Many modern communities have lost this sense of unity. There was a time when neighbors lent a hand, held potluck dinners, and worked together as part of an informal family. Sadly, for the most part, the "I'll watch your back if you'll watch mine" mentality has been replaced by "I'll leave you alone if you'll leave me alone." A preparedness network is an opportunity to reverse this trend by reestablishing a sense of working toward the common good.

Neighborhood Ready kits are available at *http://disasterpreparer.com.*

You might find the Neighborhood Ready kits (offered through *http://disasterpreparer.com*) to be helpful with establishing your DP network. Kits contain valuable worksheets to help organize your network's resources, supplies, and skills. They also include emergency contact pages for everyone in the network, as well as individual DP planning pages. Discounts are also offered on bulk purchases of this handbook.

YOUR FIRST MEETING

At your first official meeting, consider inviting everyone's families to attend. Try to make it a social gathering: share food, let the kids play; and get to know one another. The goal is to introduce everyone, talk about the basic purpose of the group, share war stories, and in general, ensure that everyone has compatible goals and agendas.

Have everyone tell a little about themselves, describing their jobs, hobbies, families, pets, and personal concerns. This will help the group to better understand the strengths that each person brings to the team. Keep politics and religion out of it unless these are intentionally going to be a cornerstone of your group.

Some people may come well versed on disaster preparedness, while others may be new to the subject—perhaps affected by a recent crisis. Try to make everyone feel that their participation is important, regardless of how much they appear willing to contribute to the group at this initial meeting.

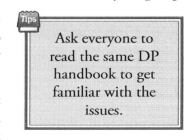

Ask everyone to read the same DP handbook to get familiar with the issues.

Discuss the basic goals of the network, but don't make any specific assignments at this stage. Rather, suggest that everyone buy or borrow a copy of your favorite disaster preparedness book—this or another one. Ask them to read it by the next meeting and jot down suggestions and concerns. By doing this, everyone will come to the second meeting with a similar understanding of the specific preparations that need to be addressed. Also, ask each person to consider what contributions they would be willing to make to the group.

THE FOLLOW-UP MEETING

Hold a follow-up meeting—perhaps a few weeks later, giving everyone time to read the handbook and think about their respective roles on the team. The follow-up meeting can either be another large family event or something a little quieter with just the adults. The goal of the second meeting is to begin to outline emergency plans. Keep things simple. The simpler the plans, the more likely they are to actually work.

Collect and combine everyone's contact information, and arrange to distribute the list to every member— perhaps through email. Much like a book club, discuss the disaster preparedness book you will be using. If members read different (or additional) books, this is a good time to swap with one another.

Begin working through the list of 14 basic needs (see *Chapter 2: Staying Alive*), prioritizing those that should be met first. Determine who has interests in specific areas, and let them take the lead on those. People tend

Working together *(FEMA photo/Bob McMillan)*

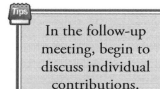
In the follow-up meeting, begin to discuss individual contributions.

to volunteer solutions to things. If you bring up the concern that the group needs some sort of water storage, a member may quickly step forward with a solution that he already has in hand. Make sure that everyone has something to do. Don't expect to find good solutions to all the needs in this initial meeting. Preparing takes time.

KEEPING THINGS MOVING

An effective network is constantly evolving. Members join, others may drop out or move away. As the group changes, so will its needs and capabilities. The network should keep a list of what every member will provide. Careful notetaking will help to identify shortages or overlaps as the group evolves as well as make responsibilities clear to its members.

Have each member identify someone in the group to share a house key with. There are many situations that require someone to have access to your house, perhaps to care for a pet, cut off utilities, or retrieve emergency supplies, not to mention when you simply lock yourself out and just need

No matter how mild or severe the crisis, work together as a group to meet your members' needs.

someone to let you in. But don't push the key sharing too quickly. Give everyone a chance to get to know and trust one another first.

When a disaster event does occur, reconnect as a group and take action to ensure that everyone's needs are being met. If the situation is mild, perhaps the power goes out for a few hours one evening, have everyone check in on each other to see how they are doing. It is this sense of camaraderie that goes a long way when things become more challenging.

If the event is more serious, have the group gather together and determine how best to handle the situation. Your network's response will depend on the particular crisis. For example, if the community gets snowed in, perhaps one member who owns a four-wheel drive vehicle can help distribute supplies.

If an ice storm brings trees down in a member's yard, the group can gather to help with cleanup. If heavy rains flood one member's home, another member can take that family in until the water recedes. Every situation will be different and thus require different responses. Being part of the preparedness network is making a commitment to help one another through difficult times. Remember the simple Neighborhood Ready motto: "We stick together."

PREPPER RESOURCES

The internet does an excellent job of bringing together people with shared interests. From bunny farmers to one-legged tap dancers, the web interconnects nearly everyone. The internet can also be useful for connecting preppers. The American Preppers Network serves as an online forum for people interested in disaster preparedness. At the website, *www.americanpreppersnetwork.com*, you will find blog postings, chat rooms, podcasts, "survival" products, and a host of prepper-related resources. The network also provides links to affiliates in every state as well as in Canada.

The internet is also alive with online broadcasts on every imaginable subject. The Preparedness Radio Network, hosted by "Dr. Prepper" (my friend, James Talmage Stevens), is the largest disaster preparedness broadcast and does an excellent job of spreading the word about the importance of getting better prepared. Topics include what-to-do-next preparedness, survivalism, self-reliance, and homesteading. Dr. Prepper also regularly hosts authors of preparedness books, industry professionals, and vendors of related food and water storage products. A schedule of shows is available at *www.bepreparedradio.com*.

Whatever method you choose to connect with others, whether it be by talking to your neighbors or exchanging email with people half way across the country, find ways to come together. Disaster preparedness is an excellent centerpiece to developing long-lasting friendships that will withstand not only the test of time but also incredible hardships.

Doctor Prepper hosts the Preparedness Radio Network

Quick Summary - Creating a DP Network

➢ You are safer and more likely to survive a serious disaster if part of a group of like-minded individuals.

➢ A DP network is most effective when members strive to pool resources, share supplies, and leverage skills.

➢ DP networks can include anyone but often consist of friends and family, neighbors, service providers, and members from church or civic organizations.

➢ Hold an initial social gathering so that everyone can get to know one another. Identify a DP handbook that everyone can read to get a clearer understanding of the needed preparations.

➢ At the follow-up meeting, begin to discuss the group's needs, allowing members to volunteer taking the lead in different areas. Also, gather and distribute contact information.

➢ When a disaster event occurs, no matter how small, connect with your network members to make sure that everyone's needs are being met.

➢ Consider becoming active in online disaster preparedness communities, such as the American Preppers Network or the Preparedness Radio Network.

CHAPTER 18

FIVE HORSEMEN OF DEATH

Some threats are so pervasive that they warrant special attention beyond just considering the 14 basic and supporting needs. To emphasize their destructive nature, these dangers are collectively referred to in this book as the "Five Horsemen of Death." Together, these natural disasters make headlines almost daily as they kill people and destroy property around the globe.

This chapter focuses on understanding and preparing for five types of natural disasters. This material is meant to supplement, not replace, the broader DP plan discussions throughout the book. Your priorities should be to first develop a practical DP plan that will meet your family's needs through nearly any disaster, and then consult this chapter for additional information and preparations regarding these specific threats.

Five Horsemen of Death
- Earthquakes
- Tsunamis
- Hurricanes
- Tornadoes
- Floods

EARTHQUAKES

Earthquakes are the result of the sudden movement of a portion of the earth's crust. At the surface, they are felt as shaking and sometimes ground displacement. Earthquakes may last anywhere from a few seconds to several minutes and can occur day or night at any time of the year.

Most earthquakes are due to ruptures along thin bands of crushed rock separating sections of the earth's crust, known as geological faults. The bands are held immobile only by the friction along the fault. When they slip with respect to one another, a rupture occurs, causing the surface to shake as the plates come to a new equilibrium. Other events can also cause earthquakes, including volcanic activity, landslides, mine blasts, and nuclear explosions. The point of initial rupture is known as the *hypocenter*, and the location directly above it at the earth's surface is referred to as the *epicenter*.

The U.S. Geological Service has created seismic maps showing the likelihood of earthquakes for various parts of the world—see illustration on the next page. Take the time to understand the earthquake risk that your community faces even if you live outside the most obvious danger zones.

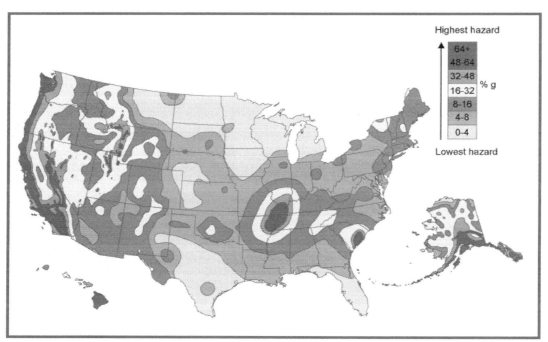

Seismic maps of the United States *(U.S. Geological Service)*[266]

DANGERS OF EARTHQUAKES

Every year, millions of earthquakes occur around the planet. Fortunately, most are small enough not to be felt, let alone cause damage. However, when a larger earthquake does occur, it can cause many serious problems, including:[264]

- **Structural damage**—Homes, businesses, bridges, roads, airports, and other structures may be damaged or collapse due to the violent shaking and surface faulting.
- **Disruption of utilities**—Damage to distribution pipes, power lines, and transportation systems can lead to disruption of services and danger to those nearby.

Devastation in Haiti from the 2010 magnitude 7.0 earthquake
(UN Photo/Logan Abassi)

- **Landslides and avalanches**—Shaking can cause rock, mud, and snow to give way and fall or slide.
- **Surface faulting**—The surface can split either vertically (resulting in a step) or horizontally (opening a gap).
- **Tsunamis**—The seabed can be displaced, resulting in a tsunami (see *Tsunami* later in this chapter).
- **Liquefaction**—Water-logged soils may lose stiffness, causing the ground to sink or slide.
- **Flash floods**—Flooding can occur due to liquefaction near rivers or lakes.

Table 18-1 Earthquake Severity and Frequency[268, 269, 270, 271]

Magnitude	Description	Severity	Approximate Frequency (# per year)
< 2.0	Micro	Not typically able to be felt by people; recorded by local seismographs	Several Million
2.0 – 2.9	Minor		365,000-1.3M
3.0 – 3.9		Often felt; rarely causes damage	49,000-130,000
4.0 – 4.9	Light	Noticeable shaking of indoor items; limited damage	6,200-13,000
5.0 – 5.9	Moderate	Heavy damage to poorly constructed buildings; minor damage to well constructed buildings	800-1,300
6.0 – 6.9	Strong	Destructive over area up to one hundred miles	120-134
7.0 – 7.9	Major	Serious damage across a very large area	15-18
8.0 – 8.9			1
9.0 – 9.9	Great	Devastating damage across a very large area	1 per 20 years
≥ 10.0		Never recorded	Extremely rare

As we all know, the net effect is terrible loss of life and destruction of property. Earthquakes have been responsible for several of our most recent large-scale disasters, including the Japanese tsunami of 2011, the devastation in Haiti and Chile in 2010, and the Indian Ocean tsunami of 2004. Together these four events have killed hundreds of thousands of people and destroyed billions of dollars of property.

MAGNITUDE SCALES

The vibrations caused by earthquakes are recorded and measured by seismographs and quantified by magnitude scale measurement systems. The most recognized scale, the Richter scale, was developed in 1935 as a method of comparing the magnitudes of earthquakes. The scale is logarithmic in nature, meaning that an increase in one number on the scale corresponds to a factor of ten increase in seismic wave magnitude. For example, an earthquake with a magnitude of 3.0 on the Richter scale is ten times stronger than one with a magnitude of 2.0. The Richter scale has largely been replaced by the moment magnitude scale (MMS), but the scales are similar enough that distinction between the two is rarely noted.

PREPARATIONS AND ACTIONS

Earthquakes occur suddenly and cannot currently be predicted with any degree of certainty. The best way to prepare is to implement a thorough DP plan for your family—discussed throughout this handbook. A broad-based plan will help you to cope with the many possible impacts of an earthquake, including disruptions in electricity, heating/cooling, water, communication, and so forth.

BEFORE AN EARTHQUAKE[265, 267]

Beyond the establishment of a thorough DP plan based around the 14 needs, consider the following additional preparations:

1. Check for hazards in and near your home:
   a. Fasten shelves to wall studs.
   b. Put heavy objects on lower shelves.
   c. Store breakables (e.g., bottles and dishes) and poisons (e.g., weed killers, pesticides, gasoline) in cabinets that latch shut.
   d. Block rollers of large appliances, such as refrigerators and washers.
   e. Install flexible gas lines to appliances.
   f. Hang heavy pictures and mirrors away from beds, couches, and chairs.
   g. Brace light fixtures.
   h. Secure water heater to wall studs using straps.
   i. Repair any cracks in the ceiling, walls, or foundations.
   j. Strengthen structure by bracing chimney, sheathing crawlspace with plywood, installing anchor bolts or plates between sill and foundation (if not already present), and adding braces between beams and posts.
2. Identify safe places in and around your home and place of work that could provide shelter. Safety might be found under a piece of sturdy furniture or against an inside wall. Stay away from windows. If outside, stay clear of buildings, trees, electrical lines, or other heavy structures that might collapse.
3. Store your important papers in a fire safe and digital copies on a secure flash drive—see the *Appendix* for a complete list of documents to store.
4. Educate your family:
   a. Teach your family about the dangers of earthquakes.
   b. Ensure that all adults know how to shut off the utilities.
   c. Have your children practice calling 911. Role-play the emergency conversation with them so they feel confident conveying important information.
   d. Practice drills with your family. Identify safe spots, note danger areas, and discuss post-earthquake hazards, such as aftershocks, tsunamis, downed electrical wires, and gas leaks.

DURING AN EARTHQUAKE[264]

When experiencing an earthquake, your immediate goal is to prevent personal injury. Most earthquake-related deaths are from collapsing walls, flying debris, and falling objects. Keep this in mind when taking shelter.

The actions you should take largely depend on your location. Consider the following three scenarios:

Indoors

- Drop to the ground, take cover under a sturdy piece of furniture, and hold on until the shaking stops.

Building destroyed in 2010 Chilean earthquake *(Wikimedia Commons/Claudio Nunez)*

- Stay away from obvious dangers, such as windows, outermost walls, large lighting fixtures, and anything that might collapse onto you.
- If you are in bed, cover your head with the pillow and hold on.
- Unless you feel your current location is particularly hazardous, don't attempt to move to another room or evacuate outdoors. It is usually safer to stay put.

Outdoors

- Move away from buildings, streetlights, trees, utility poles, and anything else that might collapse or fall on you.
- Once in the open, drop to the ground and stay put.

In a Vehicle

- Stop the vehicle quickly but safely away from buildings, trees, overpasses, and utility poles.
- Stay in the vehicle and protect yourself as best as possible—covering head, shielding face, crouching down.
- Once the tremor subsides, avoid roads, bridges, or ramps that might have been damaged.

AFTER AN EARTHQUAKE

When the tremors subside, do not immediately assume that the danger has passed. Rather, stay alert and assess your situation, taking any actions necessary to ensure your family's safety. The earthquake may have introduced numerous hazards, including live electrical wires, gas leaks, sharp objects, uneven ground, and collapsing structures. Also, be prepared for aftershocks. These secondary shockwaves are usually less violent but can still cause additional damage, especially to weakened structures. They can

Digging out survivors of collapsed hotel in Port-au-Prince *(U.S. Navy)*

occur within hours, days, or even weeks after the main quake.

Besides being aware of the immediate dangers, consider the following additional actions:

- Gather with family, friends, and neighbors to assess who is missing.
- Help those who need special assistance or are trapped or injured. Administer first aid as needed—see *Chapter 11: Medical/First Aid*.
- If possible, put out any small fires—see *Fire Extinguishers* in *Chapter 5: Shelter*.
- Listen to local emergency broadcasts using your NOAA All Hazards weather radio.
- Limit your use of the telephone (both cell and land) to keep from overloading systems.
- Stay away from the worst affected areas and out of damaged buildings.
- If you are near the coast, tune in to emergency broadcasts and be ready to move inland or to higher ground in the event of a possible tsunami—see *Tsunami*.
- Inspect your property, being careful about things that might fall from cabinets or shelves. Pay particular attention to your chimney because unnoticed damage might lead to a subsequent fire.
- Clean up any spills, such as medicines, bleaches, gasoline, etc.
- Inspect your utilities, checking for gas, water, and sewage leaks as well as electrical system damage. Call the appropriate service providers if damage is detected.
- If you must evacuate, leave a note on your door indicating where you have gone—see *Appendix*.

If you become trapped under debris, your goal is to stay alive until you can free yourself or be rescued. Some simple steps will help you hang on:

- Do not give up hope! Continually remind yourself that people are actively searching for you. It is only a matter of time before you are freed and safe.
- Do not light a match because the gas lines may have been ruptured.
- If you see a clear path out, try to extricate yourself from the rubble. Be very careful not to dislodge anything that might cause debris to fall on you.
- If you are unable to free yourself, remain still so that you don't cause the debris to settle.
- If dust is around you, cover your mouth and nose with clothing or a handkerchief.
- Make noise! Let others know that you are alive and need to be rescued. Tap on pipes or walls; blow a whistle (if you are in your emergency shelter); shout for help—but be careful not to inhale dust.

For a real-time listing of earthquake events, see: *http://earthquake.usgs.gov/earthquakes*.

TSUNAMIS

The Japanese word "tsunami" translates as "harbor wave." A tsunami (a.k.a. tidal wave) is a *series* of ocean waves caused by a large, abrupt disturbance of the sea surface. They are usually the result of earthquakes but can also be caused by underwater landslides, volcanic activity, and asteroid strikes. When a tsunami first occurs in deep water, it may only be a few feet in height but travels very rapidly. As it reaches shallower water, the propagating speed slows and the height builds (see illustration). If the disturbance occurs close to the coast, the resulting tsunami can demolish nearby communities with little warning. A tsunami can also travel across the ocean to strike distant coastal regions and claim thousands of additional lives. Two recent tsunamis, one striking Japan in 2011 and the other striking Indonesia and surrounding areas in 2004, killed over 250,000 people and disrupted the lives of countless others.

DETECTING A TSUNAMI [272]

Tsunami warning systems are designed to minimize loss of life. Predicting exactly when and where tsunamis will occur is not currently possible, however, forecasting their arrival and area of impact is done through modeling, seismometers, coastal tidal gauges, and deep ocean tsunami detectors. Following the tragic 2004 Indian Ocean tsunami, Deep-ocean Assessment and Report of Tsunami (DART) buoy stations were deployed in significant numbers. There are currently 39 U.S.-operated stations in the Pacific Ocean, Atlantic Ocean, and Caribbean Sea. The DART systems provide real-time tsunami detection by

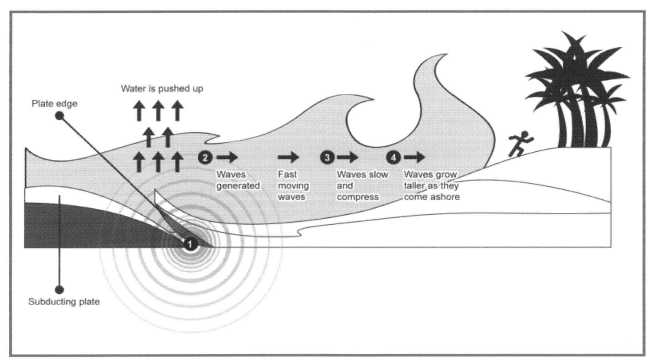

Tsunami generation and propagation[273]

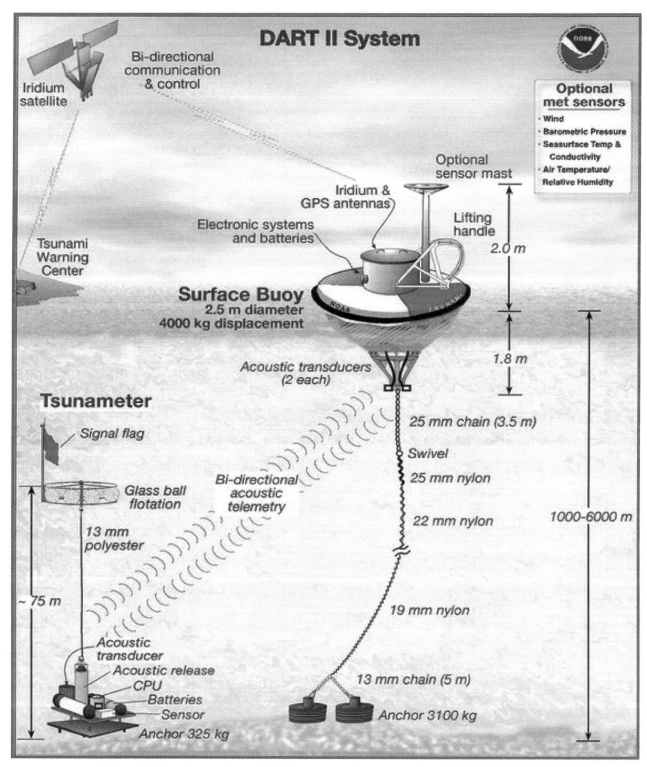

DART II System

Iridium satellite

Bi-directional communication & control

Tsunami Warning Center

Optional met sensors
- Wind
- Barometric Pressure
- Seasurface Temp & Conductivity
- Air Temperature/ Relative Humidity

Optional sensor mast

Iridium & GPS antennas

Electronic systems and batteries

Lifting handle

2.0 m

Surface Buoy
2.5 m diameter
4000 kg displacement

Acoustic transducers (2 each)

1.8 m

25 mm chain (3.5 m)

Swivel

25 mm nylon

22 mm nylon

1000-6000 m

Tsunameter

Signal flag

Glass ball flotation

Bi-directional acoustic telemetry

13 mm polyester

~ 75 m

19 mm nylon

Acoustic transducer
Acoustic release
CPU
Batteries
Sensor
Anchor 325 kg

13 mm chain (5 m)

Anchor 3100 kg

The newest DART II tsunami-detection buoy stations *(illustration by NOAA)*

> ### Tsunami-related Messages
>
> ➤ **Warning**—Danger is imminent! Immediately head for higher ground, and follow any official emergency instructions.
> ➤ **Advisory**—There is a possibility of dangerous waves or local currents. Move away from the beach, and stay out of the water. Be ready to move further inland.
> ➤ **Watch**—There exists a potential danger. Stay tuned for more information and be ready to act.
> ➤ **Information Statement**—No danger is present locally, but a distant ocean basin may be affected.

monitoring the ocean activity and then relaying it to land-based centers. The systems consist of surface buoys and seafloor sensors that detect pressure changes (see illustration on previous page).

The National Oceanic and Atmospheric Administration (NOAA) is responsible for issuing tsunami warnings to the United States. The U.S. has two warning centers: the West Coast/Alaska Tsunami Warning Center (WC/ATWC) and the Pacific Tsunami Warning Center (PTWC). The WC/ATWC is responsible for the coastal regions on the continental U.S. as well as Alaska, Puerto Rico, and the Virgin Islands. The PTWC is responsible for Hawaii and the Pacific Island Territories. The tsunami warning centers provide tsunami-related messaging to the public, which may include the use of sirens, email, faxes, radio/TV broadcasts, and texts. The messages are also relayed by the National Weather Service Forecast Offices and broadcast across NOAA All Hazards weather radios and the Emergency Alert System (AES).

Unfortunately, the current warning system cannot protect against tsunamis that occur suddenly near the coast. Some tsunamis (such as the one in Hokkaido, Japan in 1993) give only a few minutes warning, making it nearly impossible for citizens to evacuate.

What is arguably even more important than high technology detection and modeling is community preparedness. Those living anywhere near an ocean coast are at risk of a tsunami. Certainly those in the U.S. are not immune to this danger. Within the last 150 years, tsunamis have struck Hawaii, Alaska, California, Oregon, Washington, American Samoa, Puerto Rico, and the U.S. Virgin Islands.

In tsunami-prone areas, citizens should recognize the warning signs of a tsunami and have a plan for escaping from its deadly path.

> ### Warning Signs
>
> ➤ Strong ground shaking
> ➤ Loud roar from the ocean
> ➤ Water receding from the shore line unusually far, *or*
> ➤ Water level rising rapidly

Tsunami in Japan leaves behind terrible carnage *(U.S. Navy)*

PREPARATIONS AND ACTIONS

> If you suspect a tsunami, immediately rush inland or to higher ground.

Tsunamis are an excellent example of the value of the cardinal rule—some disasters can only be survived by getting out of their way. Conventional preparations will have little if any value. A tsunami is so powerful that it can destroy nearly anything in its path. You are encouraged to watch some of the videos of the 2011 Japan tsunami event to get a better understanding of the complete and utter destruction that these waves bring.

Surviving a tsunami requires two things: staying alert to the warning signs and taking action to get out of harm's way. If you suspect that a tsunami may be imminent, or if an official tsunami warning is issued, immediately seek higher ground or move inland. Do not wait for an official warning. Trust your instincts and take action. Depending on where the disturbance occurred, you may only have a few minutes to get to safety. Tsunamis occur as a series of waves, which may continue inundating the area for many hours. Do not return to coastal regions until officials declare the emergency to be over.

> For the latest tsunami warnings, watches, and advisories, see: WC/ATWC—*http://wcatwc.arh.noaa.gov*, and PTC— *http://ptwc.weather.gov*.

HURRICANES

Hurricanes are a type of tropical cyclone, a generic term for an intense tropical weather system with thunderstorms and surface circulation with sustained winds in excess of 74 mph. Ingredients for a hurricane usually include some pre-existing weather disturbance, warm tropical oceans, moisture, and light winds. If conditions persist for long enough, they may create a hurricane. These dangerous weather events are frequently accompanied by strong thunderstorms, flooding, tornadoes, and lightning. In the Northern Hemisphere, hurricanes circulate counterclockwise, and in the Southern Hemisphere, they generally circulate clockwise.

HURRICANE SEASON

The Atlantic hurricane season is from June to November, with the peak season being from mid-August through October. Each year on average, there are eleven tropical storms that develop over the Atlantic Ocean, Caribbean Sea, and Gulf of Mexico. Six of these typically become hurricanes. Every three years, approximately five hurricanes strike the U.S. coastline, killing from 50 to 100 people. Of these five, two are typically classified as *major* hurricanes (i.e., Category 3 or above according the Saffir-Simpson Scale). A NOAA illustration showing the landfalling hurricanes from 1950 to 2010 is shown on the next page.

With over 60% of our nation's population living in affected coastal states, the losses can be significant and widespread. From 1900 to the present, over 17,000 people in the U.S. have been killed by hurricanes.[274] Devastation can be catastrophic to the coastal regions, and associated flooding can reach hundreds of miles inland.

Hurricane Floyd *(NASA photo)*

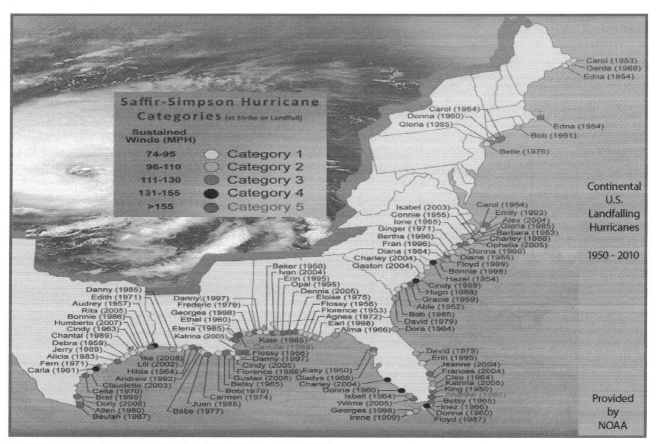

Landfalling hurricanes for Continental U.S. from 1950–2010 *(courtesy of NOAA)*

Table 18-2 Saffir-Simpson Hurricane Scale[274, 275, 277]

Scale Number (Category)	Sustained Winds (MPH)	Damage	Storm Surge
1	74-95	Minimal: Unanchored mobile homes, vegetation and signs damaged.	4-5 feet
2	96-110	Moderate: All mobile homes, roofs, and small boats damaged; flooding.	6-8 feet
3	111-130	Extensive: Small buildings damaged, roads cut off.	9-12 feet
4	131-155	Extreme: Roofs destroyed, trees down, roads cut off, mobile homes destroyed. Beach homes flooded.	13-18 feet
5	More than 155	Catastrophic: Most buildings destroyed. Vegetation destroyed. Major roads cut off. Homes flooded.	Greater than 18 feet

SEVERITY SCALE

The Saffir-Simpson Scale is used to classify hurricanes based on their sustained winds, expected damage, and storm surge. Hurricanes ranked as Category 3 or greater are considered *major* hurricanes—see Table 18-2. All hurricanes, however, are extremely dangerous and should warrant your full attention.

The category of the hurricane does not necessarily dictate the level of damage. Lower category storms may produce greater damage than higher category ones depending on where they strike and the particular hazards associated with them (such as tornadoes and flooding).

NAMING HURRICANES

Hurricanes are named according to lists originated by the National Hurricane Center. There are six lists in total, and they are rotated annually so that each list is repeated every six years—see Table 18-3 below. The lists used in 2011 will be used again in 2017. Hurricane names alternate between male and female names. If a hurricane is particularly deadly or costly, its name may be retired and a new name substituted. Examples of some names that have been retired include Katrina, Ike, Isabel, and Gustav.[283]

Table 18-3 Names for Atlantic Basin Tropical Cyclones[274]

2011	2012	2013	2014	2015	2016
Arlene	Alberto	Andrea	Arthur	Ana	Alex
Bret	Beryl	Barry	Bertha	Bill	Bonnie
Cindy	Chris	Chantal	Cristobal	Claudette	Colin
Don	Debby	Dorian	Dolly	Danny	Danielle
Emily	Ernesto	Erin	Edouard	Erika	Earl
Franklin	Florence	Fernand	Fay	Fred	Fiona
Gert	Gordon	Gabrielle	Gonzalo	Grace	Gaston
Harvey	Helene	Humberto	Hanna	Henri	Hermine
Irene	Isaac	Ingrid	Isaias	Ida	Igor
Jose	Joyce	Jerry	Josephine	Joaquin	Julia
Katia	Kirk	Karen	Kyle	Kate	Karl
Lee	Leslie	Lorenzo	Laura	Larry	Lisa
Maria	Michael	Melissa	Marco	Mindy	Matthew
Nate	Nadine	Nestor	Nana	Nicholas	Nicole
Ophelia	Oscar	Olga	Omar	Odette	Otto
Philippe	Patty	Pablo	Paulette	Peter	Paula
Rina	Rafael	Rebekah	Rene	Rose	Richard
Sean	Sandy	Sebastien	Sally	Sam	Shary
Tammy	Tony	Tanya	Teddy	Teresa	Tomas
Vince	Valerie	Van	Vicky	Victor	Virginie
Whitney	William	Wendy	Wilfred	Wanda	Walter

HAZARDS

Hurricanes are particularly dangerous because they introduce five significant hazards: storm surge, marine safety, high winds, tornadoes, and inland flooding. Any one of these can cause considerable loss of life and damage to property. Together they represent a nearly worst-case scenario that might fully test your preparations.

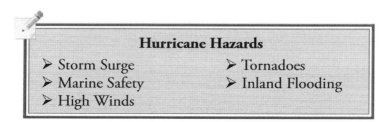

Hurricane Hazards
- Storm Surge
- Marine Safety
- High Winds
- Tornadoes
- Inland Flooding

STORM SURGE

Storm surge is the rise of water generated by a storm. This rise in water level can cause extreme flooding along coastal areas, especially when it coincides with the normal astronomical high tide level. These combined "storm tides" can reach levels of 25 feet or greater and can be up to a thousand miles wide.

Storm surge is a result of water being pushed toward the shore by the force of the winds. The level of storm surge depends on many factors, including storm intensity and size, forward speed, angle of approach to land, central pressure, and coastal and continental shelf features.

For those living on the coast, storm surge will likely be the greatest threat to life and property. Water weighs almost a ton per cubic yard, and extended pounding by waves can demolish nearly any structure. Roads and beaches can also be eroded by the water currents, and boats confined to marinas are often severely damaged.

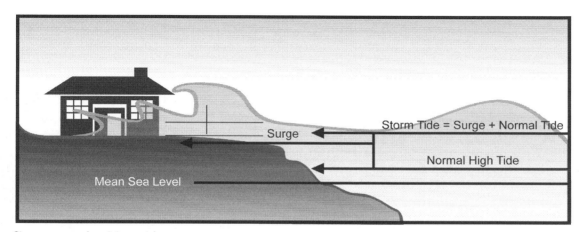

Storm surge can devastate coastal areas

House destroyed by storm surge *(FEMA photo/Dave Gatley)*

MARINE SAFETY

Hurricanes pose an obvious threat to maritime activities, including damage to vessels both at sea and in port. The best way to avoid this danger is to stay out of the path of the storm—the cardinal rule yet again. This starts with becoming familiar with favored regions and tracks for hurricane

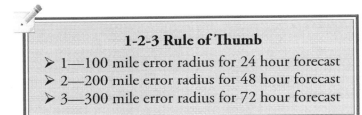

1-2-3 Rule of Thumb
➤ 1—100 mile error radius for 24 hour forecast
➤ 2—200 mile error radius for 48 hour forecast
➤ 3—300 mile error radius for 72 hour forecast

development and movement in the North Atlantic. If a storm has already formed and is approaching, boaters should make it a very high priority to stay up to date as to its projected track.

Hurricane track errors are usually smallest while the storm is moving in a west to west-northwest track, south of the Atlantic subtropical ridge. Increased uncertainty in track forecasts occurs during recurvature when the system has little environmental steering, such as when storms are in the Gulf of Mexico or Western Caribbean. Boaters are encouraged to follow the 1-2-3 rule of thumb describing hurricane path error (see tip box).[276]

HIGH WINDS

Hurricane-force winds (i.e., those in excess of 74 mph) can break out windows as well as destroy mobile homes and poorly constructed buildings. High-rise buildings are also particularly vulnerable to high winds. Occupants should stay below the tenth floor but above any floors that are at risk of flooding. Small objects such as signs, roofing material, swing sets, barbeque grills, and lawn furniture can all become dangerous airborne debris. Additionally, high winds can cause extensive damage to trees, power poles, and towers, which in turn can lead to lengthy interruptions in utility services.

Damage to home from high winds *(FEMA photo/Marvin Nauman)*

TORNADOES

More than half of all landfalling hurricanes spawn tornadoes. The tornadoes are usually produced along the right-front quadrant of the hurricane, well away from the eye. Tornadoes resulting from hurricanes are not typically accompanied by hail or significant lightning—unlike land originating tornadoes. For specific preparations, see *Tornadoes* later in this chapter.

INLAND FLOODING

Over the last 30 years, inland flooding has been responsible for more than half of the deaths associated with hurricanes. Widespread torrential rains can cause deadly flooding and mud slides hundreds of miles from the coast. The level of flooding is difficult to predict and does not correlate well with the wind speed of the hurricane. More information regarding this very serious threat is given later in this chapter under *Floods*.

Mobile home destroyed by hurricane-spawned tornado *(FEMA photo/Marvin Nauman)*

PREPARATIONS AND ACTIONS

Hurricanes routinely cause extensive property damage, shortages, and lengthy interruptions in utilities. It is therefore critical that you take the time to implement a thorough disaster preparedness plan that will meet your family's 14 needs through this type of emergency. Beyond those preparations, there are some additional steps that can be taken to minimize property losses and reduce the danger to family members.

BEFORE A HURRICANE THREATENS

Before a hurricane threatens, take steps to protect your home and property. The four areas of a home that require structural attention are the roof, windows, doors, and garage door. Improvements to protect your home against high winds were discussed in *Structural Improvements* in *Chapter 5: Shelter*.

In addition to structural improvements, there are several other preparatory steps that can help to minimize damage and injury:

- Cut back trees and bushes.
- Secure and clean out gutters and downspouts.
- Select a structurally sound room in your home to act as a wind-resistant shelter. Stock it with the basic supplies outlined in *Tornadoes*.
- Stock up on cleanup supplies, including large garbage bags, brooms, shovels, etc.
- Consider getting flood insurance. Understand that there is normally a waiting period before flood insurance becomes active, so last minute purchases won't protect you.

Inland flooding in Louisiana *(FEMA photo/Jocelyn Augustino)*

DURING A HURRICANE WATCH AND WARNING

If a hurricane is likely to affect your area, closely monitor the radio, TV, or internet for timely information regarding watches and warnings.

Hurricane Watch—Hurricane conditions are *possible* within the next 48 hours. Monitor weather events using a NOAA weather radio or local news broadcasts. Review emergency supplies, and be prepared to act if a warning is issued.

If a hurricane *watch* is announced for your area, take the following precautions:

- Keep up to date on the storm's progress and official announcements using local TV stations, a NOAA weather radio, or on the internet at *http://www.nhc.noaa.gov/*.
- Secure signs, roofing material, swing sets, barbeque grills, lawn furniture, antennas, and other outdoor objects that might be blown around by the wind.
- Double check your emergency supplies, considering all 14 basic and supporting needs.
- Secure your boat somewhere that will be unaffected by the storm surge.
- Fully fuel your automobiles, and make sure they are in good working order.
- Review possible evacuation routes.

Hurricane Warning—Hurricane conditions are *expected* within the next 36 hours. Complete your storm preparations, and be ready to evacuate if directed to do so.

If a hurricane *warning* is announced for your area, take the following additional precautions:

- Close storm shutters or board up windows; cover skylights and glass doors.
- Monitor local news broadcasts that relate to road conditions.
- If living in a mobile home, ensure that the tiedowns are secure, and then move to a sturdier structure. Don't try to ride out a hurricane in a mobile home.
- Gather keepsakes, valuables, and important papers on the highest level of your home, preferably in a waterproof container.
- Turn the refrigerator and freezer to their coldest settings in case of power outage.
- Evacuate if ordered to do so by authorities. When evacuating:
  - Turn off the electricity and main water valve to the home.
  - Pack vehicles with appropriate supplies, keepsakes, and valuables. Be prepared to become stranded on the roadway or in a shelter.
  - Inform a loved one who is away from the affected area of your travel plans.
  - If flooding is likely, elevate furniture and belongings as time permits.
  - Lock up your home securely before departing.
  - Leave a note on your door indicating that you have evacuated—see *Appendix*.
  - Avoid flooded roads or washed-out bridges.

DURING A HURRICANE

When a hurricane is directly affecting your community, your primary goal is to hunker down and stay safe. Take the following precautions:

- Stay indoors, away from windows and glass doors.
- Close curtains and blinds to help protect from flying glass and other debris.
- Secure and brace external doors. Close all interior doors.
- Keep a supply of flashlights (or other light sources) and spare batteries on hand.
- If power is lost, turn off major appliances to reduce surge currents when restored.
- Be prepared to retreat to an in-home shelter as conditions deteriorate.

AFTER A HURRICANE

After the hurricane has passed, it is time to inspect for damage and give appropriate aid to others.

Cleanup after Hurricane Ike *(FEMA photo/Robert Kaufmann)*

- Give first aid to those in need, and call emergency services. Do not move those who are seriously injured unless they are in immediate danger.
- Inspect the inside of your home thoroughly, being cautious of exposed nails, broken glass, and overturned furniture.
- Open windows and doors to ventilate and dry out your home as necessary.
- Inspect the exterior of your home, but be careful of downed power lines and debris. Report electrical hazards to the electric company, police, or fire department. Check for gas leaks. Report any hissing noises or smell of gas to the utility company. Check for water and sewage leaks. If systems appear operational, turn back on electricity and water.
- Come together as a community to meet collective needs and provide security as required.
- Beware of wild animals and snakes that may have been driven to your community due to rising flood waters.
- Inspect refrigerated foods, tossing out anything that has spoiled—see *Chapter 3: Food*.
- Take pictures of any damage to your home and contents *before* beginning cleanup operations.
- Limit telephone use to important phone calls.
- If you evacuated, return home only after authorities indicate that it is safe to do so.

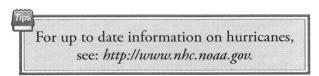

Tips

For up to date information on hurricanes, see: *http://www.nhc.noaa.gov.*

TORNADOES

> Tornadoes injure or kill more than 1,500 people annually in the U.S.

Tornadoes are violent, rotating columns of air that extend from a thunderstorm to the ground. They often form when a warm front encounters a cold front, but the specifics of why they form in some circumstances and not in others are not fully understood. Hurricanes and tropical storms can also spawn tornadoes.

The most destructive tornadoes are usually the result of supercells—rotating thunderstorms with a well-defined circulation. On average, the U.S. experiences 800 tornadoes each year, resulting in 80 deaths and 1,500 injuries. Since 1950, tornadoes have caused more than 6,000 deaths and 100,000 injuries.[275] With over 520 people killed by tornadoes in 2011 (so far), it has been the deadliest tornado season in more than sixty years.

Tornadoes can last from several seconds to more than an hour, with the majority lasting less than 10 minutes. Most tornadoes that occur in the northern hemisphere rotate counterclockwise, while those occurring south of the equator rotate clockwise—analogous to hurricanes. Tornadoes generally move from southwest to northeast, or west to east. They can change direction suddenly, even reversing direction and backtracking across already devastated areas.

Waterspouts are a special class of tornadoes that occur over water, typically along the southeast U.S. coast. They are smaller and weaker than large, land tornadoes but can still be very dangerous, capable of overturning boats, damaging larger ships, or coming ashore and affecting communities.

RATING TORNADOES

Tornadoes are rated using the Fujita scale (a.k.a. the F scale) or Operational Enhanced F-scale (a.k.a. the EF scale). The scales classify tornadoes according to their damage potential, with F0/EF0 being the

Funnel cloud *(Wikimedia Commons/Justin Hobson)*

Table 18-4 Tornado F and EF Scales[278]

Fujita Scale		Operational EF Scale	
F Number	3-second Gust (mph)	EF Number	3-second Gust (mph)
0	45-78	0	65-85
1	79-117	1	86-110
2	118-161	2	111-135
3	162-209	3	136-165
4	210-261	4	166-200
5	262-317	5	Over 200

weakest tornadoes and F5/EF5 being the most dangerous. Both scales are a set of wind estimates (not measurements) based on expected damage to different structure types. The EF scale has largely replaced the F scale in the United States.

While there is a statistical trend for wider tornadoes to cause higher damage ratings, even very narrow rope tornadoes can cause EF4 or EF5 damage. Likewise, there have been very wide tornadoes with modest damage equivalent ratings of only EF0 or EF1.

WARNING SIGNS

Even though it is true that wind, rain, and lightning are all indications of a dangerous thunderstorm, which in turn can spawn a tornado, they cannot be taken as reliable indicators of a tornado threat.

Better warning signs of a tornado are:

- Dark skies, often greenish in color
- Large hail
- A large, dark cloud, especially if rotating
- A load roaring noise, which might sound similar to a freight train

WATCHES AND WARNINGS

The Storm Prediction Center (SPC) issues watches for a variety of weather conditions, including tornadoes. Local National Weather Safety offices issue official tornado warnings. It is important to understand the difference between the two.

Tornado Watch—Conditions are favorable for a tornado.
Tornado Warning—A tornado has either been sighted or detected by radar.

Tornado Shelter Supplies

- ➤ Blankets and pillows
- ➤ Battery-operated NOAA weather radio
- ➤ Flashlights
- ➤ Spare batteries
- ➤ Cell phone
- ➤ First aid kit
- ➤ Leather gloves
- ➤ Disposable respirators
- ➤ Snacks and bottled water
- ➤ Whistle
- ➤ Some cards, books, games, or puzzles

Stairwell closets make good in-home shelters

PREPARATIONS AND ACTIONS

The time to prepare for a tornado is when the skies are clear. Start by developing a family tornado plan based on the type of dwelling that you live in. Have your family practice a simple "get to the shelter" drill at least once a year—see *Chapter 19: Trial by Fire* for this and other exercises. A reasonable goal is to require no more than thirty seconds for everyone in the family to assemble in their in-home shelter. It's also advisable to designate an after-the-storm meeting place in case the family is separated.

The optimal in-home retreat is a basement, underground storm shelter, or cellar. If your home doesn't have an underground location, then set up a tornado shelter in a small closet or bathroom, preferably one without windows and that does not share a wall with the outdoors. Stock the retreat with blankets and pillows that can be used to cover your head and body.

Tornadoes are violent, deadly events, but fortunately, they are short lived. The worst will be over quickly, so your preparations should focus on meeting immediate needs.

DURING THE TORNADO THREAT [279]

When you know that high winds are possible, secure your outdoor furniture, grills, swing sets, and garbage cans to prevent them from blowing away or becoming a hazard. If possible, park your automobiles inside the garage, and bring any pets indoors. In the event that a tornado warning sounds or you see signs of a tornado, have your family immediately retreat to the in-home shelter. Do not take time to open the windows of your home in an attempt to equalize pressures, as this is completely ineffective in reducing damage.

If you are in an office building, hospital, nursing home, or skyscraper, go to an enclosed, windowless area in the center of the building on the lowest possible level, or to an interior stairwell. Get underneath

Aftermath of 2011 Alabama tornadoes *(NOAA Photo Library)*

a piece of sturdy furniture or crouch down and cover your head with your hands. Stay off the elevators because you could become trapped if power is lost.

If you are in a mobile home, an automobile, or the open outdoors, evacuate immediately to a sturdy, permanent building. If a building is not within evacuation distance, lie flat in a ditch or on other low ground away from the mobile home or vehicle, and cover your head with your hands. Try to stay clear of things that might be blown down onto you, and watch out for rising water. Avoid seeking shelter under bridges or overpasses, which offer little protection from flying debris.

AFTER THE TORNADO

If your home is damaged by a tornado, gather your family and pets and evacuate the structure. Render aid to those in need, but be careful to avoid the many possible dangers:

- Downed power lines
- Gas leaks
- Fire hazards
- Collapsing structures
- Nails, broken glass, or other debris

Stay calm and wait for emergency personnel to arrive. Take numerous photos of the damage to your property *before* you begin cleaning up. Fill out the appropriate insurance claims, and report any substantial damage to FEMA. If the area is declared a disaster zone, FEMA often provides emergency assistances, such as food compensation and temporary housing.

For up-to-date severe weather watches and warnings, including tornadoes, go to: *http://www.spc.noaa.gov/products/wwa.*

FLOODS

Floods occur when water swells or overflows, submerging the land. Flooding can be caused by numerous events, including: extensive rainfall, hurricanes, monsoons, melting of snow or ice, tidal changes, tsunamis, accidental damage to piping, and breaks in barriers (such as levees, dams, floodwalls). Some floods occur slowly as the ground becomes saturated. Other floods can occur very quickly, leaving motorists trapped and homeowners unable to evacuate.

Floods cause extensive property damage, destroying homes, businesses, bridges, vehicles, and infrastructure items, such as sewage systems, roadways, and canals. They can contaminate the water supply, forcing services to be cut off or orders to boil water to be issued. Widespread flooding can also destroy crops and interrupt the food distribution system.

PREPARATIONS AND ACTIONS BEFORE A FLOOD

Flooding can threaten your life, compromise your ability to get food, render your water supply unsafe, leave roads impassable, damage your home and property, and force utilities to be shut off. Therefore, your most important preparation is to establish a thorough DP plan that addresses all of your family's needs—see *Chapters 3–17.*

Additional preparations for a flood might include:

- Determine the zone in which you live. Zone classifications indicate the likelihood of community flooding—see Table 18-5. Flood maps are available online through the FEMA Map Service Center.[280]
- Know the base flood elevation (BFE) of your community. This is essentially the height that a flood is expected to reach once every century. Water-resistant materials should be used on anything below the BFE level.

New Orleans resident awaiting rescue *(FEMA photo/Jocelyn Augustino)*

Table 18-5 Flood Zone Classifications[282]

Flood Zone Classification	Zones Designators	Likelihood of Flooding
Special Flood Hazard Areas	A, AO, AC, AH, A1-A30, AE, A99, AR, AR/AE, AR/AO, AR/A1-A30, AR/A, V, VE, and V1-V30	1% chance of annual flooding (i.e., once every 100 years)
Moderate Flood Hazard Areas	B or X (shaded)	0.2% chance of annual flooding (i.e., once every 500 years)
Minimal Flood Hazard Areas	C or X (unshaded)	less than 0.2% chance of annual flooding

- Elevate furnace, water heater, and other appliances above the BFE level.
- Install backflow valves in the main sewer line to prevent sewage from flowing back up into your home.
- Install flood shields or natural barriers near basement windows and doors.
- If you have a basement, install a sump pump.
- Ensure that your yard is properly graded for adequate water drainage.
- Use rain barrels or other collection devices on main gutters to prevent washout.
- Landscape using vegetation that resists soil erosion.
- Seal basement walls with waterproofing compounds to prevent seepage.
- If living in a flood zone, consider buying flood insurance.

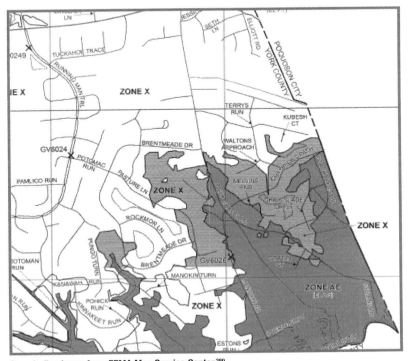

Sample flood map from FEMA Map Service Center[280]

DURING A FLOOD

If a flood threatens your area, immediately monitor NOAA weather broadcasts as well as local television stations for real-time information. If you decide to evacuate, or are directed to do so by authorities, you should first secure your home (as time allows) by doing the following:

- Unplug electrical appliances
- Bring in any outdoor furniture
- Move essential items to upper floors
- Turn off utilities at main switches and valves
- Lock your doors and windows

Poor judgment during a flood can easily lead to a deadly situation. Consider the following safety guidelines:[281]

- Do not walk through fast moving water.
- If you must walk through standing or very slow moving water, probe with a stick to test the depth.
- Do not drive into flooded areas. Cars can easily be swept away by floodwaters. If water rises around your car, abandon the vehicle and move to safety. A foot of water is usually enough to float a vehicle. Two feet of rushing water can sweep a vehicle away.

AFTER A FLOOD

After the flood water finally abates, take the following actions:[281]

- Continue monitoring news reports. Be alert for additional flooding that may force your evacuation.
- Determine whether there are any dangers to your water supply.
- Avoid floodwaters, as they may be contaminated by sewage, gasoline, or oil. Electrical hazards might also exist.

Don't put yourself in this dangerous situation *(FEMA photo/Jacinta Quesada)*

Post-flood cleanup by volunteers *(FEMA photo/Marty Bahamonde)*

- Stay clear of moving water. Six inches of moving water is often enough to sweep a person off his feet.
- Be cautious about driving across roads or bridges from which floodwaters have receded. They may have been compromised or weakened from the flow of water. Do not drive around barricades.
- Stay far away from downed power lines, and report them to the power company or emergency personnel. Electricity can arc up to twenty yards from high voltage power lines.
- Stay out of buildings that are surrounded by floodwaters. Once the water recedes, be very cautious about entering.
- Service damaged septic or sewage systems as quickly as possible.
- Clean and disinfect things that got wet. Mud left behind may contain dangerous bacteria or chemicals.
- Take photographs of any damage *before* cleaning up. Submit them as part of your insurance claim.

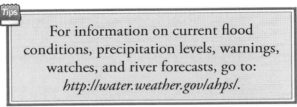

For information on current flood conditions, precipitation levels, warnings, watches, and river forecasts, go to: *http://water.weather.gov/ahps/.*

CHAPTER 19

TRIAL BY FIRE

The only way to determine if your family is truly prepared is to put your supplies and skills to the test. Too often people do a good job of stocking the basics, but miss small items that end up being critical to their survival. The classic example of this is having a pantry full of canned foods but no manual can opener. Such oversights are easily remedied by testing your preparations before a disaster strikes.

Below is a list of trials that your family should consider working through. Each exercise is designed to help identify shortcomings in your preparations. Some of the trials may not be relevant to your family, so pick and choose accordingly. Also, try to come up with your own challenges that might be germane to your family or community. The best way to do that is to identify the most likely threats, and then simulate those conditions.

> Create your own challenges by considering the most likely or most worrisome threats facing your family.

Some of the exercises outlined in this chapter should ideally be conducted on a day when everyone is off work, out of school, and available. This way the whole family can participate. Try to make this fun, but be forewarned, you are likely to hear a few complaints. No doubt you will have to remind everyone that these trials are meant to simulate hard times, so some level of discomfort is to be expected.

Of course, many disasters are more far-reaching than these short-lived experiments. However, by completing these tests, you will learn quite a bit. You will determine your family's daily food and water consumption, verify that your energy resources (batteries, flashlights, wood, generator fuel, etc.) are adequate, learn about hygiene in the absence of running water, practice evacuating and escaping from a variety of threats, and realize firsthand the importance of having a solid grab-and-go strategy.

UNPLUGGED

Scenario: A powerful storm rolls through your community overnight, causing a loss of electrical power.

As soon as you climb out of bed, turn off your main power breaker—something you should know how to do. For the next 24 hours, operate without utility power. If you have a generator and/or an inverter (see *Chapter 7: Electrical Power*), practice using them. Discover the capabilities and limitations of your equipment. Can you power your house lights? If not, do you have adequate substitute lighting for night-time? Can you stay warm (or cool)? How will you cook? Operate under the assumption that the whole community is without power. This means that you must prepare your own meals—no running out to a restaurant. Keep the power turned off until the following morning. It is advisable to try this test both in the heat of summer and the frost of winter.

BOIL ORDER

Scenario: Your county health department issues a boil order for all tap water.

Treat the water coming out of your taps as contaminated. Start by discarding all the ice made from your water supply on the assumption that it too is suspect. If you have a true purifier (see *Chapter 4: Water*), you can use the purified water from that tap without boiling it beforehand. If you do not have a purifier, then you must boil or chemically treat all water before using it. If you wash your hands in contaminated water, you must use an alcohol sanitizer afterward. Give this a try for a day and see what impact it has on your family.

DRIED UP

Scenario: Local authorities have shut off tap water because heavy rains caused sewage to contaminate the water supply.

Start by cutting off the water main coming into your home—something you should know how to do. For the next 24 hours, use stored water for your family's drinking, cooking, hygiene, and sanitation needs. Be sure to prepare for this challenge by storing adequate water. This is a great experiment to determine how much water your family really needs in an emergency. It is best to try this test in the heat of summer when your water needs are at their greatest.

RUNNING ON EMPTY

Scenario: Due to war needs, gasoline is in short supply and being rationed.

Challenge your family to cut their fuel usage by 50% for an entire week—keep track of mileage to monitor your progress. Consider carpooling with your spouse, limiting the number of shopping trips, eating at home rather than dining out, walking or riding a bike to nearby places, and a host of other fuel-saving actions.

DANGER AT THE DOOR

Scenario: Late one evening, you answer the door to find three thugs who quickly push their way into your home.

Teach everyone in your family to react to an emergency word. If a family member shouts the word, everyone should immediately attempt to escape from the home and seek help. Identify multiple escape routes from the home. Let everyone know the day when you will be conducting a practice drill, but do not announce the specific time.

LIVE FROM YOUR CUPBOARDS

Scenario: Terrorists attack the food supply system, preventing supermarkets from resupplying.

For two full weeks, have your family live off what you have in the house—no trips to grocery stores or restaurants. This is a great way to assess how much food you really need to store, as well as clear out some of the food that has cluttered up the cupboards, refrigerator, and freezer for months. It may also help you to learn to pull together some creative menus.

ESCAPE

Scenario: A dangerous event occurs at your location, such as a fire breaking out at the movie theater.

This is more of an ongoing challenge than one done over a single day or week. Over the next few months, challenge your family to identify escape routes from public locations. This could include such places as a movie theater, supermarket, restaurant, bowling alley, or shopping mall. Ask them some basic questions: If there was suddenly a fire or explosion how would they get out alive? What route would they take if the main exit was blocked? Why would some exits be better than others? Where would they go once they escaped?

FIRE IN THE HOUSE

Scenario: Your house catches fire in the middle of the night.

Before doing this exercise, discuss and practice emergency evacuations from your home. To keep panic to a minimum, let everyone know that you will be conducting an overnight fire drill, but don't tell them the specific day that it will occur.

On the night of the drill, wait until everyone is asleep and then activate a fire alarm farthest from where your family sleeps. You can activate it by holding the test button or blowing out a few matches underneath it—put in ear plugs first. See how long it takes your family to wake up and rendezvous safely at a predetermined location just outside the home. Ideally, everyone should be outside in less than one minute. If your children don't wake up quickly, investigate alternative alarm types, such as voice alarms.

Anything can happen *(FEMA photo/Michael Rieger)*

ON THE ROAD

Scenario: Due to approaching wildfires, an evacuation order has just been issued for your community.

Without prior preparation, give your family 15 minutes to grab whatever they think they might need to evacuate. Take only a single change of clothes. Use the little time you have to gather necessary supplies, irreplaceable items, and important documents. Pick a hotel some distance away, and travel to it using an alternate route—no main interstates. Spend the night at the hotel. Discuss the things you may have forgotten and how you could improve your evacuation response.

IT'S IN THE AIR

Scenario: A chemical plant has issued an urgent warning that an airborne hazard has accidentally been released. Authorities are urging residents to immediately go indoors and shelter-in-place.

Following the instructions provided in *Chapter 5: Shelter,* have your family retreat to a single room in your home. Seal the doors and windows as described. Remain in your safe room monitoring the radio for a couple hours. Discuss with your family ways to prepare the room more quickly and make it more livable for a longer stay.

GET TO THE SHELTER

Scenario: One evening during a particularly violent thunderstorm, you suddenly hear the distinctive rumble of an approaching tornado. You have only seconds to get your family to safety.

Prior to conducting this drill, set up an in-home shelter as described in *Chapter 5: Shelter* and *Chapter 18: Five Horsemen of Death*. Make sure to stock it with supplies that your family might need if they were holed up for a few hours. Supplies might include: flashlights, batteries, water, snacks, blankets, a NOAA All Hazards weather radio, a first aid kit, a whistle, a telephone, and some cards, games, or books to keep everyone entertained.

Explain to your family that you will be conducting practice drills to see how quickly everyone can get into the shelter. Without warning, and at what are obviously inopportune times, call for everyone to "get to the shelter." Use a stop watch to keep up with the response times. Determine what causes the most delays, and find ways to get everyone working together more efficiently. Once confined to the shelter, assess the sustainability of keeping everyone cooped up for hours. What could make it more tolerable? What supplies might you have overlooked?

APPENDIX

FOOD STORAGE LIST

Food Category	Type	Full Stock		Current Stock	Qty Needed
		Qty	Size		

TELEPHONE NUMBERS

Name	Telephone #
Emergency Services	
Police, Fire, Ambulance	911
Poison Control Center	1-800-222-1222
Emergency Management (FEMA)	1-800-621-FEMA
Centers for Disease Control (CDC)	1-800-311-3435
Federal Bureau of Investigation (FBI)	1-800-CALL-FBI
Local Chapter of American Red Cross	
Medical/School	
Family Doctor	
Child Doctor	
Dentist	
School	
Utilities	
Gas Company	
Electric Company	
Sewage Service Provider	
Water Service Provider	
Family/Friends	
Neighbor	
Babysitter	
Local Emergency Contact	
Out-of-town Emergency Contact	
Insurance Companies	
Other	
Local Animal Control	

IMPORTANT PAPERS

- ❑ Addresses and phone numbers of points of contact (family, friends, insurer, doctor, etc.)
- ❑ Driver's licenses
- ❑ Social Security cards
- ❑ Birth/Death certificates
- ❑ Adoption papers
- ❑ Insurance cards and policies (medical, dental, vision, auto, home, life)
- ❑ Credit cards (front and back)
- ❑ Passports
- ❑ Recent photos of family members (suitable for missing person's posters)
- ❑ Military discharge papers (e.g., DD214)
- ❑ Diplomas, certificates
- ❑ Property deeds
- ❑ Description of all vehicles (e.g., make, model, photo, VIN, and license number)
- ❑ Automobile titles
- ❑ Firearm serial numbers and photos
- ❑ Weapon permits
- ❑ Pay stub
- ❑ Marriage license
- ❑ Home inventory video or photos
- ❑ Bank/Investment account information
- ❑ Tax records
- ❑ Computer account logins and passwords
- ❑ Medical information (allergies, medicines, medical history)
- ❑ Resume (for job hunting)
- ❑ First aid information (see the Captain's Medical Guide[245])
- ❑ Survival reference information (collection of e-books, online tips, how-to manuals)
- ❑ GPS locations and driving directions of house, rally points, local medical emergency services, and other key places—see www.maps.google.com. For GPS coordinates, right click and select "What's here?"

☐ Last Will and Testament

☐ Professional licenses

☐ Bible and other e-books

☐ Relaxing music

☐ Foreign language dictionaries (e.g., Spanish)

☐ _____

☐ _____

☐ _____

☐ _____

☐ _____

☐ _____

☐ _____

☐ _____

☐ _____

☐ _____

☐ _____

☐ _____

☐ _____

☐ _____

☐ _____

☐ _____

☐ _____

☐ _____

☐ _____

☐ _____

☐ _____

☐ _____

☐ _____

☐ _____

HOME ASSESSMENT

Threat	Likelihood (1-10)	Protection (1-10)	Steps to Improve

HOME HAZARDS CHECKLISTS

Outdoor Hazards

Indoor Hazards

LIST OF MEDICATIONS AND ALLERGIES

Medications		
Family Member	**Medication**	**Dosage**

Allergies		
Family Member	**Allergy**	**Reaction**

Attention!

I have (abandoned my vehicle) / (evacuated my home) due to an emergency situation.

My Name: _____

Telephone: _____

Address: _____

My travel plans and emergency contact information are given below. I am requesting that emergency personnel confirm my safe arrival before discarding this note.

My Travel Plans:

_____

_____

_____

_____

_____

_____

_____

_____

_____

_____

_____

_____

_____

Emergency Contact Information

Name: _____

Phone Numbers: _____

Address: _____

PERSONALIZED DP PLAN

	Plan		
Dangers	Goals	Needs	Implementation

CONTACT ME

Disaster preparedness is an important subject for *every* family. If you found this book to be helpful, I would kindly ask that you do two things: (1) give a copy to your loved ones (or simply pass this one along when finished), and (2) post a review on Amazon.com to let others know that reading this handbook is time well spent.

I frequently travel the world giving disaster preparedness seminars. If you are a member of a church, business, or civic organization and would like to sponsor a disaster preparedness event, please keep me in mind.

Every author enjoys hearing from his readers, whether it be praise, criticism, or just a friendly "hello." If you would like to contact me regarding this book or any DP-related subject, please send an email to *inquiries@disasterpreparer.com*.

Best wishes to you and your family!

INDEX

A

abandoning vehicle 323–324
air pollution 189–193
air purifier 202–203
amateur radio operators 272–273
American Preppers Network 365
AM/FM radios 268
AquaStar 80

B

barbeque grill 184
batteries
    deep cycle. *See* batteries: lead-acid
    lead-acid 145, 155, 156, 157, 158, 159, 160, 161
    rechargeable 154–155
    recharging 157–160
    single use 153–155
biochemical suits 107–108, 207
biochemical threat 107, 108, 193, 195, 355
bleach 43, 67, 68, 69, 75, 76, 77, 85, 94
boiling water. *See* water: boiling
bucket kit. *See* preparedness kits

C

calling 911 220, 221, 223, 246, 271, 319, 351, 370, 404
camp stoves 185
candles 21, 22, 24, 130, 132, 133, 136, 138
carbon monoxide 110, 115, 116, 117, 126, 127, 128, 133, 138, 190, 246, 322
    alarms 116–117
CB radios. *See* citizen band radios
CBRN protective clothing. *See* biochemical suits
chemlights. *See* light sticks
children, special needs 213–215, 350–353
church or civic organizations 362
citizen band radios 273
coal-burning stoves 176–178. *See also* wood-burning stoves
commercial broadcasts 260
communication
    functions 259–260
cooking 36, 39, 40, 62, 64, 67, 87, 116, 133, 134, 139, 147, 152, 183, 183–185, 184, 185, 186, 187, 220, 265, 396

D

deadly force 334
debt 280, 281, 282, 283, 284, 285, 286, 287, 288, 289, 291, 292, 293, 296, 297, 299, 300, 302, 303, 304, 305, 307, 308, 309, 310
deep cycle batteries 156. *See also* batteries: lead-acid
deficit 108, 214, 300, 302, 304, 305, 307, 308
dew collection 89
disaster
    definition 4
    man-made 8
    natural 5–7
    pandemic. *See also* pandemic
    personal 10
    ranking 10–11
    types 4–10
    war, terrorism, crime 9
disinfectant 67–68. *See also* bleach
distillation 72, 73, 77, 78, 79, 81, 82, 84, 85, 91
dogs 104. *See also* pets
DP Network. *See* Neighborhood Ready
DP Plan 15, 16, 17, 18, 20, 141, 280, 310
    sample entries 61, 92, 126, 137, 160, 186, 207, 216, 256, 276, 309, 324, 342
duct tape 21, 101, 107, 128, 194, 208, 209, 255, 317, 324, 355

E

earthquakes ix, xi, 1, 3, 4, 5, 5–6, 16, 17, 96, 110, 121, 266, 267, 269, 294, 345, 359, 367, 368, 369, 370, 371, 372, 373
effective DP networks 362–365
elderly, special needs 345–349
electrical loads 143–144
electrical power
    generators. *See* generators
    terminology 139–140
electric space heaters 179–180. *See also* space heaters
email 125, 271, 276, 299, 363, 365, 375, 412
Emergency Alert System 261–262
emergency blanket 166, 186, 318
emergency broadcasts 260–263
emergency fund 25, 279, 281, 287, 288, 289, 309, 310
EPA Guide Standard 76, 80, 83, 85, 94

12592836R00245

Made in the USA
Lexington, KY
20 December 2011